MW01519855

Apparel Manufacturing Technology

Apparel Manufacturing Technology

T. Karthik

P. Ganesan

D. Gopalakrishnan

CRC Press
Taylor & Francis Group
Boca Raton   London   New York

CRC Press is an imprint of the
Taylor & Francis Group, an **informa** business

CRC Press
Taylor & Francis Group
6000 Broken Sound Parkway NW, Suite 300
Boca Raton, FL 33487-2742

© 2017 by Taylor & Francis Group, LLC
CRC Press is an imprint of Taylor & Francis Group, an Informa business

No claim to original U.S. Government works

Printed on acid-free paper
Version Date: 20160317

International Standard Book Number-13: 978-1-4987-6375-2 (Hardback)

Library of Congress Cataloging-in-Publication Data

Names: Karthik, T., author. | Ganesan, P. (Industrial engineering professor), author. | Gopalakrishnan, D., author.
Title: Apparel manufacturing technology / T. Karthik, P. Ganesan, D. Gopalakrishnan.
Description: Boca Raton : Taylor & Francis, a CRC title, part of the Taylor & Francis imprint, a member of the Taylor & Francis Group, the academic division of T&F Informa, plc, [2016] | Includes bibliographical references.
Identifiers: LCCN 2016004249 | ISBN 9781498763752 (hardcover : acid-free paper)
Subjects: LCSH: Dressmaking. | Clothing factories. | Clothing trade. | Manufacturing processes.
Classification: LCC TT497 .K38 2016 | DDC 687--dc23
LC record available at http://lccn.loc.gov/2016004249

Visit the Taylor & Francis Web site at
http://www.taylorandfrancis.com

and the CRC Press Web site at
http://www.crcpress.com

Contents

Preface

As known to all of us, clothing is a basic necessity for humankind. It serves as a true reflection of people's economic and social status. The improvement in the standard of living of people is showcased by the variety of clothing people wear. The lifestyle of the people is depicted based on the type of apparel worn by them. Apparel manufacturing industries have gained greater popularity in recent years owing to the popularity of ready-to-wear garments. Bulk production has reduced the cost of apparel and quality is maintained. The apparel industry is very diverse in nature and along with textile industries, it forms a complex combination of performing heterogeneous functions of transforming fibre into yarn and then to fabric. The diverse nature can be attributed to the variety of products manufactured and the requirements of people related to apparel has increased manifold times. Clothing for functional purposes has been replaced by aesthetic and fashion clothing. Because of this inevitable growth, the textile and clothing sectors are gaining more relevance with respect to economic and social terms as it leads to more income, jobs in the short term and provides economic and social development for countries in the longer run. As this industry is becoming more fashion oriented with customisable clothing, the need and potential for value-added products is increasing day by day. This need has given rise to research and development in this area to stay afloat in competition. Another notable point is that the apparel industry is a labour-intensive one and hence it provides huge job opportunities at all levels of manufacturing and managing including scope for entry level unskilled labourers. This is a huge economic boost for both developing and developed nations to reduce their unemployment index levels. In this sector, technological advancements have led to further developments in the manufacturing and designing areas. The best part is that these technologies can be implemented by all countries including poorer ones as the cost required is comparatively lesser than other sectors and hence the feasibility of using new technologies is high.

The abolition of a quota system, liberalisation and globalisation have led to greater competition and changes in the world economic outlook. The success of the apparel industry has become dependent on the cost competency aspect. In the last decade or so, it has been noticed that there is an emergence of a postindustrial production system that is able to achieve the goal of mass customised, low-volume production. Supply chain efficiency has become very critical in today's scenario and hence more advanced and sophisticated technologies like JIT (just in time) delivery, vendor managed inventory and third-party logistics are being used to improve the manufacturing processes and also result in more cost efficiency. The clothing sector was initially art-based and later it evolved into technology-based after undergoing several changes. The technological advancements in the apparel industry include the use of computer-aided designing and pattern making, fit analysis software, 3D scanning technology, automation and robotics, integration of wearable technology, biomimetics, artificial intelligence and advanced material transport systems. The requirement now is that the industry should embrace all these technologies with open arms by being creative enough to design innovative products, introducing new materials and processes and having flexibility in manufacturing by following proper supply chain practices.

This book will help readers satisfy the requirements stated in the above paragraph as it provides greater knowledge in explaining the basic concepts of selection of raw material, classification of garments, various stages in manufacturing of garments and performance tools for measuring the same. Further, the book delves into the apparel engineering aspects such as production planning and control, layout of plants and costing aspects of garments. The following 20 chapters provide a critical overview of clothing manufacturing and engineering aspects in the apparel industry.

Chapter 1 reviews the structure of the garment industry and classification of garments and the selection of fibres and their significance in apparel manufacturing. The different fabric inspection systems along with their merits and limitations and fabric characteristics and their influence on sewing performance are also discussed. Chapter 2 discusses the significance of body anatomy and body measurements on garment fitting. The various aspects related to garment pattern making such as pattern making tools, types of patterns and principles of pattern making are discussed in detail. The pattern making methods, namely drafting, draping and flat pattern technique have also been elaborated. Chapter 3 converses about the various processes carried out in the cutting section such as marker making, spreading, cutting and preparation for sewing. The requirements of marker planning, its constraints, methods of spreading and cutting and equipment are discussed elaborately. The classification of sewing machines, functions of various parts of sewing machines and stitch forming mechanisms are presented in Chapter 4. Chapter 5 provides insight into the various aspects related to the classification and selection of the sewing thread and needle for construction of garments and their influence on sewing performance.

The classification of seams and stitches along with their characteristics and various sewing defects are discussed in Chapter 6. Chapter 7 deals with various kinds of sewing machine feeding mechanisms and their application and various special attachments for sewing machines along with their functions. Chapter 8 provides the different types of fusing, pressing and packing methods used in apparel industries. Chapter 9 deals with the method of construction of fullness and yoke. Chapters 10 through 12 provide the types, construction and application of collars, plackets and pockets and sleeves and cuffs, respectively. Chapter 13 discusses the various types of garment closures and their applications and various supporting materials used for construction of garments.

Production planning and control is one of the vital parts of the apparel manufacturing industry, accuracy in planning equates to timely shipment of orders, the better utilisation of operators and guarantees that proper supplies and machineries are available for each style and order and is discussed in Chapter 14. The method of handling cutting orders and planning economic cutting lays are of extreme significance for better utilisation of materials and for increasing the efficiency of the cutting process along with roll planning and fabric grouping. It is described in Chapter 15. Chapter 16 deals with the different kinds of apparel production systems with respect to an integration of materials handling, production processes, personnel and equipment that direct workflow and generate finished products.

Chapter 17 discusses the breakdown of operations for various garment styles and different control forms in production departments with their significance. Plant loading and capacity planning with respect to line balancing, determination of machinery requirement, production capacity of the industry and operator efficiency are discussed in Chapter 18. The function of merchandising differs based on whether it is performed in retail or manufacturing. It involves the conceptualisation, development, obtainment of raw materials, sourcing of production and dispatch of product to buyers. The various aspects related to merchandising activities are detailed in Chapter 19. The last chapter provides the basic information regarding the costing, pricing and determination of garment costing for various styles of garments.

This book is primarily a textbook intended for textile technology and fashion technology students in universities and colleges, researchers, industrialists and academicians, as well as professionals in the apparel and textile industry.

Authors

T. Karthik is an MTech and PhD qualified textile technologist. Currently he is working as an assistant professor (Senior Grade) in the Department of Textile Technology, PSG College of Technology, Coimbatore, India.

Dr. Karthik has 6 years of teaching experience and 5 years of industrial experience and handling various textile subjects particularly in the area of process and quality control in spinning, garment manufacture, nonwoven technology and technical textiles.

He has published more than 60 articles in reputed international and national journals. He has published five international books and contributed five book chapters. He has received 'Young Engineer Award' from Institution of Engineers India in Textile Engineering Division for the year 2015. He is a member of professional bodies such as Textile Association of India (TAI) and Member of Institution of Engineers (MIE).

P. Ganesan is an MTech and PhD qualified textile technologist. Currently he is working as an assistant professor (Senior Grade) in the Department of Textile Technology, PSG College of Technology, Coimbatore, India.

He has 8 years of teaching experience in the area of apparel manufacturing and quality control. He also has 2 years of industrial experience. He has published 43 research articles in reputed international journals and contributed three book chapters. He has received awards such as SDC-Young Talent search award and Precitex award for his academic excellence.

D. Gopalakrishnan is an MTech (Textile Chemistry) and MSc (Costume Design and Fashion) qualified Textile Technologist. He has 7 years of teaching experience and 7 years of industrial experience. The author has published more than 65 articles in reputed international and national journals and presented more than 20 research papers in conferences. He has published two books.

1

Introduction to Apparel Industry

The textile and fashion industry is a major contributor to several national economies, including both small- and large-scale processes globally. With concern to the employment as well as production, the textile sector is one of the prime industries in the world (Abernathy 2004). The garment sector is a labour-oriented one and provides enormous job opportunities at the entry level for unskilled labour in developed as well as developing nations. Further, it is a sector where comparatively modern technologies could be implemented even in poor countries at moderately low investment costs (Ashdown 1998).

The textile and clothing sector also has the high potential market segment for value added products where design and research and development (R&D) are key competitive factors. The luxury fashion industry utilises higher labour in design and marketing segments. The same applies to market sectors like sportswear where both design and material technology are vital (Ashdown 1998).

1.1 Structure of Textiles and Clothing Industry

The clothing industry is a labour-oriented, low wage industry but a vibrant, innovative sector, depending on the type of market segments upon which the industry focuses. The high-end fashion sector is considered modern technology, with comparatively well-paid workers and designers and a high degree of flexibility (Bailetti and Litva 1995).

The core operations of industries servicing this market sector are mostly situated in developed nations and often in certain geographical locations within these nations. The other kind of major market sector is bulk production of standard products like t-shirts, uniforms, underwear, etc. Manufacturers for this type of standard product market sector are mostly seen in developing countries (Abernathy 2004). For lower- to medium-priced products in the market, the responsibility of the retailer has become more and more important in the organisation of the supply chain. The retail market sector has turned out to be more intense, leaving more market power to multinational retailers (Ashdown 1998).

Textiles are responsible for the key raw material input to the garment industry, developing vertical supply chain relationships between the two containing sales and distribution functions (Bailetti and Litva 1995). The textile and clothing sectors involve

- Acquiring and processing raw materials, that is, the preparation and production of textile fibres.
- Manufacturing of textile yarns and fabrics.
- Dyeing and finishing of textile materials, which provide visual, physical and aesthetic properties that consumers demand, such as bleaching, printing, dyeing and coating.
- Conversion of textiles into garments that can be either fashion or functional garments.

1.1.1 Clothing

The fundamental manufacturing process of the apparel industry has not undergone much change over the past century, and is considered by the progressive bundle system. Work or operation is planned in a manner that each operator is specialised in one or a few operations (Ashdown 1998). The fabric is first cut into various garment panels and then grouped by components of the garment, tied into bundles and sent to an assembling (sewing) section for making a garment. An operator receives a bundle of cut garment panels and executes his or her single operation and keeps the bundle in a buffer. A buffer of about one day's work is common at each operation. It takes about 40 operations to finish a pair of pants, which entails about 40 days of in-process inventory. Though numerous advances in the industrial engineering segment for systematising the operations and reducing the production time of each individual operation have taken place over a period of time, the basic method has remained the same (Ashdown 1998; Abernathy 2004).

The new technologies, systems and innovations in the clothing sector have improved efficiency at each production stage and enhanced the harmonisation between stages and provided a more seamless interface between them (Bheda et al. 2003). The major breakthrough innovation was the use of computers in clothing manufacturing in areas like pattern making, marker planning and computerised automatic cutting machine. This machine has made it possible to cut increasingly thick layers of cloth accurately (Tyler 1992; Chuter 1995; Fairhurst 2008). These advancements are mainly associated with the preassembly phase of production, where technological developments have been more important than at the assembly stage. The organisation structure of a medium-sized garment industry is shown in Figure 1.1.

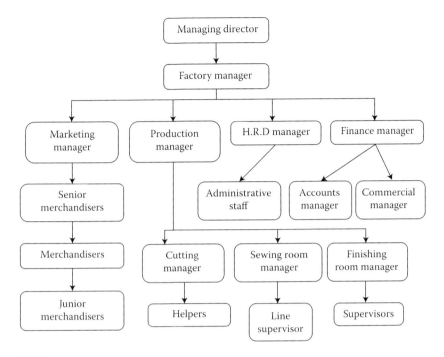

FIGURE 1.1
Organisation of an apparel industry.

1.1.2 Textiles

The textile business generally needs more investment compared to the garment sector and it is an extremely automated area. It comprises yarn manufacturing, fabric manufacturing and dyeing and finishing, and these three functions could be carried out in integrated plants. On the other hand, the textile sector suffers from the higher lead time as well as high investment cost, which results in relatively large minimum orders (Chuter 1995).

1.2 Various Departments in the Garment Industry

The various departments or sections in an apparel industry are given below.

1. Merchandising
2. Sampling department
3. Fabric sourcing
4. Purchasing department

5. Fabric inspection department
6. Accessory stores department
7. Planning department
8. Laboratory department
9. Machine maintenance
10. CAD section
11. Cutting section
12. Production department
13. Industrial engineering section (IE)
14. Embroidery department
15. Fabric washing section
16. Quality assurance department
17. Finishing department

1.2.1 Merchandising

It is a vital process that involves planning, developing, executing and dispatching the order (product) to the buyer. The merchandising process comprises guiding and supervising for the successful processing of an order. The types of merchandising done in a garment unit are marketing merchandising and product merchandising (Tyler 1992; Banumathi and Nasira 2012). The main objective of marketing merchandising is development of product, costing and ordering, and it has direct contact with the buyer. Product merchandising is carried out in the respective apparel unit and involves all the responsibilities starting from sourcing to finishing (Chuter 1995).

1.2.2 Sampling Department

The sampling department coordinates with the merchandising and production department. It is carried out to foresee finished product appearance and fit when produced in bulk and to confirm whether there are any inconsistencies in the pattern according to the buyer's specification (Banumathi and Nasira 2012). It also aids to determine the fabric consumption along with that of thread and other accessories used.

1.2.3 Fabric Sourcing

Fabric sourcing is mainly engaged in deciding where and how the fabrics have to be procured. It works in conjunction with the merchandising department and looks after the delivery of the required garments within the scheduled time and cost (Jacob 1988).

1.2.4 Purchasing Department

The main difference between the sourcing and purchasing department is that the sourcing section works for sourcing the fabrics alone whereas the nature of the work of the purchasing department comprises sourcing of accessories and trims as well.

1.2.5 Fabric Inspection Department

The main aims of fabric inspection team are

- Identification and analysis of fabric defects using various standard methods.
- Selection of fabric according to AQL (accepted quality level) 1.5.

1.2.6 Accessory Stores Department

The receipt of the raw materials or the accessories is normally completed in terms of documents that are received from the merchants.

1.2.7 Production Planning Department

Upon receipt of the orders from the merchants, preproduction meetings with the departments have to be done. After that, the production department will assign the style to the specific line that has the capacity to complete it on time. The planning section then carries out the estimation and planning of order quantity, plan cut date (PCD), breakup of order, operation breakdown, etc. based on the particular unit (Gilmore and Gomory 1961).

1.2.8 Laboratory Department

The laboratory or testing centre in the industry should be equipped with all the essential instruments that are mandatory for the testing of fabric and accessories. If the facility for specific tests mentioned by the buyer is not available in the industry, it should be sent to external laboratories that are authorised by the buyers.

1.2.9 Machine Maintenance

Undesirable quality of garments mostly results from ill-maintained machines. Breakdown and preventive maintenance is mainly aimed toward reducing the downtime and increasing lifetime, respectively.

1.2.10 CAD Section

Normally, large-scale garment industries have their own designing department for various garment styles. The CAD department is accountable for the
following functions:

- Determining cutting average for costing
- Making the most efficient cutting marker
- Development and alteration of patterns
- Development of size set pattern by grading
- Digitising the pattern

1.2.11 Cutting Section

The cutting department normally receives the order from the production manager who has approved the cutting order to cut a given quantity
of garment styles. The cutting order sheet contains the following
information:

- Sampling average, garment weight and averages of other trims
- Measurement sheet
- Design of the garment
- Purchase order
- Fabric request sheet
- Marker planning – length of lay, etc., size ratio and colours in which
  the patterns are to be cut

1.2.12 Production Department

The production department will obtain the details like

- The garment style
- Number of operators required
- The batch for which the style has to be installed
- Target for each day
- Breakup of the production quantity

After receipt of all of the above details, the production department sends
a request from the cutting section for the cut parts. After assembling of the
components, a line check has to be done where the shade matching and the
measurements are checked.

1.2.13 Industrial Engineering Section

It coordinates with several departments since this department provides the entire plan of the garment manufacturing and the thread and trims consumption criteria, operator's skill level categorisation and other related aspects.

1.2.14 Embroidery Department

It comes into play only when the particular garment style demands. It receives the garment panel, style and the embroidery details from the merchandisers and they will also get a sample of the garment on which the embroidery has been already done and it will be used as a reference sample.

1.2.15 Fabric Washing Section

After the completion of assembling and inspection process, the garments are sent to the washing department for the washing or finishing that is required for the particular style according to the specification sheet.

1.2.16 Quality Assurance Department

To maintain and control the quality, the quality assurance department divides the work into different stages of manufacturing, which are categorised into three major groups such as preproduction unit, cutting audit and sewing unit.

1.2.17 Finishing Department

The finishing department is the last section in the garment production prior to packing and dispatch and it plays a significant role in the final garment appearance (Mehta and Bhardwaj 1998). It involves the following processes.

- *Trimming:* It removes the extra threads from the garment at the stitched areas.
- *Inspection:* The inspection is done as per the AQL 2.5 system and mainly depends on the buyer requirements.
- *Pressing:* This is carried out after the garment has been inspected completely and the garments are pressed or finished based on the method of their folding during packing.
- *Tagging section:* After the completion of fabric inspection and pressing, they are sent for labelling, which includes the size labels, price tags and miscellaneous labels if any are mentioned in the specification sheet.
- *Packing:* The packing is done in the carton boxes. Individual packing of garments in the poly bag and folding the garments and organising

them in the carton boxes without placing them in the poly bag are
the two types of packing followed in the garment industry.

1.3 Classification of Garments

Garments could be classified based on several aspects as there is no standard classification system available. However, the garments could be classified based on the gender as male or female, or age as children's garments.
Generally, based on use, style and material, different varieties of garments
show different styles (David Rigby Associates 2002; Fan et al. 2004). Presently,
the garments are classified based on the following aspects.

1. Type of fabric:
    a. Knit (T-shirt, sweater)
    b. Woven (shirt, suitings and denim)
    c. Nonwoven (diaper, socks)
2. Season:
    a. Winter (jacket)
    b. Summer (tank top)
    c. Spring (singlet)
    d. Autumn (shirt)
    e. Late Autumn (shirt (design))
3. Events:
    a. Party (fashion wear)
    b. Active (regular wear)
    c. Evening gown (outfit)
    d. Night (soft fit)
4. Application:
    a. Formal (collar shirt)
    b. Swimwear (bikini, cover ups)
    c. Sportswear (trouser)
    d. Lingerie (inner wear, sleep wear)
5. Method of manufacture:
    a. Readymade (complete)
    b. Tailored (measurement)
    c. Furnishing (automated)

6. Source:
   a. Leather (leather)
   b. Natural (leaf)
   c. Artificial (fur)
7. Gender and age:
   a. Women's (skirts)
   b. Men's (tongo)
   c. Kid's (toga)
   d. Toddler (bibs)
8. Shape and styling:
   a. Dresses (sari)
   b. Shirts (neck wear)
   c. Skirts (elastic and stitches)
   d. Suits (official outfit)
9. Length of garment:
   a. Shorts (panty wear)
   b. Three quarters (cargo pant)
   c. Full wear (pant)
   d. Bermuda wear (thigh wear)
   e. Pullover (stockings)

1.3.1 Harmonised System

A harmonised system (HS) was established under the support of the Customs Cooperation Council (CCC) on 14 June 1983. The purposes of the HS are to

1. Support international trade
2. Facilitate the collection and comparison of statistics
3. Facilitate the standardisation of the trade documentation and transmission of data
4. Promote a close correlation among import and export trade statistics and production statistics

1.3.1.1 Classification and Categories of Apparel under Harmonised System

The classification of readymade garments is highly complicated due to diverse assortments of fashions and at the same time highly sensitive because of the imposition of quantitative restraint under the Multi Fibre Agreement (Nordås 2004). The classification of garments under HS is shown in Figure 1.2.

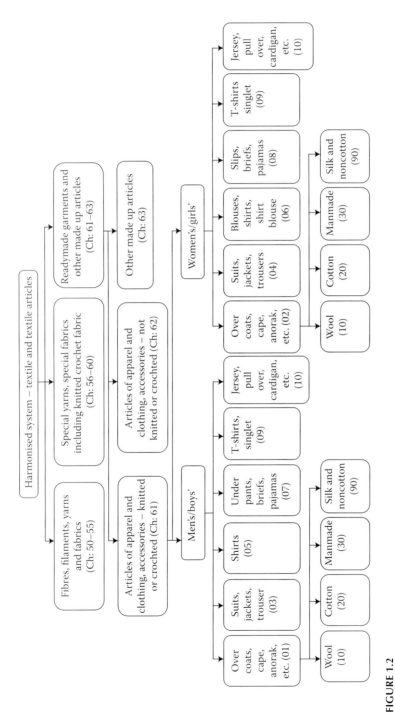

FIGURE 1.2

Garment classifications under harmonised system.

1.4 Raw Material for Garment Manufacturing

1.4.1 Fibre Selection in Garment Manufacturing

Fabric requirements can be classified into four categories: aesthetic (handle, drape, lustre, etc.); performance in use (easy-care, stretch, comfort, pilling tendency, abrasion resistance, etc.); image and cost, which can be subdivided into the fibre or yarn cost and the finished fabric processing cost. All of these factors have an influence based on the type of garment and its market position or price point. For example, a fabric to be used in a formal ladies' suit for a highly regarded brand house will have a high value placed on aesthetics and image, be less sensitive to performance in use, and will be largely insensitive to cost. On the other hand, a department store's own-brand jeans will be worth highly with respect to its durability and a low cost more than its aesthetics value and brand image (Rana 2012).

Fabrics are the centre of the analysis rather than fibres, yarns or garments because fabrics act as the crossroads in the apparel market. Yarns and fibres have a comparatively low range of variety but are difficult to interpret until they are transformed into fabric. Garments are an intricate mix of design and shape, which disguises the role played by the component materials. By contrast, fabrics are finite in number, visible from both ends of the supply chain and recognisable by all (Rana 2012). The fibre performance in the clothing market is basically decided by three factors:

- Inherent characteristics of the fibres itself matches with the aesthetic, cost and other needs of each fabric.
- How easily and economically the fibre's properties could be improved by processing in yarn or fabric form.
- How well the fibre blends with other fibres to enhance the overall fabric properties.

1.4.2 Yarns

Yarns are the immediate strand elements used to make woven and knitted fabrics. A yarn is a strand made from spun or twisted fibres or twisted filaments. Fibres are short lengths varying from 1/2 to 20". The length and diameter of a fibre depends on its natural type and source.

1.4.2.1 Yarn Specifications

Yarns that are spun (staple) or twisted (filament) are specified with respect to twist and size. There are two major types of twist, S and Z, as shown in Figure 1.3.

FIGURE 1.3
Direction of yarn twist.

Yarns sizes are designated with terms referring to yards of yarn per pound. Cotton, spun rayon and spun silk yarn sizes are numbered with the same system. A 1s yarn has 840 yards to a pound; a 2s yarn has 1680 yards, a 3s has 2520 yards, etc. If a yarn weight is 16,800 yards to a pound, the yarn is a 20s yarn. Wool yarns are numbered by three different systems. Worsted yarn measures 560 yards for a 1s. Woollen-cut yarn is 300 yards per pound for a 1s; woollen-run yarn is 1600 yards for a 1s. Woollen yarns differ from worsted yarns in structure. The fibres in woollen yarns are intermixed, whereas the fibres in a worsted yarn are long fibres that are parallel (Laing and Webster 1998).

Filament yarns (rayon, silk, etc.) are numbered with a denier count. The denier number is the number of 0.05 g units per 450 lengths. If a 450 length of filament yarn weighs 3 g, the yarn is a 60 denier yarn. A multifilament yarn is a strand composed of a group of filaments twisted into one strand, whereas a monofilament yarn consists of only one filament.

Plied yarns, in fabric construction, are fabric strands consisting of two or more yarns. Cabled yarns are fabric strands composed of two or more plied yarns as shown in Figure 1.4.

In the early 1960s, the Tex System for stipulating yarn sizes was introduced (sponsored by the ASTM Committee D-13 on Textiles). The various yarn sizes defined previously in this section assigned each a yarn size number for a given length and weight relationship in the yarn. The Tex System seeks to use the same length and weight relationship as the size system for all yarns. In the Tex System, the yarn size is equal to the gram weight of 1000 m of yarn; that is, a No. 1 yarn is 1000 m whose weight equals 1 g. Under the International Metric Count, the decitex size of a yarn is equal to the gram weight of 10,000 m of the yarn (Hollander 1993).

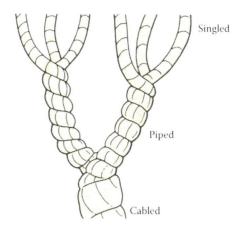

FIGURE 1.4
Construction of single, ply and cable yarns.

1.4.3 Fabric

Specifications for fabrics, and other raw materials used in apparel manufacturing, can be categorised into two groups: properties of fabrics and fabric characteristics. A fabric property represents physical dimensions like yards, pounds, etc., whereas a fabric characteristic refers to the response of the fabric when an external force is applied to it like elongation, elasticity, shrinkage, seam strength, etc. (Geršak 1996). These are measures of reactions to dynamic conditions. Characteristics are physical or chemical changes in the fabric resulting from the application of outside forces. Stress and strain properties are another term used to denote characteristics (Beazley and Bond 2006). There are three perspectives for specifying the fabric requirements:

1. The consumer's viewpoint
2. The fabric producer's viewpoint
3. The garment producer's viewpoint

The consumer's importance lies exclusively in the visual appearance, aesthetics and wearability properties of the fabric; the durability, utility and style values. The garment manufacturer is concerned with the garment production working characteristics of the fabric, and the cost of manufacturing a garment (Gupta 2011; Carr and Latham 2006). If the garment manufacturer is a job worker or manufacturer who retails the garment directly or indirectly to consumers, then he will be concerned with all the consumer values (Dockery 2001). If the garment manufacturer is a contractor, then he is only concerned with the production cost. In case of a fabric manufacturer, he is concerned with the garment production work characteristics.

1.4.3.1 Woven Fabrics

Woven fabrics are constructed by intertwining two groups of yarns perpendicular to one another. Weave constructions are classified in relation to the manner in which the warp and weft yarns intertwine. The primary weave classes are shown in Figure 1.5.

- *Plain weave:* Every filling (or warp) yarn passes alternately over and under consecutive warp (or filling) yarns.
- *Twill weave:* Every weft yarn passes over (or under) two or more warp yarns, after passing under (or over) one or more warp yarns in staggered fashion, so as to produce a diagonal line on one or both sides of the fabric.
- *Satin weave:* Filling yarns pass over (or under) enough warp yarns after passing under (or over) a warp yarn so as to give the fabric a smooth glass-like surface when the float process is staggered. The satin float is the yarn (filling or warp), which passes over many of its complimentary yarns before going under a complimentary yarn.
- *Basket weave:* This is similar to the plain weave but with a multiple yarn grouping. Two or more yarns travel as a set (Figure 1.6).
- *Jacquard weave:* Any combination of plain, twill, satin and basket weave counts used to give a complex configuration with a bias-relief effect.
- *Lappet weave:* A weave that has two superimposed warp layers in sections of the fabric.
- *Leno weave:* A weave with an open-space effect (Figure 1.7). Each filling yarn passes through the ellipse formed when two adjacent warp yarns cross over each other in reciprocal fashion from filling to filling. These warp yarn amplitudes pass over or under each other before and after encompassing the filling yarn.
- *Pile weave:* A weave that has the end of looped or cut yarns protruding out of one fabric surface (Figure 1.8). A double pile weave has yarn stubs protruding out of both surfaces.

Plain weave   Twill weave   Satin weave

FIGURE 1.5
Primary weave constructions.

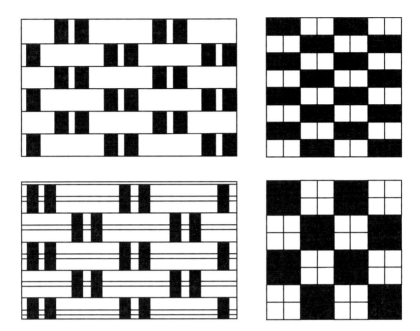

FIGURE 1.6
Basket weave.

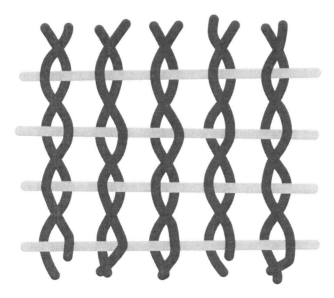

FIGURE 1.7
Leno weaves.

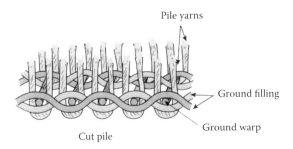

FIGURE 1.8
Pile weave.

1.4.3.2 Knitted Fabrics

Knitting is the process of constructing fabric with one or more groups of yarns by a system of interlooping loops of the yarns. The yarns are formed into rows of loops into which other yarn loop rows are interloped or interlaced. There are two basic types of knitting: weft and warp. Weft knit fabrics are manufactured by building the loops of yarn in horizontal position through the fabric width. Warp knitting constructs the fabric by making yarn loops parallel to the fabric length. Weft knit fabrics are produced in tubular- or flat-form circular knitting, whereas warp knit fabrics are made only in flat form (Anbumani 2007; Gupta 2011).

1.4.3.2.1 Weft Knitted Fabric

Weft knitted fabrics, circular and flat form (Figure 1.9) and the varieties of weft knitted fabrics are jersey, purl, rib, run-resist, tuck and interlock.

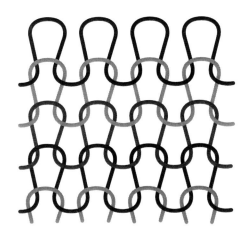

FIGURE 1.9
Weft knitted fabric.

In jersey fabric, the interlooping tie is on the same side of the fabric in all courses and wales. Purl stitching consists of changing the placement of the interloop tie from course to course. In purl stitched fabric, the interlooping ties in adjacent courses are on the other face of the fabric; alternate courses have the ties on the same face. In rib stitching, the ties in adjacent wales are on the other face of the fabric; alternate wales have like ties. Run-resist fabrics contain an alternating and staggering of course ties. A tuck stitch is an arrangement of tying two consecutive course loops in one wale structure. The interlock structure is essentially a double thickness rib knit. It consists of an interlooping of two adjacent layers of a rib knit (Anbumani 2007).

1.4.3.2.2 Warp Knitted Fabrics

In warp knit fabrics, the yarn forms successive wale loops instead of successive course loops as in weft knitting. The successive course loops in warp knitting are in different courses, whereas in weft knitting the successive course loops are in different wales (Figure 1.10). The varieties of weft knitted fabrics are single warp tricot (one bar tricot), double warp tricot (two bar tricot), Milanese, Raschel and simplex (Anbumani 2007).

Single warp tricot (one bar tricot) is made with one set of yarns and double warp tricot (two bar tricot) is made with two sets of yarn which form loops in opposite directions. Milanese is made with two or more sets of warp yarns which form loops across the fabric width in the same direction. Milanese is characterised by small diamond-shaped parallelograms which form fine rib lines diagonally through the width of the fabric on one side of the fabric. The basis of Raschel knitting is an interloop structure similar to that of a chain of slip-knots in which the single and double strands of the knot change sides in adjacent stitches (Fan et al. 2004).

FIGURE 1.10
Warp knitted fabric.

All yarn-constructed fabrics have three basic grain lines: straight, cross and bias. The straight grain in woven fabrics is the grain parallel to the warp yarns; in knitted fabrics it is parallel to the wales. Warp grain or length grain are other trade terms used for this grain. The cross grain is parallel to the weft in the case of woven and the course in the case of knitted fabric. The bias grain is parallel to the bisector of the right angle formed by the intersecting straight and cross grains. All other grain lines passing through the right angle are off-bias grains (Rana 2012).

1.4.3.3 Matted Fabrics (Felted and Nonwoven)

Felts are produced directly from fibres by matting of fibres in a sheet form. This is accomplished with heat, moisture and pressure. Many felts are isoelastic; the elongation is alike in all directions on the fabric. Such felts have no grain from this standpoint. A felt has a grain from the design viewpoint when its surface has a definite repetitive surface contour or line design or if it is not isoelastic (Geršak and Zavec Pavlini 2000).

1.4.3.4 Leather and Furs

Leather and furs have restricted sizes because they come from hides and skins. The outer surface of the hide or skin is the grain side of leather. The inner surface, the area inside the animal, is the flesh side. Flesh finished leather is leather whose flesh side has been treated in order to use the flesh side as the face side. The grain side of leather is usually used on the face side. The grain side is treated with various processes to give it the desired colour and surface values (Fletcher 2013). Natural surface structures may be enhanced or eliminated and substituted with surface markings such as the popular pebbly surface used for shoes which is known as scotch grain.

1.5 Fabric Characteristics for Apparel Manufacturing

Physical properties are generally the static physical dimensions of a fabric. The physical properties used for describing a fabric are given next.

1. Fiber or filament – type, size and length
2. Yarn – linear density, diameter, twist and number of ply
3. Weight – grams per metre or yards per pound
4. Thickness

5. Fabric structure – for woven fabrics: type of weave, count of war and weft, ends per inch (EPI), picks per inch (PPI). For knitted fabrics: type of knit, wales per inch (WPI), course per inch (CPI) and loop length

6. Non-fibrous matter – residual chemicals left over the fabric

7. Finishes – chemicals and mechanical finishes applied to the woven fabric to improve the durability, and/or utility values

8. Fabric width – the length of the filling or course

9. Colour, hue, value and intensity – hue in colour refers to the type of spectrum such as red, green, blue, yellow, etc. Value refers to the shade of spectrum such as light blue or dark blue. Intensity refers to the degree of brilliance such as bright light blue or dull light blue

10. Fabric density – weight per unit of volume

11. Surface contour – the geometric dimension of the surface plane

Six major categories of fabric characteristics that are of significance for the apparel manufacturer are

1. Style characteristics
2. Hand characteristics
3. Visual characteristics
4. Utility characteristics
5. Durability characteristics
6. Product production working characteristics

1.5.1 Style Characteristics

Style characteristics generally change, which has an effect on the emotional appeal the fabric imparts to the consumer. This is validated when a customer handles a fabric and rates the fabric with adjectives like stiff, soft, hard, etc.

1.5.2 Hand Characteristics

Hand characteristics are the transforms in the fabric surface with hand manoeuvring which apply tensile, compression and moulding forces on the fabric. The hand characteristics involve few utility characteristics. The characteristics that influence the fabric hand are

1. Thickness compressibility
2. Plane compressibility
3. Elongation

 4. Elasticity

 5. Torsion

 6. Malleability

 7. Flexibility self flex, resistance flex, maintenance flex and reflex

 8. Resilience

 9. Gravity drape, gravity sag and gravity elongation

Thickness compressibility is the degree to which a fabric thickness can be compressed with a given pressure. Plane compressibility refers to the amount to which a given area of the fabric can be compressed into the lowest volume with a given spherical compression. Elongation refers to the extent to which the fabric can be stretched without breaking the fabric. Stretch and extensibility are other terms used for this concept. Elasticity is the extent to which the fabric can be extended and still have the capacity to come back to its original length after the stretching force is removed (Dockery 2001; Banumathi and Nasira 2012).

Torsion is the degree to which the groups of fabric yarns can be displaced in the fabric plane by exerting opposing force vectors parallel to the fabric plane when the fabric is on a horizontal surface. Malleability is the degree with which a fabric can be moulded into a surface by moving more than one straight line element through space (Geršak and Zavec Pavlini 2000).

Flexibility has four dimensions: self flex, resistance flex, maintenance flex and reflex. Self flex refers to the ability of cloth to support its own weight in each of two positions. In vertical self flex, the fabric supports its weight vertically from the bottom. In horizontal self flex, the fabric supports its weight horizontally. Horizontal self flex is commonly tested with the cantilever bending method. The less the fabric can support its own weight, the greater the capacity to drape softly. A soft drape is a drape with small folds that have minute amplitudes on each fold. Resistance flex is the degree with which the fabric plane can be bent in one direction with an outside force other than the gravity of its own weight. Maintenance flex is the degree with which the fabric maintains a flexed form with outside support or force. Reflex is the ability required to turn a flexed form into another flexed form (Forza 2000).

Gravity drape has two facets: gravity sag and gravity elongation. There are two kinds of gravity elongation: straight line support and curved line support. Gravity sag usually has a different value for each of the three basic grains. It is the degree to which a given size of fabric sags and forms diagonal and horizontal ripples (or folds) when it is suspended vertically from the upper end of the fabric. Resilience is the degree to which a flexed or compressed fabric returns to its original flat plane after the flexing or compressing force is removed (Gilmore and Gomory 1961).

1.5.3 Visual Characteristics

Visual characteristics are the changes in colour values when either the fabric or light is moved. Visual characteristics can be measured in all its aspects with instruments such as the Cary or Farrand spectrophotometers used for measuring static visual values. This includes measuring colour change due to either fabric or light movement (Hassler 2003).

1.5.4 Utility Characteristics

Utility characteristics refer to the comfort, fit and wearing characteristics of a garment while the fabric experiences mechanical, thermal or chemical conditions during the usage of the garment. The transmission and transformation are the two main types in this category. A transmission characteristic transmits mass or energy through the fabric. It alters physical properties of the fabric without obliterating the fabric (Glock and Kunz 2002).

1.5.4.1 Transmission Characteristics

1. Weight
2. Thickness
3. Elongation
4. Moisture transmission
5. Radioactivity transmission
6. Water permeability

1.5.4.2 Transformation Characteristics

1. Colour fastness
2. Crease resistance
3. Crease retention
4. Crack resistance
5. Dimensional stability
6. Felting (matting)
7. Fusing
8. Mildew resistance
9. Moisture absorption
10. Moisture retention (drying)
11. Pilling
12. Scorching
13. Soiling

14. Shrinkage
15. Static electricity
16. Yarn slippage

Air permeability is the rate at which air can pass through a given area of fabric when the air is impacted with air at a given pressure. Thermal conductivity is the rate at which heat passes through a given area of fabric. Fabric weight is the product of fabric thickness times fabric density (Vinkovi 1999). Light permeability is the amount of illumination that passes through a given area of fabric. Moisture transmission is the rate with which moisture travels throughout the fabric when water contacts the fabric without an impact force, gravity or otherwise. This is a measure of the fabric's ability to diffuse moisture. Water permeability is the rate with which water seeps through fabric when water contacts the fabric with angular impact force. Radioactivity transmission is the degree with which radioactive energy, such as X-ray and gamma rays, can penetrate fabrics. This is important only for special types of clothing (Nordås 2004).

Colour fastness is the capability of a fabric to maintain its original value under given conditions. Wet crocking is colour transference by a wet fabric; dry crocking is colour transference by a dry fabric. Dimensional stability is the capacity with which fabric can resist changes in physical dimensions. Fusing is the degree with which the fabric yarns or fibres melt and weld together. This characteristic applies only to thermoplastic finishes or synthetic or manufactured fibres. Pilling is the degree with which the fabric fibres are formed into minute fibre balls on the fabric surface during wear. Soiling is the degree with which a fabric gathers and retains gases, liquids or solids in a manner that changes the original colour, odour or weight of the fabric (Geršak 1996).

1.5.5 Durability Characteristics

Durability characteristics are the ability of a fabric to retain the utility and style characteristics during wear. It is an indirect measure of stress, which destroys the fabric or its capability to retain the required style or utility characteristics. The durability characteristics are

1. Abrasive strength
2. Bursting strength
3. Corrosive strength
4. Dry cleaning durability
5. Fire resistance
6. Launder ability
7. Moth resistance

8. Radiation absorption strength
9. Tearing strength
10. Tensile strength
11. Yarn severance

Abrasive strength is the measure of rubbing action necessary to disintegrate the fabric. Bursting strength is the measure of vertical pressure, against a fabric area secured in space, necessary to rupture the fabric. Corrosive strength is the measure of chemical action, acid or alkaline, necessary to disintegrate a fabric. Dry cleaning durability is the measure of dry cleaning performance that disintegrates the fabric (Nordås 2004). Fire resistance has two parameters – the ignition point and the rate with which the fabric burns. The flame size is an important dimension in each of these two parameters.

Launder ability is the measure of washings that disintegrate the fabric. Laundering is an integration of mechanical, thermal and chemical action. Moth resistance is the extent to which a fabric is disintegrated by moths and larvae. Radiation absorption strength is the rate with which radiation energy either disintegrates a fabric or destroys the utility characteristic (Jacob 1988; Sumathi 2002).

Tearing strength is the measure of torque necessary to part the yarns perpendicular to the torque vector plane. Tensile strength is the measure of a straight line pull, parallel to the fabric plane, that is needed to part the yarns receiving the pull stress. The tensile strength reading on the testing machine varies with rate of stress (Ulrich and Eppinger 2004).

1.5.6 Garment Production Working Characteristics

Garment production working characteristics affect the quality of product as well as cost of production. An example of this is the difficulty entailed in sewing some fabrics with certain types of ornamentation. Some working characteristics, such as seam strength, are measured by durability limits (Mehta 1992; Kumar and Phrommathed 2005; Thomassey and Happiette 2007). They are classified as working characteristics because the characteristics are either the reaction to, or the vital part of, an apparel production process. The working characteristics of a fabric are

1. Coefficient of friction (cutting, sewing, pressing and packaging)
2. Sewed seam strength
3. Sewed seam slippage (yarn slippage)
4. Sewing distortions
5. Yarn severage
6. Bondability strength (fused, cemented and heat-sealed seams)

7. Die mouldability
8. Pressing mouldability

The first five characteristics stated above are vital parameters for evaluating fabric sewability potential.

1.6 Fabric Inspection Systems

After the receipt of a fabric roll to the garment industry, it must be inspected to evaluate its tolerability from quality point of view or else additional cost in garment manufacturing may be incurred (Dockery 2001; Banumathi and Nasira 2012). It is normally carried out on fabric inspection machines, which are designed such that fabric rolls are mounted at the back side of the inspection table (Mehta and Bhardwaj 1998). As the fabric is moving at a slow speed and at an angle, the fabric inspector has a better view of the fabric and could identify faults easily. These machines are normally power driven or the operator has to pull the fabric over the inspection table. The fabric faults are identified, labelled and recorded in a fabric inspection form. The various fabric inspection systems used in the garment industry are given here.

1. Four-point system
2. Ten-point system
3. Graniteville "78" system
4. Dallas system
5. Textile Distributors Institute system (National Federation of Textile 1995)

1.6.1 Four-Point System

The four-point assessment method is a commonly established method of fabric inspection globally. Fabric inspection is done as per ASTM D 5430–04 standard and this system is agreed by The American Society for Quality Control, Textile and Needle Trades Division, The American Apparel Manufacturers Association and is used by the United States Government for all of their piece goods purchased (Mathews 1986; Mehta and Bhardwaj 1998; Rana 2012). The main considerations in a four-point inspection system are given here:

- The fabric has to be passed longitudinally through the inspection area at a speed approved by the customer.
- The light source should be perpendicular to the fabric surface and the fabric should run at an angle of 45° to the vertical for better vision for the operator.

- The illumination intensity in the inspection room should have a minimum of 1075 lux and the light source used should be white fluorescent lamps.
- The fabric should be checked at a distance of 1 m from the fabric inspector when it is in motion.
- Defect points should be assigned based on the length of the defect as mentioned in Table 1.1.
- Four points should be assigned to each metre of fabric where usable width is lower than the minimum specified.
- The fabric should not be penalised more than four points.
- Defects not obvious on the face side of the fabric should not be registered unless agreed between supplier and customer.

Total defect points per 100 square yards of fabric should be determined and the criterion for the acceptance of a fabric roll is generally not more than 40 penalty points. If it is more than 40 points, it will be considered 'seconds'. The formula to determine the penalty points per 100 square yards is given by

$$= \frac{\text{Total points scored in the roll} \times 3600}{\text{Fabric width in inches} \times \text{Total yards inspected}}$$

Example:
A fabric roll 160 yards long and 47″ wide contains the following defects as shown in Table 1.2.

Advantages

- Four-point system has no width limitation.
- Worker can easily understand it.

TABLE 1.1

Assignment of Penalty Points in Four-Point System

Size of the Defect	Penalty Points
Length of defects in fabric (either length or width)	
Defects up to 3″	1
Defects >3″ ≤6″	2
Defects >6″ ≤9″	3
Defects >9″	4
Holes and openings (largest dimension)	
1″ or less	2
Over 1″	4

TABLE 1.2

Example of Defective Points in Fabric

4 defects up to 3″ length	5×1	5 points
3 defects from 3 to 6″ length	2×2	4 points
2 defects from 6 to 9″ length	4×3	12 points
1 defect over 9″ length	2×4	8 points
1 hole over 1″	1×4	4 points
Total defect points		33 points
Therefore, points/100 sq. yards	$= (33 \times 3600)/(160 \times 47) = 15.79$ points	

1.6.2 Ten-Point System

The ten-point inspection system for fabric evaluation was permitted by the Textile Distributors Institute and the National Federation of Textile in 1955. It is designed to categorise the defects and to assign each defect a numerical value based on severity of defect. The system allots penalty points to each defect based on its length and whether it is in the warp or weft direction (Mehta et al. 1998; Rana 2012). Table 1.3 shows the assignment of penalty points in a ten-point system.

According to this system, the fabric roll is considered good if the total penalty points, assessed to that roll, do not exceed the length of the fabric. If the points exceed the length of fabric in a roll, then it is considered 'seconds' and may be rejected. Suppose if the fabric roll having a length of 50 yards is inspected in a ten-point system and the total penalty points are less than 50. Then the fabric roll was considered good (Mehta et al. 1998; Dockery 2001; Banumathi and Nasira 2012).

Advantages

- Oldest and most used in woven finished fabric.
- In it length of fabric is used and along the length of warp and weft defects are identified.

TABLE 1.3

Assignment of Points in Ten-Point System

Warp Defects	Points	Weft Defects	Points
Under 1″	1	Under 1″	1
1–5″	3	1–5″	3
5–10″	5	5″–1/2 width of goods	5
10–36″	10	Over 1/2 the width of goods	10

Disadvantages

- It has width limitation.
- It is difficult in practical use.

1.6.3 Graniteville "78" System

It was introduced in 1975 for the field of fabric grading. In this system, the fabric defects are categorised as major defects if they are obvious in the fabric and leads to second quality and minor defects if the severity of the fault is minor and does not lead to second quality (Mehta 1992). The assignment of penalty points in this system is shown in Table 1.4.

This system was basically established for garment cutting components, in which the short length faults less than 9″ would normally be removed. The system aims to balance the significance of longer defects (over 9″) and place less weight on 1–10″ faults such as slubs. The system recommends the viewing distance of 9′ instead of the normal 3′ distance.

Disadvantages

- As this system is used on cutting pieces, according to my point of view it also increases the cost of production. We should control problems before cutting.

1.6.4 Dallas System

The Dallas system was introduced in the 1970s and it was developed particularly for knitted fabrics. According to this inspection method, if any fault was observed on a finished garment, then the garment would be called 'seconds'. It describes the seconds as 'more than one defect per ten linear yards, determined to the nearest ten yards'. For example, one piece 60 yards long would be allowed to have six defects.

Disadvantage

- It increases the cost of production as defect is located after the garment is finished.

TABLE 1.4

Assignment of Points in Graniteville "78" System

Defect Length	Penalty Points
9″	1
9–18″	2
18–27″	3
27–36″	4

References

Abernathy, F.H., A. Volpe and D. Weil. 2004. *The Apparel and Textile Industries after 2005: Prospects and Choices.* Harvard Center for Textile and Apparel Research, Cambridge.

Anbumani, N. 2007. *Knitting-Fundamentals, Machines, Structures and Developments.* New Age International (P) Ltd, New Delhi.

Ashdown, S.P. 1998. An investigation of the structure of sizing systems—A comparison of three multidimensional optimized sizing systems generated from anthropometric data with the ASTM Standard D5585–94. *International Journal of Clothing Science and Technology* 10(5):324–34.

Bailetti, A. and P. Litva. 1995. Integrating customer requirements into product designs. *Journal of Product Innovation Management* 12(1):3–15.

Banumathi, P.N. and G.M. Nasira. 2012. Fabric inspection system using artificial neural networks. *International Journal of Computer Engineering Science* 2:20–7.

Beazley, L. and T. Bond. 2006. *Computer Aided Pattern Design and Product Development.* Blackwell Publishing, UK.

Bheda, R., A. Narag and M. Singla. 2003. Apparel manufacturing: A strategy for productivity improvement. *Journal of Fashion Marketing and Management* 7:12–22.

Carr, H. and B. Latham. 2006. *The Technology of Clothing Manufacture.* Blackwell Science, Oxford.

Chuter, A.J. 1995. *Introduction to Clothing Production Management.* Blackwell Scientific Publications, Oxford, UK.

David Rigby Associates. 2002. *Fiber Selection in Garment Markets Global Fibers and Feedstocks Report.* CMN 11:10–14

Dockery, A. 2001. *Automated Fabric Inspection: Assessing the Current State of the Art.* http://techexchangecom/thelibrary/FabricScanninghtml (accessed on November 21, 2015).

Fairhurst, C. 2008. *Advances in Apparel Production.* Woodhead Publication, Cambridge.

Fan, J., W. Yu and L. Hunter. 2004. *Clothing Appearance and Fit: Science and Technology.* Woodhead Publishing Limited, Cambridge.

Fletcher, K. 2013. *Sustainable Fashion and Textiles: Design Journeys.* Earthscan, London.

Forza, C. and A. Vinelli. 2000. Time compression in production and distribution within the textile-apparel chain. *Integrated Manufacturing System* 11:138–46.

Geršak, J. 1996. *Fabric Quality Requirements—Costs or Savings (in Slovenian).* Book of Proceedings of the Garment Engineering '96 Maribor, University of Maribor, Faculty of Mechanical Engineering, 37–46.

Geršak, J. and C.D. Zavec Pavlini. 2000. Creating a knowledge basis for investigating fabric behaviour in garment manufacturing processes. *Annals of DAAAM for 2000 and Proceedings of the 11th International DAAAM symposium 'Intelligent Manufacturing and Automation Vienna DAAAM International,* 155–6.

Gilmore, P.C. and R.E. Gomory. 1961. A linear programming approach to the cutting stock problem. *Operation Research* 9:349–59.

Glock, R.E. and G.I. Kunz. 2002. *Apparel Manufacturing—Sewn Product Analysis.* Prentice-Hall, Englewood Cliffs, NJ.

Gupta, D. 2011. Functional clothing: Definition and classification. *Indian Journal Fiber and Textile Research* 35:321–6.

Hassler, M. 2003. The global clothing production system: Commodity chains and business networks. *Global Networks* 3:513–31.

Hollander, A.L. 1993. *Seeing through Clothes.* University of California Press, Berkeley, CA.

Jacob, S. 1988. *Apparel Manufacturing Handbook—Analysis Principles and Practice.* Columbia Boblin Media Corp, USA.

Kumar, S. and P. Phrommathed. 2005. *New Product Development—An Empirical Study of the Effect of Innovation Strategy Organisation Learning and Market Conditions.* Springer, New York, 55–7.

Laing, R.M and J. Webster. 1998. *Stitches and Seams.* Textile Institute Publications, UK.

Mathews, M. 1986. *Practical Clothing Construction—Part 1 and 2.* Cosmic Press, Chennai.

Mehta, P.V. 1992. *An Introduction to Quality Control for Apparel Industry.* CRC Press, Boca Raton, FL.

Mehta, P.V. and S.K. Bhardwaj. 1998. *Managing Quality in the Apparel Industry.* New Age International, New Delhi, India.

Nordås, H.K. 2004. *The Global Textile and Clothing Industry Post the Agreement on Textiles and Clothing.* World Trade Organization Geneva, Switzerland.

Rana, N. 2012. Fabric inspection systems for apparel industry. *Indian Textile Journal,* http://wwwindiantextilejournalcom/articles/FAdetailsasp?id1/44664 (accessed on December 18, 2014).

Sumathi, G.J. 2002. *Elements of Fashion and Apparel Designing.* New Age International Publication, New Delhi, India.

Thomassey, S. and M. Happiette. 2007. A neural clustering and classification system for sales forecasting of new apparel items. *Applying Soft Computing* 7:1177–87.

Tyler, D.J. 1992. *Materials Management in Clothing Production.* Blackwell Scientific Publications, Oxford, UK.

Ulrich, K.T. and S.D. Eppinger. 2004. *Product Design and Development.* Third Edition. Prentice-Hall, Englewood Cliffs, NJ.

Vinkovi, C. M. 1999. *Garment Design I (in Croatian) First Book.* Zagreb Faculty of Textile Technology, University of Zagreb, Croatia.

2

Pattern Making

2.1 Body Measurement

Body measurement plays a vital role in better fitting of garments to the human body. Hence, it is crucial for a designer or dressmaker to have better knowledge of body anatomy as well as the correct procedure for taking body measurements. For efficient fabric utilisation in the cutting room, the designer should know the size and shape of the body for which the designing has to be done, and it begins with the eight head theory (Adu-Boakye et al. 2012).

2.1.1 Body Anatomy

The father of tailoring, Mr. Wampon, drew seven imaginary lines across the body structure for the purpose of easy measurement in tailoring. The complete body structure was lengthwise divided into eight equal parts, which is known as the eight head theory. This theory has become the foundation of all fashion drawing and for understanding the body shape and structure for fashion (Adu-Boakye et al. 2012).

2.1.1.1 Eight Head Theory

According to this theory, the normal body structure is considered 5'4" height. This body is divided into 8 parts in which each part is 8" in length. A development of this theory is the ten head theory, which is used for all fashion drawing. This figure is referred to as fashion model figure. In this system, the body structure is divided into ten equal parts or heads (Jacob 1988; Adu-Boakye et al. 2012). The bottom part of the body is longer compared to the eight head figure. The ideal height for this theory is taken to be 5'8". The division of body structure in the eight head principle is shown in Figure 2.1.

1. *Hair to chin*: The garments are generally worn on the body through the head and hence hat or cap head measurements should be taken. The right place on the neck is the chin itself. Yoke measurements are

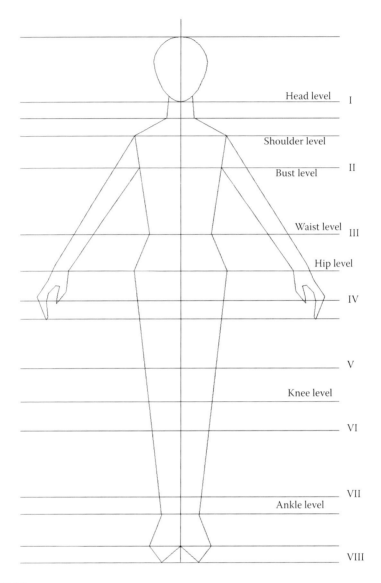

FIGURE 2.1
Eight head theory.

taken 1″ below the chin. The head is considered the first portion of the human body and the chin is considered a first imaginary line.

2. *Chin to nipple*: The upper body garments are prepared according to the size of the chest only. This is the second part and the second imaginary line passes through the nipples and the armscye. This line denotes the bust level.

3. *Nipple to navel*: The next imaginary line passes through the navel; shoulder to waist is measured up to the navel points. This level is the waist level. But for proper garment fitting, the waist measurement for ladies is taken 1″ above the waist level and for men 1/2″ below the line.

4. *Navel to pubic organs*: This part is most important for lower body garments. The lower body garments are cut based on the hip size. The hip level is usually 3–4″ above this imaginary line. This is the most heavy or fat part of the body. This is also as important as the chest measurements.

5. *Pubic organ to mid thigh*: This part is important mainly for arm measurements. The fingertips normally end near about this line. The length of the arms is measured as 3 heads.

6. *Mid thigh to small*: The part below the knee is known as small. The knee level is about 2–3″ above this imaginary line. Length of gowns is taken around this head.

7. *Small to ankle*: This head is important for full length garments like trousers. These garments usually end here. The calf level is above this head. House coats, nightgowns, etc. end at the calf level.

8. *Ankle to feet*: The eight heads are imagined on assuming a person standing on the toes. This is the last head and it comprises only the feet. This is necessary for tight fitting leggings and floor-length garments like evening gowns.

Advantages of the Eight Head Theories

1. By the knowledge of eight head theory, the observation of the body structure becomes easy.

2. It will facilitate drafting and fitting. If there happens to be any fault, then it shall be detected and rectified.

3. Knowledge of body structure shall be helpful in taking correct measurement and this will result in correct cutting and the garment shall be stitched properly.

4. Work shall be easily and speedily executed.

2.1.1.2 Ten Head Theory

The ten head figure is considered the fashion figure. This is mostly used for fashion drawing and designing. The division of body structure in the ten head principle is shown in Figure 2.2.

1. *Head to chin*: Like the eight head figure, the head is the first part. The first imaginary line is at the level of the chin.

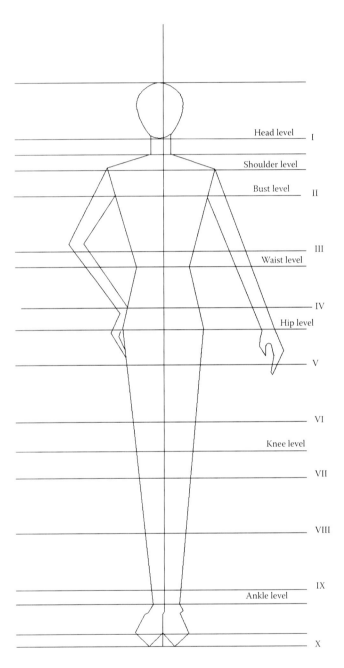

FIGURE 2.2
Ten head theory.

2. *Chin to bust*: This is the second and the most significant part of the figure. The second imaginary line is across the bust and the armscye. The shoulder level is in between this part, which is wider than the eight head figure.

3. *Waist level*: In the ten head figure, the waist level is about 2–2$^{1/2}$", below the third imaginary line.

4. *The hip level*: The hip level is also 2–3" below the fourth imaginary line. This is also the end of the torso level. The torso of the ten head figure is longer than the eight head figure. This is also a very important level for fashion figures.

5. *The end of pubic organs*: This is the position of the fifth imaginary line. The hand usually ends just below this line.

6. *Knee level*: The knee level is in between the sixth and seventh imaginary line. The sixth line signifies the end of the thigh whereas the seventh line is at the level of the small.

7. *Calf level*: The eighth line signifies the calf level. The lengths of leg are longer in the ten head figure compared to the eight head figure.

8. *Ankle level*: The ankles are at the ninth imaginary line.

9. *Feet*: The last parts of this figure are the feet. Like the eight head figure, this figure too is assumed to be standing on its toes.

2.1.2 Body Measurement

The following points have to be taken into account while taking body measurements.

1. Prior to taking the body measurements, it is advisable to understand the customer's requirements, concerning the shape, fit, and style of the garment.

2. It is important to study the human anatomy carefully and if any variation in body proportion is noticed, it has to be recorded and should be taken into account while taking measurements and pattern making.

3. While taking the measurements, the person should stand straight in front of a mirror.

4. Body measurements should be taken with tape, without keeping it too tight or loose with the body.

5. The measurements should be taken in the appropriate order and with a definite sequence.

6. All girth measurements should be taken tightly, since ease allowance is incorporated in the draft.

7. After taking all the measurements, they should be rechecked twice.

2.1.2.1 Taking Body Measurement

Bodice measurements. The various bodice measurements are shown in Figure 2.3

1. *Bust*: Measurement has to be taken about the fullest part of the chest/bust by raising the measuring tape to a level slightly below the shoulder blades at the back.

2. *Waist*: Measurement has to be taken tightly around the waist with the tape straight.

3. *Neck*: Measurement has to be taken around the neck, by keeping the tape slightly above the collar front and along the base of the neck at the back.

4. *Shoulder*: Measurement has to be taken from the neck joint to the arm joint along the middle of the shoulder (A to B in Figure 2.3).

5. *Front waist length*: Measurement has to be taken down from the high point shoulder (HPS) to waist line through the fullest part of the bust (A to C Figure 2.3).

6. *Shoulder to bust*: Measurement has to be taken down from the HPS to the tip of the bust (A to D in Figure 2.3).

7. *Separation of bust points*: Measurement has to be taken between the two bust/chest points (D to E Figure 2.3).

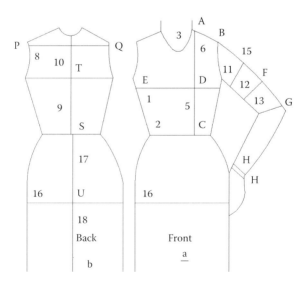

FIGURE 2.3
Body measurement.

8. *Across back measurement*: Measurement has to be taken across the back between armholes about 3″ below the base of the neck (P to Q in Figure 2.3).

9. *Back waist length*: Size has to be measured from the base of the neck at the centre back position to the waistline (R to S in the Figure 2.3).

10. *Armscye depth*: Measurement has to be taken from the base of the neck at the centre of the back to a point directly below it and in level with the bottom of the arm where it joins the body (R to T in Figure 2.3).

Sleeve measurements

11. *Upper arm circumference*: Measurement has to be taken around the fullest part of the arm.

12. *Lower arm*: For the lower arm, measurement has to be taken around the arm at the desired level corresponding to the lower edge of the sleeve.

13. *Elbow circumference*: Measurement has to be taken around the arm at the elbow.

14. *Wrist*: Measurement has to be taken around the wrist.

15. *Sleeve length*: For short sleeves, the length has to be measured from point B to F. For elbow length sleeve, measurement has to be taken from the top of the arm to the elbow point (B to G in Figure 2.3). For full length, the elbow has to bend slightly and measurement has to be taken down from the top of the arm to the back of the wrist passing the tape over the elbow point (B to H in Figure 2.3).

Skirt measurements

16. *Waist*: Measurement has to be taken tightly around the waist with the tape in a horizontal manner and parallel to the floor.

17. *Hip*: Measurement has to be taken around the fullest part of the hip horizontally (7–9″ from waist approximately).

18. *Waist to hip*: Measurement has to be taken from the waist at the centre of the back to the fullest part of the hip (S to U in Figure 2.3).

19. *Skirt length*: Measurement has to be taken at the centre of the back from the waist to length of the skirt as required (S to V in Figure 2.3). The calculation of other measurements using chest circumference is shown in Table 2.1.

After taking the measurements, compare it with the sample measurements for women, men and children garments as given in Tables 2.2 through 2.4, respectively.

TABLE 2.1

Determination of Other Dimensions from the Chest Circumference

Measurements	Women	Men
Waist	Chest – (4–5″)	Chest – (5–7″)
Hip	Chest + (1–2″)	Chest + (2–4″)
Shoulder (half)	1/4 Chest – 1/2″	1/6 Chest + (1–2″)
Armscye depth	1/8 Chest + 1″	1/8 Chest + (2–2½″)
Neck	1/3 Chest + (2–3″)	1/3 Chest + (2–2½″)

TABLE 2.2

Sample Measurements for Ladies' Garments

Bust (cir)	28	30	32	34	36	38	40	42
Waist (cir)	24	24½	25	26	28	30	32	33
Hip (cir)	30	32	34	36	38	40	42	44
Back width	14	14	14½	15	15½	15½	15¾	16
Army scye depth	6¾	7	7¼	7½	7¾	8	8¼	8½
Lower arm (cir)	9½	9¾	10	10¼	10½	11	11¼	11½
Wrist (cir)	6¼	6½	6½	6¾	6¾	7	7¼	7½
Back waist length[a]	13–16							
Shoulder to bust[a]	7½–9½							
Full sleeve length[a]	20–23							
Short sleeve length[a]	7–10							
Waist to hip[a]	7–9							
Waist to ground[a]	38–44							
Choli length[a]	12–14							
Pant top length[a]	18–23							
Kurta length[a]	38–42							
Maxi dress length	52–56							
Maxi skirt length[a]	38–44							
Middy skirt length	24–28							

Notes: All measurements are inches; cir – circumference.
[a] All the dimensions vary based on the height of the person.

2.2 Patterns

A basic or foundation pattern can be created by any of the two methods, namely, by drafting or by draping fabric on a model. Pattern drafting is defined as a technique or method of drawing patterns on brown paper with accuracy and precision, based on the body measurements or standard measurement chart. This is an efficient and economical method and can be manipulated to create the pattern for different styles by a technique known as flat pattern designing (Aldrich 2002, 2004).

TABLE 2.3

Sample Measurements for Boys' Garments

Age (in Years)	2	4	6	8	10	12	14	16
Chest	19	21	23	24	26	28	30	32
Waist	19	21	22	23	24	26	28	30
Hip (seat)	20	22	24¼	26	28	30	32	34
Neck	10	10½	11	11½	12	13	14	16
Back width	8½	9½	11	12	12½	13	14	15
Back waist length	8½	9½	10½	11½	12½	13	14	14
Short sleeve length	4	5	5	6	7	8	9	10
Full sleeve length	11	14	17	19	20	21	22	23
Cuff length	6	6½	7	7¼	7½	8	8¼	8½
Shirt length	13	15	17	18	20	22	25	28
Pant length	20	23	26	29	32	35	38	40
Half pant length	9	10	10½	11½	12	12½	13½	14½

TABLE 2.4

Sample Measurements for Children's Garments

Age (in Years)	1	2	3	4	5	6	8	10	12	14
Chest	18	19	20	21	22	23	24	25	26	28
Waist	18	19	20	21	22	22	23	23½	24	24
Hip	19	20	21	22	23	24	26	27	28	30
Back width	8	8½	9	9½	10	10½	11	12	13	14
Back waist length	8	8½	9	9½	9¾	10	10½	11	12	13
Armscye depth	4½	4¾	5	5¼	5½	5¾	6	6¼	6½	6¾
Short sleeve length	3½	4	4½	5	5½	5¾	6	6¾	7	7½
Lower arm	7	7½	8	8¼	8½	8¾	9	9¼	9½	9½
Full sleeve length	9	11	13	14	15	16	17	19	20	21
Wrist	5	5¼	5¼	5½	5½	5¾	6	6	6¼	6¼
Waist to hip	3	3½	4	4½	5	5¼	5½	6	6½	7
Waist to ankle	16	18	20	22	24	26	30	33	36	38
Maxi skirt length	16	18	20	22	24	26	30	34	37	39
Short skirt length	8	9	10	11	12	13	14	16	18	20
Frock length	15	17	18	20	22	23	24	27	30	33
Blouse length	10	10½	11	11½	12	13½	14	15	16	17

Note: All measurements are inches.

2.2.1 Types of Paper Pattern

1. *Standardised paper pattern*: Paper patterns prepared using standardised body measurements are called standardised paper patterns. This method is followed in training and tailoring schools.

2. *Individual paper pattern*: The measurement of a particular person is taken and a pattern is prepared using these individual measurements. The pattern prepared for a particular person will not suit another person. These are usually done at home and some tailor shops.

3. *Final paper patterns*: Once the individual is satisfied with the paper patterns, they are made into final paper patterns. Though, while making individual patterns all the precautions are taken, yet, there could be some minor points, which are to be considered. These minor details are corrected and finally made into permanent patterns.

4. *Block paper pattern*: Normally these are made with standard sizes with thick cardboard. These are mostly used in the garment industry. The garment made out of these block patterns will fit those who have measurements equal to that of the standardised body measurement.

5. *Readymade patterns*: These are made using a unique type of tracing paper. These can be procured from the market and are more useful for people who can do stitching, but not drafting. These can be bought readymade and can be easily used by placing on the material and cutting and stitching accordingly.

6. *Graded paper pattern*: Patterns of five consecutive sizes (e.g. 30", 32", 34", 35" and 38" chest size) are marked in one single pattern. The required size according to the individual body measurement is traced separately, cut and used.

7. *Commercial paper pattern*: The paper patterns for different designs are available in readymade forms. These patterns are called commercial patterns. These patterns are enclosed in an envelope along with an instruction sheet. The instruction sheet will provide information about selection of fabric, preparation of fabric, marking, cutting, and steps for sewing. The front side of the envelope contains the front view, side view and back view of the garment design along with the body measurements.

2.2.2 Pattern Making Tools

The tools required for pattern making are given below, based on the order of their usage (Figure 2.4).

1. Measuring devices
2. Drafting devices
3. Marking devices
4. Cutting devices
5. Sewing devices
6. Finishing or pressing devices
7. Miscellaneous or general tools

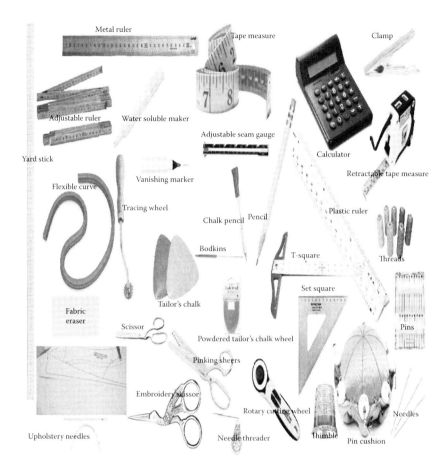

FIGURE 2.4
Pattern making tools.

2.2.2.1 Measuring Devices

Measuring tools are the most essential things in making a pattern. The key to success in garment construction lies in taking accurate measurements of the subject and by using the appropriate tool for pattern making (Aldrich 2002, 2004, 2009; Beazley and Bond 2004).

1. *Measuring tape*: It is indispensable for taking body measurements. It is 152 cm or 60″ long with measurements on both sides. Its one end is made of metal having 3″ length and the other is made of the same metal having 1/2″ length. The side with the 1/2″ length is used for measuring a circular area, while the side with 3″ length is used for a vertical area.

2. *CPG measuring tape*: This is used for taking measurements for a coat. Three measurements, that is, chest, shoulder and depth of side can be taken at a time. Apart from these, over shoulder and under shoulder measurements can be recorded with this tape.

3. *Leg measuring tape*: It is a tape used for measuring the inner part of the leg. It is made of wood in the shape of a crescent and a measuring tape is fixed at the centre of the circle. The circle is entrapped with the leg to measure the inner portion of the leg by tape.

4. *Measuring stand*: This stand is used to measure long garments such as long overcoats, frocks or gowns, as well as flare of the garment. In this stand, a rod of aluminium is fitted on the stand. The rod has a graduated scale, which gives the vertical measurement.

5. *Ruler*: It is the best device for taking long straight measurements. It is mostly used for checking grain lines and marking hems.

2.2.2.2 Drafting Devices

Drafting equipment is used for making paper patterns. This is the second stage of pattern making. Using the measurements taken, the drafting is carried out according to the design of a garment. The following drafting equipment is used:

1. *L-scale*: It is called a triscale or L-scale and is made of wood or steel. The L-scale has one arm, which measures 12″ and the other arm is 24″. It is used for drafting on brown paper to draw perpendicular lines.

2. *Leg shaper*: It is made of wood or plastic. Either 24″ or 36″ lengths are available. It is used to measure and shape the interior part of the leg.

3. *Tailor's art curve*: It is made of plastic or wood. This is used to draw curves in the drafting.

4. *French curve*: This is made of transparent plastic. It helps in marking shapes of the neck, depth of sides and bottom of the garments.

5. *Compass*: It is mainly used for making a curve for umbrella cloth.

6. *Drafting table*: It is a wooden table of 3′ height, 4′ wide and 6′ long. The surface should be smooth and firm.

7. *Milton cloth*: It is a thick, blue coloured woolen cloth used for drafting. It is mostly used for practising drafts by students. The surface can be brushed and reused until correct drafting is obtained.

8. *Brush*: A brush is used on Milton cloth to rub the mistakes while drafting.

9. *Brown paper*: It is used for drafting by placing on the drafting table. While using brown paper, a grain line should be followed.

10. *Pencil*: Pencil is used for marking on brown paper.

11. *Rubber*: It is frequently used for deleting mistakes. Good quality rubber, which does not leave black lines, should be selected.

12. *Red and blue pencils*: These are used for marking on fold (red line) and grain line (blue arrow).

2.2.2.3 Marking Devices

Marking devices are used for transferring the details of the paper draft to the fabric.

1. *Tailor's chalk*: It is made of china clay and is available in different colours. It is used for marking the paper patterns on the cloth. Alterations and construction markings are drawn using tailors' chalk.

2. *Chalk in pencil form*: This is used like a pencil and is ideal for marking thin accurate lines. This is used for marking pleats, darts and buttonholes.

3. *Tracing wheel*: It is used for transferring the pattern markings on fabrics. But for sheer fabrics and loosely woven fabrics, the tracing wheel should be used with care; otherwise, the fabric may get damaged.

4. *Dressmaker's carbon paper*: Carbon papers are mostly used for transferring patterns. In embroidery, they are used for tracing designs. They are available in several colours including white.

2.2.2.4 Cutting Devices

Cutting devices/equipment should be selected and used with maximum accuracy. A slight change in cut results in huge fitting problems. These tools must be selected and maintained properly in order to use them effectively.

1. *Cutting table and cutting board*: A cutting table is 6′ long, 4′ wide and 3′ height. People working in a standing position use the table and people who work sitting use a cutting board. A cutting board should be 6″ height.

2. *Shears*: These are typically utilised for cutting thick materials and usually 10–15″ in length.

3. *Scissors*: These are used for cutting ladies' and children's garments. They are 7–10″ in length.

4. *Paper cutting scissors*: These are small scissors available in various sizes and meant for cutting paper.

5. *Pinking shears*: This cuts the edges in a zigzag manner. It is used for finishing seams and raw edges. It gives a decorative appeal to the raw edges while at the same time avoids unravelling of yarns.

6. *Trimming scissors*: These are used for carrying out alterations, trimming seams, repairs and cutting thread while sewing.

7. *Buttonhole scissors*: These are used for making holes for buttons and eyelet holes in garments.

2.2.2.5 Sewing Devices

Sewing can be carried out either manually or by a machine. For hand sewing, the following are required:

1. *Needles*: These needles come in denominations of a 0 to 12 numbers. Based on the thickness of cloth, the needle number is used.

2. *Crewel needle or darn needle*: This is used for darning. The front side of the needle is bent.

3. *Pins*: Pins are used for fixing the pattern on the cloth. They come in different colours.

4. *Pin cushion*: It is used for keeping pins together.

5. *Needle threader*: This helps in threading the machine and hand needles.

6. *Thimble*: This is a cover that protects the finger while hand sewing. It is available in various sizes and is made of plastic or steel.

7. *Seam ripper*: It has a sharp curved edge for opening and cutting seams. It can also be used for slashing machine work buttonholes.

2.2.2.6 Finishing or Pressing Devices

The following equipment are needed for pressing:

1. *Iron*: A good brand with after sale services should be chosen. A steam iron with a thermostat regulator is preferred.

2. *Ironing board*: For ironing clothes, a table or ironing board can be used. An ironing board is 36″ long and 12″ wide. Six inches are left on its right side to keep the iron box. The left side of the board is angular and is suitable for ironing dart edges and sleeve darts while stitching. The table or ironing board should have proper stuffed backing.

3. *Sleeve board*: It is in the shape of a sleeve. This board is 30″ long and 3/4″ thick.

2.2.2.7 Miscellaneous or General Tools

Often, a few more tools and equipment may be required other than the above-mentioned items, in making the pattern and constructing the garment. These can be termed miscellaneous tools.

1. *Sponge*: While pressing, a sponge is used to wet the fabric pieces to smooth the surfaces.

2. *Water container*: A container with water, which will accommodate the sponge, should be selected. While ironing, water is sprinkled to remove wrinkles.

3. *Damp cloth*: If a steam iron is not available, a damp cloth can be used. Any rectangular absorbable cloth can be chosen for this purpose.

4. *Hole maker*: It is a sharp-edged instrument with a handle. This is used to make buttonholes.

5. *Orange stick*: This is a long tool with a pointed edge. This is inserted into the collars or seams to get pointed edges.

2.2.3 Principles of Pattern Drafting

Pattern drafting can be carried out on an ordinary brown sheet paper which is not too thin. To achieve an accurate and precise pattern draft, use of appropriate tools should be practiced, for example, for drawing a straight line a sharp pencil and a ruler have to be used and to draw right angle lines, an 'L' square or set square can be utilised. Prior to pattern drafting, it is essential to know the procedures and instructions (Beazley and Bond 2004; ISO 3635 1981; ISO 3636 1977; ISO 3637 1977). The basic principles of pattern drafting are given below.

1. Patterns must be created larger than actual body measurements to permit free body movements, ease of action and comfort in wearing. Normally used ease allowance for various parts of the body are as follows.

   a. Bust – 3–5" (3" for a tight fitting garment and 5" for loose fitting one).

   b. Waist – 1/2".

   c. Hips – 3–5".

   d. Upper arm – 3–4".

   e. Arm hole depth – 1".

   The ease allowance must be incorporated in the pattern drafting before cutting out the pattern.

2. For a symmetric garment (the right and left sides of the garment panels are similar), the paper pattern could be made only for half front and half back. But for the sleeve part, a full pattern must be made.

3. It is better to draft the basic pattern blocks such as plain bodice, plain sleeve, and plain skirt without including seam allowances. However, while marker planning or keeping the patterns directly on the fabric for cutting, adequate seam allowances have to be ensured between

the patterns before cutting. Otherwise, to avoid the risk of cutting without seam allowance, it is better to add seam allowances in the paper pattern itself after completing the draft.

4. The following construction detailed information should be recorded and marked clearly on the pattern after drafting to aid in further processes.

   a. Identification mark of every pattern piece by its name (bodice front, bodice back, sleeve, etc.).

   b. Number of pattern pieces to be cut with each pattern piece.

   c. If seam allowances are not included in the draft, this should be pointed out in the pattern. If it is included, then seam and cutting lines should be clearly drawn on the pattern.

   d. Length grain line should be marked in a different colour pencil on every pattern piece.

   e. Notches should be provided for easy matching of components while sewing.

   f. Centre front (CF) as well as centre back (CB) lines should be marked in the block pattern.

   g. Fold lines in the pattern should be clearly marked and should be visible to show the location where the material should be folded.

   h. Dart and pleat markings, etc. should also be marked clearly on the pattern.

2.2.3.1 Advantages of Paper Pattern

1. A better pattern of the appropriate size manipulated to individual requirements results in a better fit.

2. A pattern made in a thick paper or cardboard shall be maintained for a longer period of time and can be reused several times.

3. By modifying the basic pattern pieces using the flat pattern technique, it is feasible to make patterns for intricate and original designs.

4. A paper pattern of a specific size can be used to produce patterns of other sizes by means of a grading process.

5. The errors that occur during pattern drafting can be corrected in the pattern itself.

6. Patterns can be changed/modified according to the latest fashion trend.

2.2.4 Commercial Pattern

These are generally made on tissue paper as it permits compact packing of many pattern pieces in an envelope. Normally, in commercial patterns seam allowances are included for safety purpose (Aldrich 2002, 2004, 2009). It normally comprises all the pattern constructional information such as grain lines, seam lines, cutting lines, darts, centre lines, etc. and common information like name of the pattern piece, pattern size, number of pieces to be on each pattern piece, etc.

2.2.4.1 Merits

If the personal measurement is closer to the standard measurement sizes, then a commercial pattern can be procured from the market to draft on our own. It saves time and gives a better fit than a homemade pattern.

2.2.4.2 Demerits

Commercial patterns are normally costlier compared to drafted patterns and patterns for various styles of garments are not available.

2.2.5 Steps in Pattern Drafting

The sample measurements (7 years old): Chest 24″, waist length 10½″, waist 23″, back width 11″ and sleeve length 5″.

2.2.5.1 Basic Front Bodice and Back Bodice Pattern (Figure 2.5)

For children, back and front patterns can be drafted within the same rectangle as it is not necessary to make the front larger than the back.

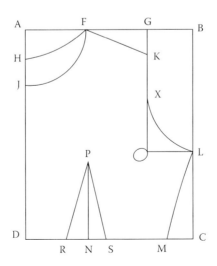

FIGURE 2.5
Pattern drafting of front and back bodice.

First, the rectangle ABCD has to be constructed with the following measurements:

- AB = 1/4 (bust + 5″ ease allowance)
- AD = BC = back waist length + 1/2″
- Mark AG = 1/2 back width
- AF = 1/12 chest
- AH = 1″
- AJ = 1/12 chest + 1/4″ = AF + 1/4″ and GK = 1″
- Connect points H and F with a bold line, which is referred to as a back neck line
- Connect points J and F with a dotted line, which is referred to as a front neck line
- Connect points F and K with a straight line, which is referred to as a shoulder seam
- Mark BL = 1/4 chest
- Draw GO parallel to and equal to BL
- Mark KX = 1/3 KO and XY = 1/2″
- Connect points K, X, and L with a bold line, which is referred to as the back armscye line
- Connect points K, Y, and L with a dotted line, which is referred to as the front armscye line
- Mark CM = 1/2″. Connect LM. This is the side seam

For a dart, mark DN = 1/2 DM – 1/2″ and NP = CL –1″. Mark R and S 1/2″ on either side of N and connect RP and SP.

2.2.5.2 Basic Sleeve Pattern (Figure 2.6)

In Figure 2.6, AD is on fold and is equal to sleeve length. AB = 1/4 bust – 1/4″ (for adults, this was 1/4 bust – 1½″). Mark BE = 1/2 AB and DF = 1/2 lower arm + 1/4″. Connect AE. Divide it into four equal parts and mark a, b, c. Mark CG = 1/2″, BF = 1/4″, AE = 1/4″ and AD – 1/2″. Connect AGFE (back armscye line) and AGBDE (front armscye line).

2.2.6 Pattern Draping

Pattern draping is otherwise known as toiling or modelling. Pattern draping is the manipulation of two-dimensional fabrics on a three-dimensional torso or body form to get a perfect fit. The dress form generally used for draping is a muslin padded dress form, positioned in an adjustable stand that duplicates the human body structure. The dress form should be firm, yet resilient

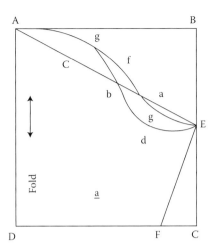

FIGURE 2.6
Pattern drafting of sleeve.

and should not resist pins (Aldrich 2002, 2004, 2009; Anon 2015). A range of dress forms exist in standard sizes for children's and men's figures. The steps for the preparation of a dress form are given below.

1. The dummy should be checked for both bust and hip measurement.
2. It should be padded to get the required measurement by using high density foam. The foam is adjusted in such a way that it assumes the shape of the human breast.
3. A square piece of quilt batting is pinned onto the formed breast. While pinning the batting, care is taken to see that it is slightly larger than the required size at the sides. When it is covered with muslin, it gets compressed to the right size and looks more natural.
4. Batting should be done equally on both sides.

2.2.6.1 Draping an Adhesive Paper Dress Form

Since it is hard for a person to fit him or herself for a long time, a dress form – a duplicate of the figure – is essential (Cooklin 1999; Crawford 2005; Ashdow 2007). One that is made on the individual is more adequate and less costly than a commercially made form.

2.2.6.1.1 Materials Needed

- Two shirts of thin knitted fabric like a T-shirt.
- Thin muslin fabric which is cut in the bias direction or a dress form kit, which will contain all the required material.

- Medium weight 1″ wide adhesive papers.
- Coloured scotch tape, two sponges, two small basins for water, needle, thread, sharp scissors, pencil, rule, tapeline, and a sharp razor blade.

2.2.6.1.2 *Method of Draping*

Generally four persons are required to build the body form quickly. Two persons are needed for moistening the cloth strips and two persons for pasting the fabric strips to a person who should wear a tight garment which gives the needed style lines. All strips should be cut prior to creating the form on the person concerned (Yarwood 1978; Armstrong 2006; Anon 2015). The various steps in the pattern draping method are shown in Figure 2.7.

a. The figure shows the depth that each group of strips is cut from the two rolls. Each batch should be labelled as it is cut.
   1. The first batch of strips has to be cut 3/4″ from outside of both fabric rolls and is roughly 15–12″ long. These fabrics are used for draping from shoulder to waist on the first layer, and on bias manner from the neck down front and back on the second layer.
   2. The second batch of fabric strips of about 12–9″ has to be cut 3/4″ from outside of each roll. These strips are utilised for draping on the first layer from waist to hip (lower edge) as well as on the second layer down from the neck in front and back.
   3. The third batch of strips of about 9–5″ has to be cut 3/4″ from the outside edge of the roll.

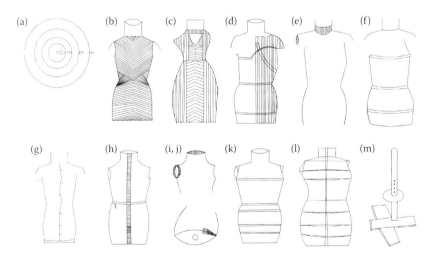

FIGURE 2.7
Draping in adhesive paper dress form.

4. About 250 of 3″ strips have to be cut for finishing edges of the form and for joining the two sections.

b. The shirt has to be put on the person concerned and then the edges are sewn together; hence, it will fit firmly on the figure. Afterward, the shirt has to be pulled down from the shoulder line and a strip of moistened tape should be pasted carefully around the waistline. Another strip of tape has to be pasted below the widest hip and under each bust to maintain contour.

c. The first layer of strips, which is about 12–15″, has to be pasted on the figure from neck and shoulder edges to taped waistline, overlapping 3/4″ of the width of the strip. Further pasting of strips has to be done for the armscye edge and overlapping of strips at the top of the shoulder. The same method has to be repeated for another side of the front and back. After pasting the strip at right angles to the waist, the tape ends should be joined to those above the waist and another strip of about 5–9″ in bias way from underarm to lower hip edge. The procedure has to be repeated for the other side also.

d. For pasting the second layer of tape, strips of about 9–12″ have to be used. Starting from below the neck, the strips have to be pasted in the bias way across the front, alternating strips from the right and left side. The crossed strips should be continued down the body to lower hip line edge, using strips of about 12–15″ below the waistline. The second layer of strips from the underarm to the hipline should be reinforced with strips of about 5–9″ in a similar manner as in c.

e. The armscye area should be strengthened to form a good structure by using strips of about 3–5″. The neck area has to be finished with a layer of strips about 5–9″ around the neck to maintain a good shape.

f. Before removing the neck, bust and waist, measurement have to be taken from the floor up to 2 or 3″ below the widest hip and it is marked at the base of the form.

g. Centre front and back with ruled line and across the front and back have to be marked before removing the form from the person, for matching when joining the half sectional. The difference between the person with and without form at the neck, bust, waist and hip has to be noted and necessary calculations have to be done. Then the form has to be cut along the hipline which is parallel to the floor.

h. Both sections are joined using strips of about 3″ wide in front as well as back.

i. The neck and armscye edges have to be trimmed and reinforced with 3″ strips around the cut edge.

j. The bottom of the form has to be trimmed until it stands evenly on the table. Cardboard has to be cut from paper patterns and a 1¼″

hole in the centre also has to be cut. Then the cardboard flush is fitted onto the inside edges of the hip and neck and joined to the form with 3″ strips.

k. and l. Outer covering – The form should be shellacked and dried to avoid curling of paper. A top shirt should be tightly and smoothly fitted over the form, and sewing can be done wherever necessary. The taping should be done at the armscye, neck and under the lower edge of the form.

m. A stand made of two 2 × 4 × 18″ pieces of wood for a base, and a pole 5–5½′ tall and 1¼″ in diameter with nail holes bored up the length to adjust the height can be used to keep the form.

2.2.6.2 *Draping on the Stand*

One of the main advantages of this method is building up a desired effect before cutting into the fabric through experimenting (Chen-Yoon and Jasper 1993; Cooklin 1999; Crawford 2005; Ashdow 2007; Anon 2015). The various stages of draping on the stand are given below.

Stage 1: Draping of uncut length of dress fabric over the stand – The fabric should be draped on the stand for analysing the overall effect by observing its natural characteristics such as handle, texture and weight and it has to be evaluated.

Stage 2: Substitution of dress fabrics – Modelling the whole garment using actual fabric is ideal but it makes experimentation more costly. Hence, it is advisable to utilise a fabric having similar properties as that of the actual fabric which has been left over from a previous collection.

Stage 3: Taping of stand – Centre front, centre back shoulders, seam lines, style line, neck lines, waist, hip and bust line and position and direction of drapes.

Stage 4: Selection and preparation of material – Prepare the garment material and since the whole garment is cut, allow enough material to cover both sides of the stand for each section. The draping quality of the warp and weft grain should be the same in order to match both sides of a drape. Allow plenty of excess material beyond the outer edges of the stand and mark in the centre vertical line and the warp grain with a contrasting thread.

Stage 5: Placing and pinning of material onto the stand – A full toile or torso is normally required; however, one side only is needed to model, except for asymmetric designs. Pining of excess fabric has to be done temporarily to the side of the stand and then the fabric is moulded around the stand as desired, allowing the extra fabric to fall freely into the area where the fullness is desired. The techniques used for controlling the drape are given below.

- By mounting the drapes on a fitted section
- By weights places inside the drapes
- By taping

Stage 6: All the details should be indicated with pins rather than chalk and it should follow the direction of any darts, tucks, seams, etc.

Stage 7: The fabric should be removed from the stand.

Stage 8: Then pressing of the fabric (except pins) has to be carried out.

Stage 9: Trueing of rough design needs to be carried out in order to establish the correct grain line and to ensure that the armhole, underarm seams and shoulder are the same length.

Stage 10: Seam allowances have to be checked.

Stage 11: Make up and press.

Stage 12: The model has to be checked for any discrepancies.

2.2.7 Flat Pattern Technique

The flat pattern technique is a method of manipulating the pattern while the pattern is laid flat on the table (Aldrich 1999). Pattern manipulation is a common word applied to the act of slashing and spreading or pivoting a pattern section to alter its original shape. Darts play an important role in the flat pattern technique (Chen-Yoon and Jasper 1993; Cooklin 1999; Crawford 2005). The darts can be shifted to any location around the pattern's outline from the pivot point without affecting the size and fit of the garment. There are three methods of flat pattern technique, which are as follows:

- *Pivot method*: By this method, darts can be moved from one point to another. For this, a thick cardboard, which is firm and sturdy, is required. Seam allowance is not added.
- *Slash and spread method*: In this method, darts are shifted by cutting and spreading the pattern along the dart to the desired position. This is a relatively easy method provided the slashes are made correctly. Care is taken to see that the cuts are not made through the pivot point.
- *Measurement method*: This method is commonly used when the darts have to remain in the same seam line and the width of the darts can be divided into two or three darts.

In all three methods, the darts on the bodice play an important role in creating the different patterns.

2.2.7.1 Types of Darts

A dart is a wedge-shaped cut out in a pattern used as a means of controlling the fit of the garment. A dart is a fold of a fabric stitched to taper gradually to a point (Eberle et al. 2002; Sumathi 2002; Fairhurst 2008). The location, length and width at the base vary according to the style. Wider darts provide better shape to the garment. These are used as the basic pattern in all positions where a bulge or hollow occurs in the figure. Darts can be single pointed or double pointed (Figure 2.8). Single pointed darts are mostly used for saree blouses and plain skirts to give shape and fit. Double pointed darts are mostly used for tops and long blouses, cholies and kameezes to give shape at the waist (Jacob 1988; Cooklin 1999; Crawford 2005).

There are two terms that are used in relation to darts – fitting darts and decorative darts. Decorative darts do not have any functional purpose in a garment and are used only for decorative purposes. Fitting darts are functional darts, which are triangular folds in a cloth making the flat fabric fit to the curves of the body (Pheasant 1986; Cooklin 1999; Crawford 2005).

2.2.7.2 Locating the Dart Point

The basic of the flat pattern work is locating the pivotal or the dart point. The dart point, also called pivotal or apex point of the front bodice, is a place on the pattern from which the darts radiate. Two darts – one near the shoulder and another at the waistline – could be found in the back bodice. Each of these darts has its own pivot point (Taylor and Shoben 1990; Fan 2004; Glock and Kunz 2004).

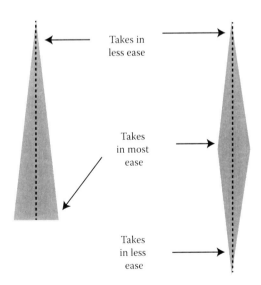

FIGURE 2.8
Types of dart.

2.2.7.2.1 *Method of Locating the Dart Point in the Back Bodice*

In the back bodice there is no well-defined location for the common pivot point to be located. The pivot point is at about $1^{1/2}''$ away from the tip of the dart. Figure 2.9 shows the position of the pivot point in the back bodice.

2.2.7.2.2 *Method of Locating the Dart Point in the Front Bodice*

This method is used for the pattern having two darts in the front bodice. The two darts are the bust fitting dart and the waist fitting dart. The bust fitting dart originates from the side seam and moves toward the bust point. The waist fitting dart originates from the waistline and moves toward the bust point (Gillian Holman 1997; Gupta and Gangadhar 2004).

For locating the dart point, a line has to be drawn from the middle of both bust and waist fitting darts and it is extended until they intersect. The point of juncture of these lines is the dart point in the front bodice. In Figure 2.10, point A and B are the centre line of the waist fitting dart and bust fitting dart, respectively. The lines are extended and the intersecting point C is labelled as the pivot point. The other vital aspect is the drawing of bust circle, which encloses the bust area in the pattern. The bust circle is generally drawn around the bust point with varying radius, which depends upon different sizes. For instance, 1 1/2″ of radius is used to draw the bust circle for sizes 8, 10 and 12, and 2″ radius is used to draw the bust circle for the sizes above 12.

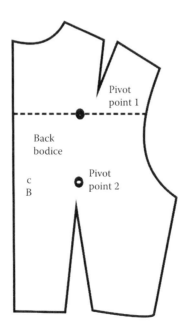

FIGURE 2.9
Pivot points in the back bodice.

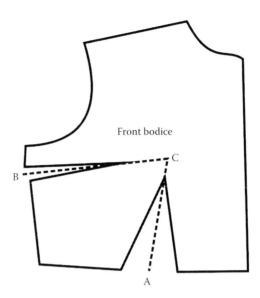

FIGURE 2.10
Pivot point in the front bodice.

Figure 2.11 shows the pattern with the bust circle (Cooklin 1999; Natalie Bray 2004; Crawford 2005; Gupta et al. 2006).

2.2.7.2.3 Rules for Dart Location

1. Minimum length of the darts – The fitting darts of the front bodice must extend to the bust circle. This is the minimum length.
2. Maximum length of the darts – All the fitting darts must extend to the bust circle but should not lengthen outside the bust point. This is considered the maximum length of the darts. In some of the patterns, there would be one larger dart that would be extended until the bust point for proper fitting.
3. If both fitting darts are equal in size, both darts will end at the bust circle.
4. Darts may point away from the bust point for certain design effects but they must not point outside the bust circle.
5. A decorative dart, which does not help in fitting, does not point toward the bust circle. It should be kept small in angle so it does not create a 'bulge'.

2.2.7.3 Pivot Method

The pivotal point is the designated point on the pattern that is used as a basis for the slash and spread method and the pivot method. The pivot point on

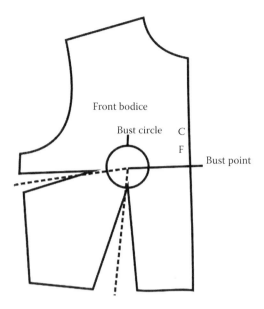

FIGURE 2.11
Front bodice with the bust circle.

the front pattern is the bust (Cooklin 1999; Crawford 2005; Beazley and Bond 2006; Gupta et al. 2006).

Front Bodice. The example of shifting of the waistline dart to the neckline dart using the pivot point method is shown in Figure 2.12.

- The dart leg AB has to be marked on the front bodice pattern.
- The new position of dart C has to be marked as indicated in the figure, at the neckline to which the dart needs to be shifted.
- Tracing of the pattern from point C to point A has to be done so that the dart can be moved as shown in the figure.
- Thumbtack the pattern at the bust point and slowly the pattern has to be moved from point B to A, thus closing the dart at the waistline.
- Tracing of the pattern from point B to point C has to be continued and then the block bodice can be removed. Now an opening at the neckline can be observed, which can be marked as DE. This is the new dart located at the neckline.
- Label the pivot point. This would be a guide for locating the new dart.
- The midpoint of ED should be marked and a dotted line is drawn until the bust point.

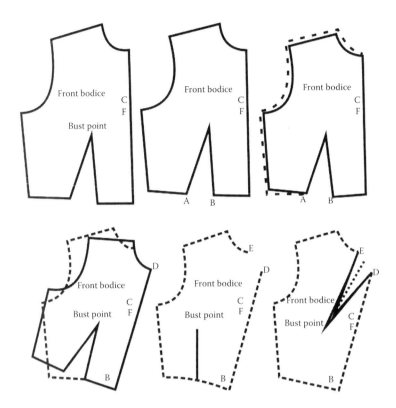

FIGURE 2.12
Pivot point method for front bodice.

- The dart legs are drawn by joining E and D to the bust point.
- Point F is located at 1/2″ above the bust point and the dart legs are completed as shown in the figure.

Back Bodice. The back bodice has two darts, namely, shoulder dart and the waistline dart. The shoulder dart is often used in creating new designs at the back (Cooklin 1999; Crawford 2005; Carr and Latham 2006; Gupta et al. 2006). The steps followed are shown in Figure 2.13.

- Take the back bodice block with the pivot points located on it.
- Mark the dart legs as AB and CD as shown in Figure 2.13.
- Mark the location of the new dart E on the neckline.
- Starting from point E, trace the pattern toward the centre back, then to the waistline.

  Trace the dart AB at the waistline and proceed to tracing the side seam, armhole and the shoulder until point D as shown in the figure. The dotted line shows the traced pattern.

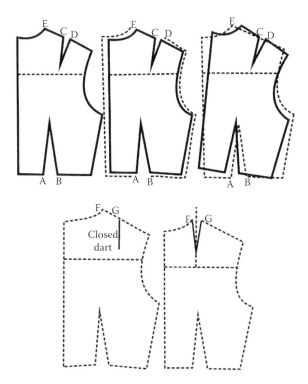

FIGURE 2.13
Pivot point method for front bodice.

- Thumbtack the pivot point corresponding to the shoulder dart and pivot the pattern thus closing the dart. See Figure 4.15.
- Remove the pattern. You would notice the dart opening at the neckline. Mark the new dart opening as FG.
- Locate the centre point of the dart opening FG and mark the point H.
- The length of the dart is the same as the length of the shoulder dart. Measure the shoulder dart. Draw a line from point H parallel to the centre back line or slanted slightly away from the centre back.
- Measure the length of the shoulder dart and mark the point on the line drawn from H. Draw the dart legs and complete the dart.

2.2.7.4 Slash and Spread Method

The sloper or block pattern has to be traced on a separate paper for moving the darts in front and the back bodice using the slash and spread method and the pivot points on the back bodice have to be marked as well (Cooklin 1999; Crawford 2005; Gupta et al. 2006). The factors that need to be considered during the slash and spread method are given below.

- The slash should always be made along the lower line of the horizontal darts.
- For vertical darts, the slash needs to be made along the line nearer to the centre front or centre back.
- Make all the slashes go to but not through the pivot point.

Front Bodice. The steps involved in creation of a dart in the front bodice using the slash and spread method are shown in Figure 2.14.

- Take the front bodice pattern.
- Locate the pivot point on the front bodice.
- Label the dart legs as AB.
- Mark the position of the new dart point C at the neckline to which the dart needs to be shifted as indicated in Figure 2.14.
- Draw a slash line from point C to the bust point.
- Slash along the line marked from point C to the bust point but not through the bust point.
- Similarly slash along line B to the bust point but not through the bust point as shown in the figure.
- Close the dart AB by overlapping the dart legs.
- Trace this on another sheet of paper.
- You would now find an opening at point C.
- Mark the opening as DE.
- Mark the centre line as F.

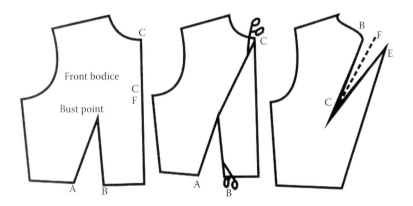

FIGURE 2.14
Slash and spread method for front bodice.

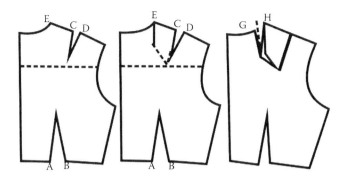

FIGURE 2.15
Slash and spread method for back bodice.

- A line has to be drawn toward the bust point and then the dart legs are drawn.
- Locate a point G 1/2″ above the bust point and redraw the dart legs.

Back Bodice. The steps involved in shifting of the shoulder dart of the back bodice to any part of the seam line in the upper portion of the pattern are shown in Figure 2.15.

- The shoulder dart of the bodice back may be shifted.
- The back bodice paper pattern is prepared and the pivot point is marked.
- Mark the dart legs at the waistline as AB and at the shoulder as CD – the dart to be moved.
- Mark the position to which the dart needs to be shifted as point E at the neckline.
- A line is drawn at the point E to indicate the full length of the new dart. The line is extended to the pivot point and indicated as a dotted line.
- Cuts are made from the shoulder dart to the pivot point and along the new dart line to the pivot point.
- The shoulder dart is closed and pinned. The neck line opening is filled for the new dart by pinning the pattern to another piece of pattern.
- A dotted line is drawn to indicate the centre of the new dart. The new dart can be parallel to the centre back or it could be a slant line from the centre.

- The new dart lines, that is, GH is drawn in such a way that the length of the new dart coincides with the original shoulder dart.

2.2.7.5 Measurement Method

This method is commonly used when the darts have to remain in the same seam line and the width of the darts can be divided into two or three darts (Cooklin 1999; Le Pechoux and Ghosh 2004; Crawford 2005; Gupta et al. 2006). The method of shifting the darts is described wherein the darts at the waistline are manipulated.

- The front bodice pattern is traced onto a new paper leaving the space of the dart which is to be divided as shown in Figure 2.16.
- Measure the gap left for the dart and divide into two halves by using dressmaker's tape. Dressmaker's tape is a strip of paper folded to give a firm, straight edge.
- The two halves would give the dart space on each dart.
- Locate the position for two darts on the waistline.
- Now, keep the section between the dart and the centre front of the original sloper unchanged. This part is called the centre panel or centre section.
- The two dart points need to be located so that the distance between the two darts is 1″.
- Draw the guidelines for darts as indicated by a dotted line toward the bust circle.

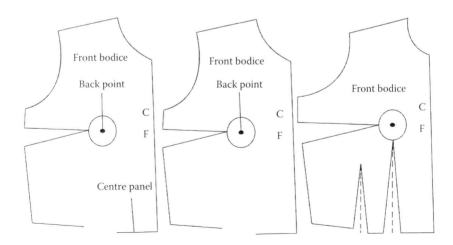

FIGURE 2.16
Measurement method.

- The tips of the darts are located toward the bust circle. Draw the dart legs and complete the dart on the pattern.

2.2.8 Pattern Grading

Grading is a technique used either to maximise or minimise the size of a pattern. This becomes necessary when large numbers of different sized garments have to be produced in a relatively shorter time as is done in the garment industry (Hulme 1944; Knez 1994; Joseph-Armstrong 2004). The different terms associated with grading are as follows:

1. *Suppression grading*: The controlling features of the garment like darts, pleats and gathers when decreased in size undergo suppression. To suppress the girth measurement, of say No 28″ size in relation to the girth of 26″ waist size, only a tuck has to be placed at the waist. This has nothing to do with styling.
2. *Three-dimensional grading*: This technique is commonly used for tight fitting and knitted garments. This involves not only suppression, but also the changes in girth and height.
3. *Two-dimensional grading*: In this simple and easy method, only girth and height measurements are changed without altering the shape.
4. *Cardinal points*: The points where the grading increments are applied are called cardinal points.
5. *Balance*: This refers to the perfect relationship between the units as explained earlier. It explains that when the increase is done in the front, then care should be taken to increase at the back also.
6. *Nested or stacked grading*: In this method, the difference in the increase in size is made visible by superimposing one size to another. The progression of sizes can be more noticeable.

2.2.8.1 Types of Grading Systems

There are two types of grading systems commonly used.

1. Two-dimensional grading
2. Three-dimensional grading

2.2.8.1.1 Two-Dimensional Grading

The two-dimensional grading could be done using two techniques:

1. *Draft technique*: This involves the increments being applied to the actual pattern draft. For example, if you are grading for one size up at the front bodice, the sloper is taken and the measurements to be

added at the different cardinal points like shoulder, armscye, centre front, etc. are added simultaneously.

2. *Track technique*: This involves applying grade increments to individual pieces of pattern by moving the base pattern piece along predetermined tracks. In this method, the pattern is altered section by section along the predetermined tracks. Let us take the example of grading the pattern for one size up at the shoulder, neckline, centre front, etc. The tracks are drawn on a separate sheet of paper and the pattern is moved as follows:

   a. *Shoulder*: The pattern is moved along the track for grading the shoulder, and then returned to the original track.

   b. *Neckline*: From the original track, again the pattern is moved for grading the neckline, then again returned to the original track. The process is continued until all the sections of the pattern are graded.

2.2.8.1.2 Three-Dimensional Grading

Three-dimensional grading is used not only to increase a pattern for size, but also to increase or decrease suppression in the following areas:

1. Bust to shoulder
2. Hip to waist
3. Elbow to wrist

References

Adu-Boakye, S., J. Power, T. Wallace and Z. Chen. 2012. Development of a sizing system for Ghanaian women for the production of ready-to-wear clothing. In: *The 88th Textile Institute World Conference*. Selangor, Malaysia.

Aldrich, W. 1999. *Metric Pattern Cutting*. Blackwell Science Ltd, UK.

Aldrich, W. 2002. *Pattern Cutting for Women's Tailored Jackets: Classic and Contemporary*. Blackwell Science Ltd, Oxford.

Aldrich, W. 2004. *Metric Pattern Cutting*. Fourth Edition. Blackwell Publishing Ltd, Oxford.

Aldrich, W. 2009. *Metric Pattern Cutting for Women's Wear*. Wiley, Chichester.

Anon. 2015. The Basic Figure Types, Chapter V. http://haabetdk/patent/Corset_fitting_in_the_retail_store/5.html (accessed on December 21, 2015).

Armstrong, H.J. 2006. *Patternmaking for Fashion Design*. Pearson Prentice Hall, Upper Saddle River, NJ.

Ashdow, S.P. 2007. *Sizing in Clothing—Developing Effective Sizing Systems for Ready to Wear Clothing*. CRC Press, Textile Institute & Wood Head Publishers, UK.

Beazley, A. and T. Bond. 2004. *Computer Aided Pattern Design and Product Development*. Blackwell Publishers, UK.

Beazley, L. and T. Bond. 2006. *Computer Aided Pattern Design and Product Development.* Blackwell Publishing, UK.

Bray, N. 2004. *Dress Fitting.* Blackwell Publishing Company, UK.

Carr, H. and B. Latham. 2006. *The Technology of Clothing Manufacture.* Blackwell Science, Oxford.

Chen-Yoon, J. and C. Jasper. 1993. Garment sizing systems: An international comparison. *International Journal of Clothing Science and Technology* 5(5):28–37.

Cooklin, G. 1999. *Pattern Grading for Women's Clothing.* Wiley Blackwell, UK.

Crawford, C.A. 2005. *The Art of Fashion Draping.* Fairchild Publications, New York.

Eberle, H., H. Hermeling, M. Hornberger, R. Kilgus, D. Menzer and W. Ring. 2002. *Clothing Technology.* Third English Edition. Haan-Gruiten Verlag Europa-Lehrmittel, Nourney.

Fairhurst, C. 2008. *Advances in Apparel Production.* The Textile Institute, Woodhead Publication, Cambridge.

Fan, J., W. Yu and L. Hunter. 2004. *Clothing Appearance and Fit: Science and Technology.* Woodhead Publishing Limited, Cambridge.

Gupta, D. and B.R. Gangadhar. 2004. A statistical model to for establishing body size standards for garments. *International Journal of Clothing Science and Technology* 16:459–69.

Gupta, D, N. Garg, K. Arora and N. Priyadarshini. 2006. Developing body measurement charts for garment manufacture based on a linear programming approach. *Journal of Textile and Apparel Technology Management* 5(1):1–13.

Glock, R.E. and G.I. Kunz. 2004. *Apparel Manufacturing—Sewn Product Analysis.* Prentice Hall, Englewood Cliffs, NJ.

Holman, G. 1997. *Pattern Cutting Made Easy.* B T Batsford Ltd., London.

Hulme, W.H. 1944. *The Theory of Garment-Pattern Making Textbook for Clothing Designers Teachers of Clothing Technology and Senior Students.* The National Trade Press Ltd, London.

ISO 3635. 1981. Size designation of clothes—Definitions and body measurement procedure.

ISO 3636. 1977. Size designation of clothes—Men's and boys' outerwear garments.

ISO 3637. 1977. Size designation of clothes—Women's and girls' outerwear garments.

Jacob, S. 1988. *Apparel Manufacturing Hand Book—Analysis Principles and Practice.* Columbia Boblin Media Corp, New York, USA.

Joseph-Armstrong, H. 2004. *Pattern Making for Fashion Designing.* Prentice Hall, New York.

Knez, B. 1994. *Construction Preparation in Garment Industry.* Zagreb Faculty of Textile Technology University of Zagreb, Croatia.

Le Pechoux, B. and T.K. Ghosh. 2004. *Apparel Sizing and Fit.* Textile Progress 32. Textile Institute, Manchester.

Pheasant, S. 1986. *Body Space: Anthropometry Ergonomics and Design.* Taylor & Francis, London, Philadelphia.

Sumathi, G.J. 2002. *Elements of Fashion and Apparel Designing.* New Age International Publication, New Delhi, India.

Taylor, P. and M.M. Shoben. 1990. *Grading for the Fashion Industry.* Stanley Thomas Publishers Ltd, London, UK.

Yarwood, D. 1978. History of brassieres. *The Encyclopedia of World Costumes.* The Anchor Press Ltd, UK.

3

Fabric Spreading and Cutting

In most circumstances (except one-piece knitted garments), garments are assembled from several components. This is essential to establish profile/ shape to the garments, to overcome the constraints of fabric width. The cutting is the process of reproduction of shape of pattern pieces in the fabric during the production of garments. The cut fabric panels are then joined using a series of stitches and seams to produce three-dimensional garments (Aldrich 2002a).

3.1 Cutting Department

The main purpose of cutting section involves cutting of garment panels precisely, consistent with the pattern shape and size as well as economically and in a necessary volume to keep the sewing department supplied with work. The process flow in the cutting department is shown in Figure 3.1.

The four main operations or processes involved in the cutting section are

- Marker planning
- Fabric spreading
- Fabric cutting
- Preparation for the assembling process

3.2 Marker

It is an illustration of accurate and precise planning of patterns for a particular style of garment and the sizes to be cut from a single spread on a marker paper. To prepare an efficient marker, the width of the fabric to be spread in a lay as well as the number of pattern pieces to be included in the marker plan for all the required sizes should be known prior to it. Marker width that is less than fabric width leads to more fabric wastage while marker width that

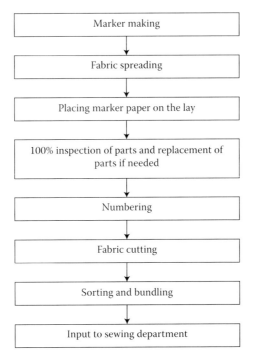

FIGURE 3.1
Process flow in cutting section.

is wider than fabric results in incomplete cut components. The individual marker has to be prepared for linings and interfacing materials (Anon 1993).

3.2.1 Marker Parameters

Markers are generally defined by two parameters at the beginning of the marker, namely, the relation to the relative symmetry of the garment and mode.

3.2.1.1 Relation to the Relative Symmetry of Garment

It is related to the technique of utilisation of patterns concerned with the relative symmetry of garments (Beazley and Bond 2003).

Mixed marker: It is a kind of marker that is created for attaining higher marker efficiency. It is mainly utilised while processing asymmetric garments or when the fabric is spread in all face up mode.

Open marker: This kind of marker is prepared while processing asymmetric garments with the objective of getting better spreading and cutting quality. It generally keeps a set of garment panels, that is, left and right panels close to each other along the marker length and is utilised when the fabric is in open condition and spread in all face up mode.

Closed marker: In this marker, only one half of the pattern set is generally used for symmetric garments and the fabric is folded lengthwise on the table after spreading it in face-to-face mode. Subsequent to the cutting operation, one pattern could produce both sets of pattern pieces, that is, right and left garment panel when choosing a pair of successive plies.

Closed-on-open marker: Though it is like a closed marker, fabrics are not folded as the patterns should be either kept at the left or right. The main limitation in this kind of marker is that single piece garment panels like back bodice could not be prepared without the blocking and re-laying process.

Blocking and relaying: This type of marker is normally used along with the closed-on-open marker. In this process, after the garment panel is cut, the cut block for the pattern is separated in half, and the smaller size pattern of that component is placed on the second half. This is then recut to the smaller size.

Single section marker: In this type of marker, different sizes of patterns are distributed all over the length of the marker. It gives higher marker efficiency as a huge quantity of patterns could be accommodated in the marker.

Section marker: This kind of marker is utilised when the ratios of different garment sizes are known in advance. Step markers are used when the order ratios of different sizes are not known in advance. It aids in processing small orders of different sizes in an efficient manner by means of a stepped lay.

Grain: The patterns should be positioned in the marker in such a way that the grain line in the pattern piece should be parallel to the selvedge of the fabric, which has a direct influence on the draping quality of the garments.

3.2.1.2 Mode

Generally, markers are made in several modes. The direction of the nap on the fabric is used to define the mode of spreading (Burbidge 1991; Beazley and Bond 2003).

Nap/One/Way marker: The Nap/One/Way marker (N/O/W) is produced with every pattern placed in the 'down' direction of the pattern in the identical direction. This kind of mode is required for asymmetric fabrics.

Nap/Either/Way marker: When there is no restriction of orientation of pattern placement this kind of marker could be used. The patterns could be placed in 'up' or 'down' direction but as parallel to the fabric grain. It is the most efficient marker yielding the highest fabric utilisation.

Nap/Up/and Down marker: The Nap/Up/and Down marker (N/U/D) is more effective than the Nap/One/Way marker; however, it is not as efficient as the Nap/Either/Way. In order to get a better fit between the patterns, alternating sizes of patterns are oriented in opposite directions.

3.2.2 Types of Markers

3.2.2.1 Sectioned Markers

Markers can be created as sections or a continuous one. Sectioned markers comprise all of the patterns of a particular garment size and style. They are easier to imagine and handle; however, they may not give the maximum marker efficiency. High-volume blocks could be kept on one side and low-volume blocks at the opposite side of the marker; hence, the fabric can be spread to correspond with the volume needed for each block. These kinds of markers are beneficial if there is an end-to-end shade difference in the fabric (Sumathi 2002).

3.2.2.2 Continuous Markers

Continuous markers comprise patterns of all garment sizes for a particular garment style. It could be lengthier and need reorganising of pattern pieces frequently. These markers normally have higher marker efficiency due to its flexibility in grouping and manipulating larger as well as smaller pattern pieces. A splicing process where the fabrics can be cut across the width and overlapped with the next fabric piece in case of occurrence of fabric fault to keep a continuous spread could be carried out in this type of marker to minimise fabric wastage (Tyler 1992; Anon 1993; Shaeffer 2000).

3.2.3 Marker Planning

Marker making is the process of finding out the most proficient arrangement of pattern pieces for a particular garment style, fabric and range of sizes. This process requires skill, time and concentration to get the maximum efficiency (Chuter 1995; Aldrich 2002a,b). The process of marker making can be explained in two aspects:

1. *Marker planning*: It is the placement of patterns in a paper to meet the technical requirements as well to minimise wastage of fabric.

2. *Marker production/marker utilisation*: This involves drawing of a marker plan directly on fabric or creating it on a paper marker, or copying information related to the pattern piece on the fabric without drawing any pattern lines over it as in computerised cutting.

3.2.3.1 Requirements of Marker Planning

Marker planning is more of a creative, intuitive and conceptualising process rather than a technical one and there is no final result for a marker planning. The main purpose is to produce a shortest marker by considering all the practical and technical constraints. The constraints in making a shortest marker are related to

- Fabric characteristics and the design requirement in the finished garment
- Cutting quality
- Production planning

3.2.3.1.1 Fabric Characteristic and the Design Requirement in Finished Garment

1. Alignment of patterns with respect to fabric grain – All the patterns in the marker plan should be kept such that the grain line in the pattern should be parallel to the fabric selvedge for better hanging and draping of garments (Figure 3.2) (Chuter 1995; Aldrich 2002a,b).
2. Fabric symmetry and asymmetry – If the face and back side of the fabrics have a similar appearance, then they are called 'two-way' or 'symmetrical' fabrics and it does not warrant any special requirement while marker planning. The asymmetric fabric where the face

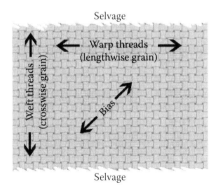

FIGURE 3.2
Grain line in fabric.

and back are dissimilar needs some attention during marker planning. Examples of asymmetric fabrics are those having a nap or pile. More complicated fabrics are 'one way' or 'asymmetrical'. These kinds of fabrics have a surface pile or a print design that has a recognisable object which can only be used one way.

3. Design requirements in final garment – Design aspects of final garments also have to be considered while marker planning to get a better visual appearance of the finished garment. For instance, if a vertical stripe in a garment does not exhibit a complete replica of a repeat on the right and left panels of garment it looks awkward (Chuter 1995; Aldrich 2002a,b).

3.2.3.1.2 Cutting Quality

Since most of the garment units utilise a vertical blade for cutting of fabric panels, the marker plan should take into consideration the space required for movement of the knife blade especially while cutting in curved areas. The space required between the patterns in the marker plan depends on the cutting method used.

A counting of number of patterns in the marker should be carried out to verify the complete set of patterns has been incorporated in the marker plan. After cutting of fabric panels, sorting of pattern pieces as per the size, bundling of cut fabric panels as per the colour and size and finally pattern count in each size should be done to confirm that all the patterns are available for the assembling process. The marker planner should give coding to all patterns with its size during the preparation of marker planning (Tyler 1992; Anon 1993).

3.2.3.1.3 Production Planning

Each order is specified by a certain quantity with respect to size and colour. For example, an order for 12,000 trousers may include 4800 blue, 4800 green and 2400 red, across sizes 30, 32, 34 and 36 in the ratio 2:4:4:2. The production planning and control department have to ensure adequate supply of cut components to the sewing room at regular intervals (Eberle et al. 2002).

3.2.4 Construction of Markers

A marker is generally made by keeping the patterns one after the other in the length of the marker. The marker length states the length of fabric that could be used in a lay (Tyler 1992; Anon 1993; Fairhurst 2008). The marker length is defined by the following components, which are common to all markers.

1. *Selvedge lines*: The two parallel lines should be drawn parallel to the edge of the cutting table. The gap or distance between the selvedges represents the maximum fabric width that could be used during marker planning.

2. *Beginning line*: The beginning line is at the left side of the marker as seen by the marker planner and is perpendicular to the selvedge lines and is considered a beginning position of the marker.

3. *End line*: The end line is marked at the end/right side of the marker (opposite the beginning line) which is located after the extent of the last pattern and is drawn parallel to the beginning line joining the selvedge line.

4. *Splice marks*: Splice marks represent the area in the cutting table where the fabrics are overlapped during the run out of fabric rolls or elimination of fabric defects during spreading. These marks are placed along the control selvedge.

5. *Legend*: The legend is used to give the key about the marker and normally consists of reference information about the marker.

6. *Placement rules*: The marker planner has to consider the following general rules while marker planning:

    a. The grain line in the pattern should be parallel to the fabric selvedge.

    b. The patterns should be placed on the marker by considering the grain line in it.

    c. The patterns should be kept as close as possible to minimise fabric wastage.

    d. The patterns can be placed from largest to smallest, to get higher marker efficiency, leading to the least amount of pattern manipulation as necessary.

3.2.5 Methods of Marker Planning

3.2.5.1 Manual Marker Planning

It is the conventional marker planning method and is still used by the garment industries where they make single garment markers. The marker planner works easily by moving around the full-size patterns until an acceptable marker plan is obtained. Multiple copies of the marker are usually required, which can be done by reproducing the master marker with a range of duplicating methods (Fredrerick 1999; Eberle et al. 2002).

1. Carbon duplicating – This method is utilised when very few numbers of copies are needed. Double-sided carbon paper or special NCR-type (no carbon required) paper can be used for duplicating the master marker. In this method, only six to eight copies of master marker can be made without much deterioration in the line.

2. Spirit duplicating – In this system of duplicating, a special hectograph sheet is placed underneath the marker. The hectograph paper

transfers a blue line on the back side of the master marker as it is drawn. A master marker is then utilised to produce multiple copies one at a time in a duplicating machine where the master marker along with the white paper wetted in alcohol is moved through the rollers which transfer the line onto the copy.

3. Diazo photographic method – This technique can be used to make copies as required, one at a time. Here, both the light-sensitive paper and a marker are passed through a UV light source, where the light-sensitive paper can be developed by ammonia vapour, which produces a copy.

3.2.5.2 Computerised Marker Planning

This method is generally a part of an integrated system that comprises digitising of full-size patterns into the computer, conveniences for pattern alteration, and by inputting suitable grading rules to create all the required sizes (Knez 1994). The various components involved are visual display unit with keyboard, tablet, data pen and mouse (Figure 3.3).

The marker planner indicates the precise make-up of the marker plan such as fabric width, the pattern pieces to be utilised, and product sizes to be included in the marker and the constraints to be considered including any matching of checks. Then the system generates a marker plan automatically or interactively. The automatic marker planning needs data defining the placement of pattern pieces in markers previously planned, and selection of a suitable marker which gives the highest marker efficiency (Carr and Latham, 2006). In the interactive method, the marker planning was done by interaction of the marker planner with the system. All the available patterns will be exhibited in miniature form at the top right of the screen. For manoeuvring the patterns data pen, the mouse and the keyboard can be utilised. The system finally positions the pattern pieces accurately based on marking rules specified. Subsequent to selection of an economical marker plan, the computer will also give a pattern count, marker efficiency and total

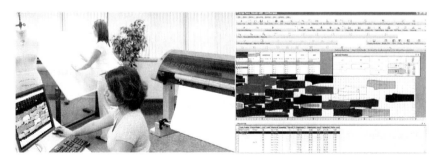

FIGURE 3.3
Computerised marker planning system.

marker length at the bottom of the screen. The computerised planning provides a pattern grading facility as well and allows the reproduction of as many copies of a marker as are necessary (Knez 1994; Carr and Latham 2006).

3.2.6 Marker Efficiency

Marker efficiency refers to fabric utilisation and is defined as the percentage of the total fabric that is actually utilised in garment components. It depends on how closely the patterns are arranged in the marker; that is, length of marker (Laing and Webster 1998). The marker efficiency is defined by the formula as given below:

$$\text{Marker Efficiency (\%)} = \frac{\text{Area of patterns in the marker plan}}{\text{Total area of the marker}}$$

The influencing factors for the marker efficiency are characteristics of fabric, profile/shape of the pattern pieces and grain requirements (Workman 1991).

3.3 Spreading

The main aim of the spreading process is to lay the several fabric plies essential for the production process to the marker length without any tension on the fabric. The lay height depends on order size, fabric characteristic, capacity of the spreader, cutting method and equipment used. The preference of mode of spreading will influence the cost of spreading as well as finished garment quality (Aldrich 2002a,b; Clayton 2008).

3.3.1 Types of Spreads

The spreads can be categorised into two basic types, namely, flat spread and stepped spread.

- *Flat spreads (scrambled spread)* – It is the economical method of spreading where a single section maker comprises patterns in the ratio that the style is ordered. Fabric is normally spread in multiples of the ratio of the marker. In this type of spread, all plies are of the same length.
- *Stepped spreads (section spreads)* – In this method, the spread is normally built like small steps, with all the fabric plies in a step having the same length. It is commonly used when the order needs

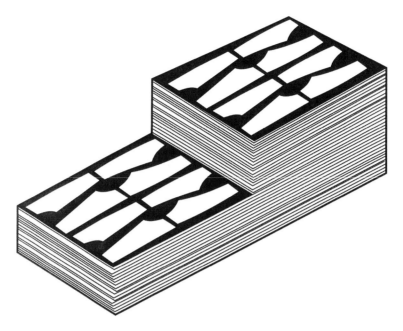

FIGURE 3.4
A stepped lay with markers on top.

to cut the imbalance between the quantities to be cut, which prevents the use of the flat spread. In most circumstances, the marker section with the need for the greatest number of plies is situated closest to the left of the spread (Burbidge 1991; Aldrich 2002a,b). Then each section in the order of decreasing numbers of plies is located after the first going down the table (Figure 3.4).

3.3.2 Objectives of the Spreading Process

3.3.2.1 Shade Sorting of Fabric Rolls

Generally one spread requires more than one fabric roll; several rolls are necessary to produce the required order quantity. Hence there is a chance for roll-to-roll shade variation. A garment assembled from components cut from these different fabric rolls could exhibit a shade variation between its different panels. While spreading fabrics of more than one roll in a spread, they have to be separated by means of interleaving paper, which aids in easy identification and separation of the plies for bundling (Burbidge 1991; Aldrich 2002a,b).

3.3.2.2 Ply Direction and Lay Stability

It is influenced by the type of fabric, shape of the pattern and the spreading equipment. For example, the fabric could be spread either face up or face-to-face manner for symmetrical patterns. On the other hand, the fabric can be spread only in face-up or face-down manner for asymmetrical patterns (Knez 1994).

3.3.2.3 Alignment of Plies

Each ply of the spread must have the length and width of the marker and also the minimum possible extra outside those measurements especially in width due to the possibilities of width variation between fabric rolls as well as within the roll to a lesser extent. By considering this, generally the marker plan is created to the narrowest width of fabric. The excess fabric width could be dispersed outside the marker plan at the opposite end of an operator by aligning the fabric edges at the end or the fabric could be aligned centrally by distributing the extra width equally on both sides.

3.3.2.4 Correct Ply Tension

It is very crucial to spread the fabrics with adequate tension during spreading. Suppose if the fabric is spread with low tension, then the fabric will form ridges with irregular fullness. Conversely, if the fabric is spread with high tension, they will maintain their tension while held in the lay, however, it will contract after cutting or during sewing, leading to a smaller size of the garment component. The normally recommended methods for removing tension in the lay are relaxing the fabric overnight, beating the lay and positive fabric feed.

3.3.2.5 Elimination of Fabric Faults

A plastic tag is normally pasted on the fabric edge in line with the fault during fabric inspection. The fundamental ways of taking action to localised faults are make-through system, cut out at the lay and the sort and recut system.

- In the make-through system, the fabric faults are left in the garment as it is and it is inspected at the final stage of manufacturing. This option can be utilised when fabric faults are relatively lower and the market is available for 'seconds'.
- Cutting out at the lay uses 'splicing' during spreading. In this method, the fabric is cut across the fabric ply at the point where the fault is located and overlays it as far back as the next splice mark, which is adequate to allow a complete garment panel rather than sections only to be cut.
- In the case of the sort and recut method, the fabric faults are marked with a strip of contrasting fabric; however, no action is taken at the spreading stage. After the cutting process is completed, the cut components are inspected for faults and the defective panels are recut

from the remnant fabric. This is a cost-effective method and is particularly used when the cost of fabric is high, the garment pieces are
large and the fault rate is high.

3.3.2.6 Elimination of Static Electricity, Fusion and Tight Selvedge in Cutting

- Static electricity can build up within a lay in the case of synthetic
  fibres particularly on dry days. In such circumstances, the spreading
  process will be more difficult.
- Cut edges of thermoplastic fibre fabrics could fuse together during
  cutting due to heat generation in the knife blade. Generally, anti-fusion
  paper comprising a lubricant could be used that lubricates the knife
  blade, therefore reducing the heat generation in the cutting knife.
- Tight selvedges generally lead to fullness in the central area of the
  spread. They can be corrected by cutting into the selvedge to release
  the tightness.

3.3.2.7 Fabric Control during Spreading

Preferably, each ply in the lay should be spread by superimposing the fabrics
one above another with their ends aligned.

- Smoothing fabric – During spreading it is important to open out any
  unnecessary folds, and to avoid 'bubbles' caused by uneven tension
  in softer fabrics.
- Skewing – Skewing is a condition where the fabric is angled across
  the course.
- Bowing – Bowing is created when the cross-grain weft bends additionally down the table in the centre of the fabric which is difficult
  to minimise.

3.3.2.8 Avoidance of Distortion in the Spread

Spreaders are vital to lay up the fabric without any tension. Therefore, the
garment panels do not shrink after cutting. Normally, a glazed paper with its
glossy side kept down is put at the top of the spreading table before spreading to avoid disturbance of lower plies of fabric while the base plate of a
straight knife cutting equipment passes beneath it.

3.3.3 Method of Spreading

Generally, the spreading process can be done manually or by computer
controlled machines. One or two person, based on the fabric type and width

of fabric, type of spreading machine and size of spread, can be involved in the spreading process. In case of the manual spreading process, two persons are normally required except when the spread is too short. One person on each side of the spreading table could work during spreading to keep the fabric flat, smooth and tension-free. With the automatic spreading process, the equipment itself controls the fabric tension, fabric placement and rate of travel (Mathews 1986; Mehta 1992; Bhuiyan 2015).

3.3.3.1 Spreading Table

Spreading normally requires a flat, smooth surface. Spreading and cutting tables are available in standard widths. A spreading table should be about 10″ wider than the fabric width (Figure 3.5). It may have rails fixed on the top of a spreading table to guide and control the spreader as it moves along the length of the table. With modern high speed spreading machines, all the drives are synchronised to control the fabric tension.

Specialised spreading tables are also available depending on the type of fabric and cutting. One kind of spreading has an option of a row of pins that are placed below the table surface and can be drawn-out above the table through slots for better gripping of fabric in an accurate position for getting an accurate pattern matching in case of repeats. Vacuum tables are also available to compress the lay and prevent sliding of plies during cutting.

The fabric can be spread on one table and then transferred to the cutting table. With the air flotation facility in spreading tables, a lay can be transported easily to the adjacent cutting table. A layer of air between the top of the table and the bottom layer of paper reduces friction and allows a lay to be moved easily. Spreading tables with a conveyor arrangement move the fabric from the spreading table to the cutting table with ease to minimise the handling and transportation time (Ukponmwan et al. 2001).

FIGURE 3.5
A spreading table. (Reproduced by permission of IMA spa, Italy.)

3.3.3.2 Solid Bar

Even though this method is apparently unsound, this kind of spreading by two workers is still used. There is no tension control in this type of spreading and hypothetically can be used for any mode of spreading.

3.3.3.3 Stationary Rack

This machine has basically two uprights fixed at the end of the table. A steel bar is passed through the fabric roll and two spreaders, one on each side of the table, pulls the fabric from one end of the table to the end of the spread length. After the fabric ply is aligned and weighted at the end, the spreaders then smooth out any ridges or wrinkles in the fabric and align the fabrics with respect to any one side of the fabric. These kinds of spreading are preferably used for F/O/W, N/O/W fabrics but are not suited for N/U/D modes of spreading.

3.3.3.4 Drop-In Unwinder

This equipment has a cradle with rollers that enclose the fabric roll. It is most commonly used when the tube in the fabric roll is crushed or too small for the steel bar in a stationary rack.

3.3.3.5 Rolling Rack

In the case of a rolling rack machine, it is understood that it rolls down the table length with the fabric roll kept on it. The entire arrangement sits on a rail mounted on both edges of the table and the wheels roll over it. Similar to other rolling machines, the wheels on the far side of the machine ride on the top of the opposite edge of the table. The rolling rack is preferably used for F/O/W, N/O/W, F/F and N/U/D spreading. This method has no tension control on the fabric, hence apart from smoothing out wrinkles and aligning the fabric edges, the spreader must cautiously unroll the fabric slightly ahead of the speed that the machine is advanced.

3.3.3.6 Turntable

It is another manually operated spreading machine. Similar to the rolling rack, it is manually pushed down the table, and there is no control over fabric tension like the rolling rack. Conversely, as the fabric roll is fixed on a rack that can easily be rotated, the turntable is ultimate for F/F, N/O/W, F/O/W and N/U/D modes of spreading.

3.3.3.7 Semi-Automatic Rolling Rack with Electric Eye and Catchers

Semi-automatic spreading machines are designed with electric eye edge sensors that use a servo-motor to move the rack side to side to align the fabric selvedge on the control side of the table. Moreover, for F/O/W, N/O/W spreading at one end, and for F/F, N/O/W spreading at both ends, a mechanical catcher device is used.

3.3.3.8 Automatic Rolling Rack

It has a drive motor and end switches that allow the machine to automatically drive itself from beginning to the end of the table and back. The machine can be fitted with an end cutter that would also automatically cut off the end for F/O/W, N/O/W spreading (Gardiner 2003). The fundamental components in this machine consist of carriage, wheels travelling in guide rails at the top edge of the table, a fabric support and guide collars to aid the perfect unrolling of the fabric as shown in Figure 3.6. An advanced version of the spreading machines consists of a platform for the spreader to walk, a motor to drive the carriage, an automatic ply catcher and cutting device, a ply counting device, automatic ply alignment system using photoelectric sensors and alignment shifters, a turntable and a positive drive for the fabric support which is synchronised with the spreading speed to minimise the fabric tension during spreading (Rogale and Polanovi 1996; Leaf 1999; Nayak and Khandual 2010).

FIGURE 3.6
Fully automatic spreading machine. (Reproduced by permission of IMA spa, Italy.)

3.3.3.9 Automatic Turntable

The common form of turntable has all the features of the automatic rolling rack, may use a cut-off knife mechanism, and is self-powered. It requires the spreader to manually rotate the fabric turntable rack when spreading F/F, N/O/W and F/O/W, and N/U/D modes of spreading (Leaf 1999). The fully automatic turntable spreader is also capable of rotating the fabric as well (Figure 3.7).

3.3.3.10 Tubular Knit Fabric Spreader

Tubular knit fabrics create an exceptional challenge during spreading. Since two fabric layers are being kept on the table from the roll concurrently, a frame is inserted inside the fabric tube to control both layers as they traverse along the machine and avoid folding of fabric. Positive feed roll arrangement minimises the tension in the fabric during spreading (Leaf 1999; Glock and Kunz 2004; Ukponmwan et al. 2001).

3.3.4 Nature of Fabric Packages

The option of fabric packages to be delivered to a cutting department depends on the characteristics of fabric and the spreading method employed

FIGURE 3.7
Automatic turntable spreader. (Reproduced by permission of IMA spa, Italy.)

(Mehta 1992). The most commonly used types of fabric packages are given below:

- *Open fabric – rolled:* The majority of fabrics are supplied in a rolled form as a single layer wound directly onto a tubular cardboard about 7–8 cm in diameter. The width of the open fabric may vary from less than 75 cm to over 3 m, particularly in case of knitted fabrics.
- *Tubular knitted fabric – rolled and plaited:* This kind of form is utilised for the manufacture of garments like underwear, sports shirts and t-shirts. Plaiting, which is presenting the fabric in width-wise folds, facilitates preventing tension in the fabric.
- *Folded fabric – rolled and plaited:* This is common for woollen and woollen mixture fabrics used in tailored garments. The fabric is rolled onto a flat board and the width of the fabric varies from about 70–80 cm folded.
- *Plaited folded fabric:* This is used more commonly for checks and a few tubular knitted fabrics to avoid the distortions due to tight rolling.
- *Velvet – hanging:* More rarely, velvets may be delivered wound on specially constructed frames to prevent the pile from becoming crushed.

3.3.5 Advancements in Spreading

- *Fabric defect marking sensors:* In this system, a reflective label is normally fixed at the selvedge of the fabric during the fabric pre-inspection. Automatic spreading machines having this sensor detect the label as it crosses the electric edge control eye and stops the machine and allows spreader to trace the defect (Mehta 1992).
- *Air flotation tables:* The cutting tables having air jets fitted at the bottom of the table facilitates moving the entire spread down the table.
- *Vacuum table:* The vacuum is applied from the bottom of the table through small holes in the table after a polythene cover is spread over the top of the entire lay to compress the lay and stabilise it.
- *Heavy roll loaders:* It is used when roll weights surpass 200 lb per roll. It has the capacity to manage rolls weighing over 1200 lb and these are used to lift the fabric rolls from the floor to the spreading machine.
- *Automated panel cutting systems:* In this system, the fabric is pulled automatically from the roll by an exact measured distance and then is cut off squarely and accurately. These are used for cutting home textile products such as table cloths, mattress pads, sheets, napkins, bedding and curtains.

3.3.6 Evaluation of Spreading Cost

The two cost determinations related to spreading are

- The labour cost for the time to spread
- The cost of fabric engaged in the spreading of garments as well as the fabric cost of ends and damages

3.3.6.1 Spreading Labour Cost

Labour cost is determined by the following formula:

$$\text{Spreading labour cost} = \frac{\text{Labour cost per hour} \times \text{Spreading time}}{\text{Garments/marker}}$$

3.3.6.2 Spreading and Deadheading

While the spreader is deadheading (reverse movement without spreading the fabric) as in the F/O/W–N/O/W and F/O/W–N/O/W spreading modes, the spreading cost will be double the cost of the F/F spreading mode. As the labour time is directly influencing the spreading cost and quality, the utilisation of labour saving devices is important to spreading (Leaf 1998; Gardiner 2003).

3.3.6.3 The Cost of Ends and Damages

The cost of ends and damages is given by the following formula:

$$\text{Cost of damages and ends} = \frac{\text{Yards lost to damages and ends} \times \text{Fabric cost per yard}}{\text{Garments/marker}}$$

3.4 Cutting

Cutting is the process of separating a spread into garment components as a replica of pattern pieces on a marker. It also involves transferring marks and notches from the marker to garment components to facilitate sewing. The cutting process is frequently done in two stages: rough cutting and the final accurate cutting (Aldrich 2002a,b; Vilumsone-Nemes 2012).

3.4.1 Objectives of Cutting

The main purpose of cutting is to separate fabric plies as replicas of the patterns in the marker plan. In attaining this objective, certain requirements must be fulfilled.

3.4.1.1 Accuracy of Cut

The garment components have to be cut accurately and precisely as per the shape of the pattern to facilitate assembling process and for better fitting of garments. The effortlessness in achieving this accuracy is based on the cutting method engaged and on the marker.

3.4.1.2 Clean Edges

The fabric edges after cutting should not show fraying or snagging. These defects are due to an imperfectly sharpened knife, which could result in heat generation due to friction with fabric which leads to fabric damage. The heat generation during cutting with knives could be reduced by means of using sharpened knife blades, serrated or wavy edge knife, utilisation of anti-fusion paper between fabric, spraying of lubricant over the blades and reducing the lay height and blade.

3.4.1.3 Support of the Lay

The cutting method should provide the support for the fabric in addition to allow the blade to pierce the lowest ply of a spread and separate all the plies.

3.4.1.4 Consistent Cutting

Based on the method of cutting employed, the lay height will vary. To get a consistent quality of cutting, the lay height should be as low as possible without affecting the production planning and quality of cutting.

3.4.2 Preparation for Cutting

After the laying process has been completed, the spreader has to recount the numbers of plies as in the cutting ticket. Then the following additional steps have to done prior to cutting.

3.4.2.1 Moving the Spreading Machine Aside

The spreading operator will place the spreading machine aside and remove catchers if they were used. The spreading machine must be placed back far enough from the lay to permit the cutter to work.

3.4.2.2 Facilitating Shrinkage of the Lay

If the lay is knitted fabric, then the lay should be cut into sections and left on the spreading table overnight to relax. These sections are cut at natural splice sections in the lay. The cutter would cut between the components through the fabric width to release the tension in the plies nearby the table.

3.4.2.3 Rechecking the Marker

After the spreading process is completed, the marker is kept on top of the spread. The beginning line in the marker is aligned at the starting point of the spread. The spreader has to ensure that the length and width of the spread matches with the length and width of the marker.

3.4.2.4 Fastening the Marker to the Spread

The methods for fastening the marker to the lay of fabric are given below.

- Cloth weights – Cloth weights made of metal about 2–10 lb can be used to hold the marker down on the lay.
- Lay tacks (sharp staples) – In this method a lay tacker, similar to a stapler, is utilised to hold the marker by pressing them with the top layer of fabric to keep the marker in place and stabilise the spread.
- Straight T-pins – Straight T-pins of 1 ½″ to 3″ long are used on softer woven fabrics such as wools and wool blends, and terry cloth.
- Light spray adhesive – In this method, the bottom portion of the marker is covered with a rubber type adhesive to hold the marker to the top layer of the fabric and it can be easily separated after cutting.

3.4.3 Methods of Cutting

In most of the cutting methods, a sharp blade is pressed against the fibres of the fabric to separate them. The cutting knife has to present a very thin edge to the fibres, to shear the fibres without exerting a force that will deform the fabric. The act of cutting desharpens the blade, which should be sharpened frequently (Rogale and Polanovi 1996).

3.4.3.1 Fully Manual Methods

1. Hand shears – Hand shears are commonly utilised for cutting single or double fabric plies. The lower blade passes under the plies; however, the consequent distortion of the fabric is temporary and accurate cutting to the line can be attained only with practice.

The major drawback in this method is that it is a time intensive one and incurs a higher labour cost per garment.

2. Short knife – It pierces through the fabric; 10 to 12 fabric layers could be accurately cut. Heavy weight or denser fabrics have to be used for cutting using this short knife as it distorts several fabric layers while cutting through the fabric.

3.4.3.2 Manually Operated Power Knives

Portable power knives are normally moved manually through a lay by means of an operator. Two main kinds of power knives are vertical straight knives and round knives. Construction-wise, both the knives have a base plate, power system, handle, cutting blade, sharpening device and blade guard. The round knives operate with a single force as the circular blade makes contact with the fabric, but the vertical knives cut with an up-and-down action. A straight blade will always maintain a perpendicular contact with the lay (90°) so that all the fabric plies in a spread could be cut at the same time. However, this will not be the case for a rotary knife blade as it contacts the spread at a certain angle. In both cases, the fabric that has to be cut is kept stationary and the knife blade fixed on the machine is moved by an operator to cut the fabric. The basic elements of manually operated power knives are given below:

- Knife blades – Knife blades have a major influence on the quality of the cut. The performance of the knife blades are influenced by factors such as the blade edge, surface texture of the blade, fineness of the blade edge and blade composition. Blade edges may be straight with a flat surface, saw-toothed, serrated or wavy surface. Straight edge blades are used for general-purpose, serrated blades to reduce heat generation during cutting, wavy edges for cutting plastics and vinyl, and saw-type blades for cutting canvas.

- Base plate – It supports and balances the equipment. It guides the knife along the cutting table and raises the spread off the table for contact with the blade. It is normally supported by bearing rollers at the bottom to facilitate easy movement of the base plate.

- Power system – The power required to cut a lay depends on the lay height and fabric weight (grams per square metre, GSM). The motor horsepower determines the cutting power of the blade; higher horsepower increases machine power as well as the motor weight.

- Sharpening devices – Blades become blunt very quickly while cutting higher spread height or heavy weight fabrics which leads to frayed or fused edges. Sharpening devices such as emery wheels, abrasive belt sharpeners or stone could be used on the machine.

- Handle – It is used to guide and impel the knife through the spread. The operator stabilises the fabric plies on the other hand, which is ahead of the knife to prevent bunching of the fabric.

The different types of power knives are described below:

1. *Straight knife:* This is the most frequently used equipment for cutting garments in bulk. It comprises a base plate, vertically moving blade, an upright, a motor for providing the power for cutting the fabric plies, a handle for the cutter to direct the blade, and a sharpening device as shown in Figure 3.8.

   Typically, the height of the knife blade varies from 10 to 33 cm and strokes vary from 2.5 to 4.5 cm. The straight knife is versatile, portable, cheaper than a band knife, more accurate on curves than a round knife, and relatively reliable and easy to maintain. In a few cases, a straight knife system is used as the preliminary process to cut the lay and then a band knife is used for accurate cutting as the final process.

2. *Servo assisted straight knife:* A development from a straight knife machine has a travelling suspension system which supports the knife from the top, hence heavy base plate and rollers could be

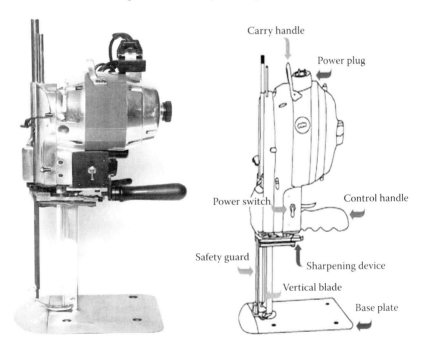

FIGURE 3.8
Straight knife cutting machine.

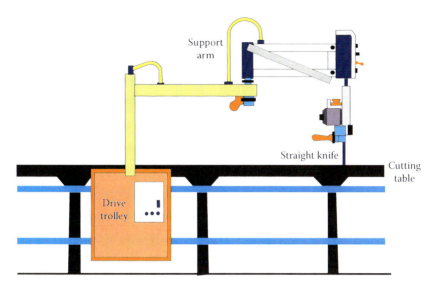

FIGURE 3.9
Straight knife with servo assisted arm support.

changed with a small, flat base (Figure 3.9). These servo knife systems provide a higher degree of cutting precision than the previous
version of unsupported straight knife systems, with the requirement
of less operator skill.

3. *Round knife:* The basic elements of a round knife are analogous to a
   straight knife except it has a round blade as shown in Figure 3.10.
   The blade diameter varies from 6 to 20 cm. Round knives are not

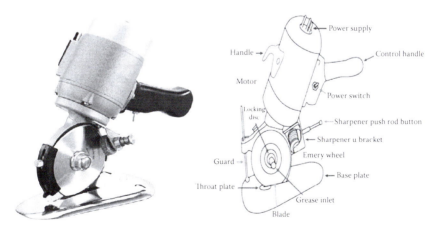

FIGURE 3.10
Round knife cutting machine.

appropriate for cutting curved lines especially in high lays as the circular blade could not cut all the plies at the same point as well as the same time as in a vertical blade. Hence, it could be utilised only for cutting straight lines rather than curved ones.

4. *Band knife:* It is normally engaged for accurate cutting of garment components. It consists of an electrically powered motor and a constantly rotating steel blade mounted over it (Figure 3.11). In this cutting system, the knife is stationary which moves through a small slot provided in the table and the fabric has to be moved manually to the blade area for accurate cutting.

5. *Die cutting:* The die is a knife blade in the profile/shape of a pattern margin, including notches (Figure 3.12). It involves forcing a firm blade through a fabric lay. Free-standing dies normally have two categories. One kind is a strip steel, which cannot be sharpened and must be replaced when worn and another one is forged dies, which can be resharpened but the cost is five times higher than strip steel. The position of the tie bars, which hold the die, determines the depth of the cut. Free-standing gives higher accuracy of cutting and is used for cutting the small components of larger garments like collars and pockets (Fairhurst 2008).

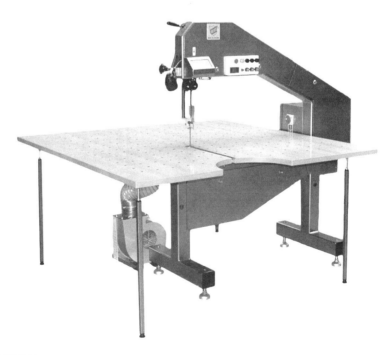

FIGURE 3.11
Band knife cutting machine.

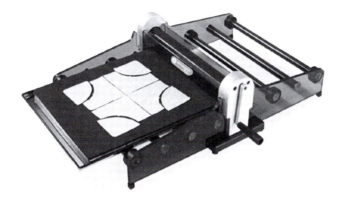

FIGURE 3.12
Die cutting machine.

3.4.3.3 Computerised Methods of Cutting

1. *Computer controlled knife cutting:* This method gives the most precise and accurate cutting at high speed. The complete setup of a computerised cutting system is shown in Figure 3.13. A characteristic computerised cutting system has nylon bristles at the top of the cutting table to support the fabric lay, which is flexible enough to allow penetration and movement of the blade through it. It also allows passage of air through the table to produce a vacuum for decreasing the lay height. The frame/carriage supporting the cutting head has two synchronised servo-motors, which drive it on tracks on the edges of the table. A third servo-motor keeps the cutting head at an accurate position on a beam through the width of the carriage. The cutting

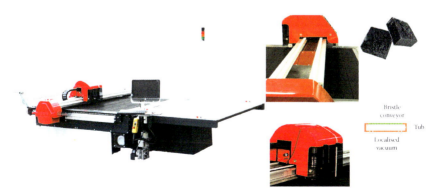

FIGURE 3.13
Computer controlled knife cutting and table. (Reproduced by permission of IMA spa, Italy.)

head includes a knife, sharpener and a servo-motor to rotate the knife to position it at a tangent to the line of the cut on curves. An airtight polyethylene sheet could be spread over the top of the lay to facilitate vacuum creation in the lay to reduce the lay height. A control cabinet houses the computer and the electrical components required to drive the cutter, its carriage and the vacuum motor (Beazley and Bond 2003; Rogale and Polanovi 1996).

An operator spreads the fabric lay on a conventional cutting table or cutting table equipped with air flotation or conveyorised cutting table. Perforated paper is spread below the bottom fabric ply to support it during cutting as well to avoid distortion during moving to the cutting table. After loading the disc having the marker plan into the computer, the operator positions the cutting head's origin light over the corner of the spread (reference point). A motorised drill at the back of the cutting head provides drill holes as required and facilities are available to cut the notches as well. The maximum height is usually 7.5 cm when compressed, with the height before compression, and hence the number of plies, being based on the nature of the fabric.

As the computerised cutting system works on the predetermined instructions from the computer/disc, markers are not compulsory for this type of system. However, to identify the cut garment panels for sorting and bundling, labelling of garment components that are to be cut is required.

2. *Laser cutting:* A laser produces a beam of light that could be focused into a very small point (0.25 mm) to produce high energy density and result in localised increase in temperature. In this system, cutting takes place by way of burning, melting and vaporisation. The limited depth of fabric cutting (single or two plies) is the major drawback of this system.

The cutting system comprises a stationary gas laser, a cutting head carrying a system of mirrors to reflect the laser beam to the cutting line, a computer which operates the entire system and a system for removing cut parts from the conveyor carrying the single ply of fabric (Figure 3.14).

An automatic, single ply, laser cutting system is speedy compared with automatic multiple ply knife cutters, with speeds of 30–40 m/min being realised compared with 5–12 m/min for knife cutting. The main hindrances to utilising laser cutters in the garment industry are the quality of the cut edge (which may become charred and, with thermoplastics, may affect the feel of the edge), the possibility of less than 100% efficient cutting and the requirements to maintain the equipment.

FIGURE 3.14
Laser cutting machine.

3. *Plasma cutting:* The plasma cutting process was developed to satisfy a demand for high quality accurate cutting on stainless steel and aluminium; however, it could also be utilised to cut textile materials. In this system, cutting is accomplished through a high velocity jet of high temperature ionised gas (argon). This method has the potential to become the faster cutter of single plies, but the cutting method has similar issues as in laser cutting related to quality of cutting.

4. *Water jet cutting:* A high velocity, small diameter stream of water is generated by applying high pressure water to a nozzle (Figure 3.15). The high pressure water jet acts as a means to cut the fabric, tearing the fibres on impact. As the water jet penetrates succeeding plies in a spread, the energy decreases and cutting capability is also reduced. The water jet spreads out and the cutting point becomes wider at the

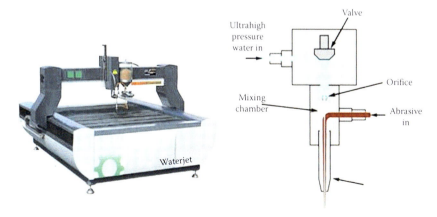

FIGURE 3.15
Plasma cutting machine.

bottom of the lay. There is a problem of water spotting, wet edges and inconsistent cutting quality.

5. *Ultrasonic cutting:* In this cutting system, vibration frequencies in the 20 kHz range are used to produce 1/20 mm movement in the blade, small enough to remove the need for a bristle base to the cutting table. Disposable knife blades save sharpening time and last for 10–14 days. Single ply and very low lays can be cut and low vacuum only is needed.

3.4.3.4 Auxiliary Devices

- *Notchers*: Notchers are machines used to create notches in the edge of cut components.
  - *Cold notcher* – The cold notcher is a spring-loaded device with a small blade fitted on a plunger. For making a notch in the fabric panels, it is kept at the edge of the panel where the notch has to be produced and by a single downward stroke the notch is cut into the edge of fabric plies (Sumathi 2002).
  - *Hot notcher* – In loosely constructed woven or knitted fabrics, the cut notch will vanish in the edge fraying during handling each component. To make a permanent notch, a hot notcher (Figure 3.16) is utilised. It uses a vertical heated edge to burn a

FIGURE 3.16
Hot notcher.

notch without the danger of melting or scorching into the edge of the bundle.

- *Ink notcher* – It is analogous to the hot notcher except after burning a notch it leaves a drop of UV marking ink that is visible under UV light.

- *Cloth drills*: Cloth drills are utilised when an identification mark is required inside the body of a panel to illustrate the dart point, pocket location, or location of an inner element such as a pocket or appliqué.
  - *Cold drill* – It cuts a tiny circle of fabric plies as it drills down through the fabric lay.
  - *Hot drill* – It utilises an electrically heated solid shaft for drilling, which leaves a burn mark to create a permanent identification on loosely constructed woven and knitted fabrics. The hot drill machine is shown in Figure 3.17.
  - *Thread markers* – It uses a needle that penetrates all the fabric plies in the lay (Figure 3.18). The thread carried by a needle is left in the fabric, which shows a location of a drill hole. This is suitable on loosely constructed woven and knits where use of a hot notch could lead to fabric damage.
- *Inside slasher* – It is a device used to cut the inside slashes for interior 'slash' pockets. The cut is completely inside of the component, thus cutting from the fabric edge becomes impossible. The device has a double edge blade that reciprocates and is inserted from above the part bundle, where the part bundle is moved under the knife.

3.4.4 Preparation of Cut Work for Sewing Room

The essential preparatory activities for sewing are bundling, shade separation, indicating the face side of the fabrics and work ticketing (Jacob 1988; Wong 2000, 2011; Ruth and Knuz 2004).

3.4.4.1 Bundling

Most of the sewing rooms use the bundling system, where small batches of garments move from one workstation to another in a controlled manner. In order to prepare the cut work, it is essential for operators to be able to identify each pile. This is the function of the marker, if used, as the style number, the size and the part identification will be part of the plot. If markers are not used, a top-ply labelling system is required.

3.4.4.2 Shade Separation

Shade variation in fabric roll is common. However, within the batch of cut components, there are likely to be shade differences. It can be ensured in

FIGURE 3.17
Hot drill machine.

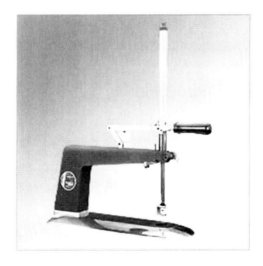

FIGURE 3.18
Thread markers.

cutting sections by inserting tissue paper between every piece. With quality outerwear garments, it is quite common to give every garment piece a pressure-sensitive adhesive ticket with a ply number known as soabaring.

3.4.4.3 Indication of the Face Side of Fabrics

Few fabrics have a noticeable difference between the face and back, which does not pose any problem for machinists to identify it. However, the fabrics that are identical on both sides pose a problem. The need for identification of face side becomes crucial when there is a close resemblance between the face and the back side of the fabric. Right side identification may use soabar tickets, whereby the ply number is always positioned on the fabric face.

3.4.4.4 Work Ticketing

Whenever the bundling system is used, it should be accompanied by work tickets or bundle tickets. It gives fundamental information about the work such as the style number, the size of the garment, the number of garments in the bundle and the date issued. Work tickets are usually created on site once the outcome of spreading/cutting is known.

References

Abernathy, F.H. and J.T. Dunlop. 1999. *A Stitch in Time – Apparel Industry.* Blackwell Scientific Publications, Oxford, UK.

Aldrich, W. 2002a. *Metric Pattern Cutting.* Blackwell Science Ltd, Oxford.

Aldrich, W. 2002b. *Pattern Cutting for Women's Tailored Jackets: Classic and Contemporary.* Blackwell Science Ltd, Oxford.

Anon. 1993. A system for made-to-measure garments by Telmat Informatique France. *Journal of SN International* 12(93):31–4.

Beazley, A. and T. Bond. 2003. *Computer-Aided Pattern Design and Product Development.* Blackwell Publishing, Oxford.

Bhuiyan, Md.T.H. 2015. Marker Making Planning Efficiency and Use of CAD. http://textilelearnerblogspot.com/2014/06/an-overview-of-garments-marker-making.html (accessed on March 10, 2015).

Burbidge, G.M. 1991. Production flow analysis for planning group technology. *Journal of Operation Management* 10(1):5–27.

Carr, H. and B. Latham. 2006. *The Technology of Clothing Manufacture.* Blackwell Science, Oxford.

Chuter, A.J. 1995. *Introduction to Clothing Production Management.* Blackwell Scientific Publications, Oxford, UK.

Clayton, M. 2008. *Ultimate Sewing Bible – A Complete Reference with Step-by-Step Techniques.* Collins & Brown, London.

Eberle, H., H. Hermeling, M. Hornberger, R. Kilgus, D. Menzer and W. Ring. 2002. *Clothing Technology*. Haan-Gruiten Verlag Europa-Lehrmittel Vollmer GmbH & Co, Nourney.

Fairhurst, C. 2008. *Advances in Apparel Production*. The Textile Institute, Woodhead Publication, Cambridge.

Gardiner, W. 2003. *Sewing Basics*. Sally Milner Publishing, Australia.

Glock, R.E. and G.I. Kunz. 2004. *Apparel Manufacturing – Sewn Product Analysis*. Prentice Hall, Englewood Cliffs, NJ.

Jacob, S. 1988. *Apparel Manufacturing Hand Book – Analysis Principles and Practice*. Columbia Boblin Media Corp, New York, USA.

Knez, B. 1994. *Construction Preparation in Garment Industry*. Zagreb Faculty of Textile Technology, University of Zagreb, Zagreb, Croatia.

Laing, R.M. and J. Webster. 1998. *Stitches and Seams*. Textile Institute Publications, UK.

Mathews, M. 1986. *Practical Clothing Construction – Part 1 & 2*. Cosmic Press, Chennai.

Mehta, P. V. 1992. *An Introduction to Quality Control for Apparel Industry*. CRC Press, Boca Raton, FL.

Nayak, R. and A. Khandual. 2010. Application of laser in apparel industry. *Colourage* 57(2):85–90.

Ng, S.F., C.L. Hui and G.A.V. Leaf. 1998. Fabric loss during spreading: A theoretical analysis and its implications. *Journal of Textile Institute* 89(1):686–95.

Ng, S.F., C.L. Hui and G.A.V. Leaf. 1999. A mathematical model for predicting fabric loss during spreading. *International Journal of Clothing Science and Technology* 11(2/3):76–83.

Rogale, D. and C.S. Polanovi. 1996. *Computerised System of Construction Preparation in Garment Industry (in Croatian)*. University of Zagreb, Faculty of Textile Technology, Zagreb, Croatia.

Ruth, E.G. and G.I. Kunz. 2004. *Apparel Manufacturing – Sewn Product Analysis*. Prentice Hall, Englewood Cliffs, NJ.

Shaeffer, C. 2000. *Sewing for the Apparel Industry*. Woodhead Publication, Cambridge.

Sumathi, G.J. 2002. *Elements of Fashion and Apparel Designing*. New Age International Publication, New Delhi, India.

Tyler, D.J. 1992. *Materials Management in Clothing Production*. Blackwell Scientific Publications, Oxford, UK.

Ukponmwan, J.O., K.N. Chatterjee and A. Mukhopadhyay. 2001. *Sewing Threads*. Textile Progress, The Textile Institute, Manchester.

Vilumsone-Nemes, I. 2012. *Industrial Cutting of Textile Materials*. Woodhead Publications, Cambridge.

Wong, W.K., C.K. Chan and W.H. Ip. 2000. Optimization of spreading and cutting sequencing model in garment manufacturing. *Computer Industry* 43(1):1–10.

Wong, W.K, Z.X. Guo and S.Y.S. Leung. 2011. Applications of artificial intelligence in the apparel industry: A review. *Textile Research Journal* 81(18):1871–92.

Workman, J.E. 1991. Body measurement specifications for fit models as a factor in clothing size variation. *Clothing Textile Research Journal* 10(1):31–6.

4

Sewing Machine

4.1 Classification of Sewing Machine

Sewing machines are normally considered general working and from the perspective of its technical features sewing machines could be classified as shown in Figure 4.1.

1. *Basic sewing machines* comprise machines that sew with a lock stitch and multi-thread chain stitch. These are mainly intended for attaching garment components that are not exposed to high amounts of loads during wear using a lock stitch. Garment components that undergo higher load during wear require basic sewing machines that utilise multi-thread chain stitch (Araujo 1992; Anon 1993; American Efird 2014).

2. *Special sewing machines* are proposed for specific technological operations and they can be categorised by (i) function, such as machines for pocket piping, sewing zips, knitted fabrics, etc. and (ii) stitch class and types except 301 and 401 stitch types, with blind stitch, zigzag stitch, for attaching pocket bags, etc.

3. *Sewing automata* are sophisticated specialised machines. The key features of these kinds of machines are its capability to perform automatic sewing when the fabric is positioned and the machine is actuated, cut the thread after sewing, release the fabric, etc.

4. *Sewing systems* are under advanced sewing machines. They have all the features of sewing automata and facilitate the automatic performances of two or more sub-operations.

5. *Numerical-controlled sewing machines* are machines where the fabric is guided automatically as in the case of sewing automata and sewing systems, but which follows a predetermined seam contour line. The numerical data are stored in the memory of a computer.

6. *Robotic sewing systems* have a multifunctional manipulator, where reprogramming and designing possibilities can be enabled for easy sewing.

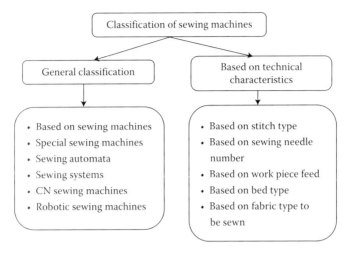

FIGURE 4.1
Classification of sewing machines.

4.1.1 Sewing Machine Classification Based on Its Bed Type

The machine bed is the lower part of the machine and feed dog and loopers are positioned beneath it. Table 4.1 shows the various kinds of machine beds along with their uses (Beazley and Bond 2003; Gersak 2013).

4.1.2 Sewing Machine Classification Based on Machine Type

The sewing machine classification based on the machine type is given in Table 4.2.

4.2 Sewing Machine Parts and Functions

The basic sewing machine components are shown in Figure 4.2 and their functions are listed in Table 4.3.

4.3 Single Needle Lock Stitch Machine

These kinds of sewing machines are generally used for sewing fabric, leather, etc. particularly one that uses two threads such as an upper and a lower thread. Figure 4.2 shows the single needle lock stitch machine. The one side

TABLE 4.1

Classification of Sewing Machine Based on Bed Type

Sewing Machine	Features and Applications	Stitch Type
Flat-bed machine	A vast working space permits a wide range of sewing applications and is utilised for all types of flat sewing work.	Lock and chain stitch
Raised bed machine	The machine bed is in the form of a pedestal which helps in assembling of presewn parts. This is specifically used for attachment of accessories and special attachments.	Lock and chain stitch
Post bed machine	It has a raised working machine bed and is used for stitching of three-dimensional products such as shoes and bags.	Lock and chain stitch
Cylinder bed machine	It has a horizontal arm-shaped bed as well as increased working height. It is most suited for sewing tubular components like sleeves, cuffs and trouser legs, and can also be utilised for button sewing and bar tacking.	Lock and chain stitch
Side bed machine	These are dedicated for edge sewing and requires a lesser working space	Chain and over-edge stitch

TABLE 4.2

Classification of Sewing Machine Based on Machine Type

Machine Types	Applications
1. Lock stitch machine	Straight and zigzag seams
2. Chain stitch machine	
3. Double chain stitch machine	
1. Blind stitch machine	Blind stitch and hemming
2. Linking machine	Linking machine attaching trimming and cuff of knitted fabrics
1. Over-edge machine	Edge neatening and seam closing
2. Safety stitch machine	Safety stitching
1. Buttonhole machine	Specific sewing operation
2. Button sewing machine	
3. Bar tack machine	
1. Profile sewer	Automatic, complex sewing operation
2. Pocket sewer	

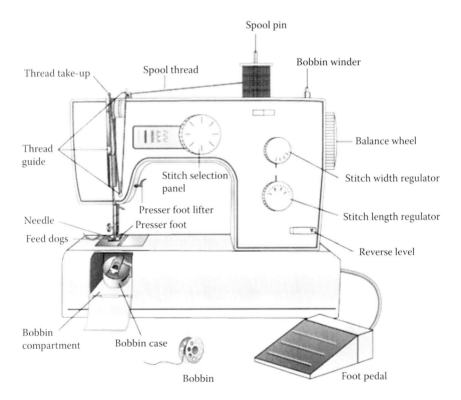

FIGURE 4.2
Sewing machine components.

TABLE 4.3

Sewing Machine Parts and Their Functions

Sl. No	Name of the Component	Description
1	Foot pedal 	It controls the speed of the machine which depends on the force exerted on it. But it is not an essential part of high-speed sewing machines as the machine speed can be set by one single adjustment and start and stop of the sewing machine is then controlled with the push of a button.
2	Power cord and switch 	The electricity for the machine is supplied by the power cord which has to be connected tightly to the machine for constant supply of power. The power switch is used for switching ON and OFF of the sewing machines electrically.
3	Hand wheel 	It is used for slowly raising and lowering the sewing needle manually to provide better control to position fabric under the needle. The clutch knob positioned inside the wheel acts as a safety feature, that is, when the knob is pulled out, it avoids the needle from jabbing up and down while winding a bobbin.
4	Reverse lever 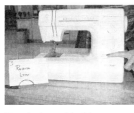	It is situated on the front side of the machine. This is used for making reverse stitching while sewing at the end of every seam to secure it.
5	Spool pin and holder 	It holds the sewing thread besides controls the sewing thread direction as it goes through the machine.  Spool pin/creel

(Continued)

TABLE 4.3 (Continued)

Sewing Machine Parts and Their Functions

Sl. No	Name of the Component	Description
6	Bobbin winder 	It is used to wind the bobbin thread on the empty bobbin. Bobbin winders can be located at the top or right side of the machine.
7	Pattern selector 	It is used to decide the kind of stitch to be sewn on the fabric, such as straight stitches or zigzag or an embroidery stitch. Based on the machine type, a variety of stitches can be selected beside straight stitches.
8	Stitch length adjustment 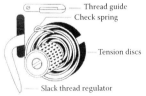	• Stitch length determines the length of the stitch • The range on the machine is from 0 to 4. 0 is the shortest stitch, 4 is the longest. • The stitch length adjustment adjusts the length of stitches the sewing machine makes. The adjustment takes place at the feed dog not the machine needle. • Shortening the stitch length shortens the amount of fabric that is fed under the presser foot before the needle comes down and vice versa.
9	Tension disks	Thread tension determines the looseness or firmness of the stitch. Tensions disks control the pressure applied to the thread for uniform feed to the machine needle. The main functions of tension device is to 1. Position the thread to needle 2. Regulate the flow of the thread 3. Maintain the smoothness in stitching 4. Control the thread passage precisely There are two kinds of tension device, such as direct tension device and indirect tension device. Both types have parts like (a) pressure disk, (b) tension spring, (c) thumb nut, (d) tension mounting bar and (e) pressure releasing unit. On high speed and modern machines, the tension dial with numbers graduated on it is used for

(Continued)

TABLE 4.3 (Continued)

Sewing Machine Parts and Their Functions

Sl. No	Name of the Component	Description
		varying the tension. The higher the number, the greater the tension and vice versa. When the tension is adjusted correctly, the stitch line will be straight and even on either side of the fabric.
10	Needle and needle clamp 	The needle fits into the needle bar, which holds it in place with a small screw. The needle clamp is used to fix the needle in place.
11	Take-up lever 	The take-up lever moves up and down during the stitch formation to provide the extra thread while forming the loop and takes back the needle thread after each stitching to set the stitch. It is used to regulate the needle thread tension at an optimum level.
12	Presser foot 	It is used to grip the fabric from the top counter to the feed dog; therefore, the feed dog can move the fabric through the machine. It applies downward pressure on the material as it is fed under the needle. 
13	Presser dial 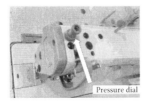	The presser dial determines the quantity of pressure to be exerted on the fabric through the presser foot. Lighter weight fabrics necessitate higher pressure for better control of fabric during stitching and vice versa.

(Continued)

TABLE 4.3 (Continued)

Sewing Machine Parts and Their Functions

Sl. No	Name of the Component	Description
14	Feed dog	Feed dogs are a 'teeth-like' component that combines with the presser foot to transport the fabric by one stitch. It also regulates the stitch length by adjusting the fabric movement per stitch.
15	Face plate	It is a cover that conceals all the internal working elements of the machine.
16	Throat plate	It has a hole for the needle to go through to the bobbin casing, a pair of slots for the feed dog to move and stitching guide lines. It is a removable part, which covers the bobbin and bottom of the sewing machine.
17	Sewing light	It aids in threading the needle and allows you to see the stitching in both day and night.
18	Presser foot lever	It is used to engage and disengage the presser foot on the fabric against the feed dogs gently. When it is in the upward position, the tension disks are disengaged and vice versa.

(Continued)

TABLE 4.3 (Continued)

Sewing Machine Parts and Their Functions

Sl. No	Name of the Component	Description
19	Thread cutter	Sewing machine thread cutters are usually located behind the needle of the sewing machine, so that it is convenient while the fabric is moved to the back of the machine, the sewing thread can be cut using the thread cutter.
20	Slide plate	It is a plastic cover that protects the bobbin case from the dirt and dust. It also gives the open space for accessing the bobbin zone under the sewing machine for changing the bobbins and other maintenance work to be carried out in this area.
21	Bobbin case	It is the case where the bobbin has to be fixed. This can be found under the needle plate and usually has a piece of plastic that flips up to cover the bobbin case when not sewing. Bobbin cases are not exchangeable in different sewing machines.
22	Bobbin	A bobbin is a small package that carries the bottom sewing thread and is fitted onto the bobbin case. Bobbins are filled on the bobbin winder and the thread should be evenly distributed on the bobbin.

of the needle has the eye and a sharp tip while the other side is attached to a needle bar that moves up and down to form a stitch. The arm also secures a presser foot which presses the fabric while sewing. The needle penetrates into the fabric from the top, to bring the needle thread through the fabric to the bottom to form a stitch (Chmielowiec and Lloyd 1995; Carvalho et al. 2012).

The needle thread in a spool is threaded through the thread guides, tensioning mechanism and into the sewing needle eye. The bottom thread (bobbin thread) is wound onto the small package called a bobbin before sewing and is secured inside a bobbin case. For stitch formation, the needle brings the needle thread loop to the bottom of the fabric where a small hook on the bobbin case catches it. As the bobbin hook rotates, it takes the needle thread around the bobbin to produce a stitch and then the take-up lever takes back the extra loop of top thread to tighten the stitch. As the needle is out of action of fabric, the feed dog comes above the throat plate, and pushes or moves the fabric up against the presser foot for next stitch (Chuter 1995).

4.4 Double Needle Lock Stitch Machine

This machine (Figure 4.3) is similar to a single needle lock stitch machine; however, here all items have two sets, that is, two sets of bobbin case, tensioner, take up lever, thread guides, spool pin, needle holders, etc. In this machine, the bobbin case is fixed compared to a removable one in the case of a single needle lock stitch machine. It has a synchronous tooth belt for driving, a knob-type stitch regulator and a plunger pump for lubricating. These kinds of machines are suitable for stitching shirts, uniforms, jeans, overcoats or similar clothing (Chuter 1995; Clapp et al. 1999; Shaeffer 2000).

4.5 Special Sewing Machines

4.5.1 Overlock Machine

It sews over the edge of the fabric plies for edge neatening, hemming or seaming. Generally, an overlock machine called a 'sergers' (Figure 4.4) will trim the fabric edges during sewing using cutters. The addition of automated cutters ensures these machines to produce a finished seam easily and quickly (Coats 2003; Cooklin et al. 2006).

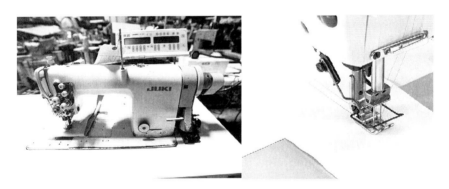

FIGURE 4.3
Double needle lock stitch machine. (Courtesy of Juki Corporation, Japan.)

FIGURE 4.4
Overlock machine. (Courtesy of Juki Corporation, Japan.)

It uses a bottom thread known as a looper thread fed by larger size cones compared to smaller size bobbins in the case of lock stitch machines. The loopers provide thread loops that pass from the needle thread to the fabric edges thereby raw edges are enclosed within the seam. Overlock sewing machines are normally run at high speeds, from 1000 to 9000 rpm.

4.5.2 Bar Tacking Machine

The bar tack machine shown in Figure 4.5 is used for sewing dense tack around the open end of the buttonhole. These machines are used to make more tight stitches across the point to be reinforced and then sew covering stitches at a right angle over the first stitches (Tyler 1992; Fairhurst 2008).

The applications of the bar tack machine in the garment industry are given below.

- Closing of buttonhole end.
- Reinforcing the ends of pocket opening.
- Sewing on belt loops.

FIGURE 4.5
Bartack machine. (Courtesy of Juki Corporation, Japan.)

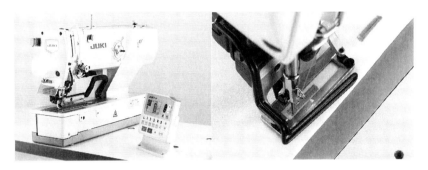

FIGURE 4.6
Buttonhole sewing machine. (Courtesy of Juki Corporation, Japan.)

4.5.3 Buttonhole Sewing Machines

Buttonhole machines are used for creating buttonholes in the garment and to finish the edges. A buttonhole machine as shown in Figure 4.6 may form a simple circle where the stitches radiate from the centre of an eyelet home, two legs on either side of a straight out with a bar tack on both ends as in a shirt, a continuous line of sewing up one leg, round the end and down the other without the cut as in a shank (Fredrerick 1999; Fairhurst 2008; Coats 2014).

The selection between lock stitch and chain stitch is based on seam security requirements and the requirements of edge finishing. In general, a buttonhole on a tailor outward utilises two thread chain stitches; the chain affect gives an attractive purl appearance to the buttonhole. The simpler shape of a buttonhole on shirts and other lightweight garments is frequently used with a single thread chain stitch and in some cases the sewing is done inside the garment.

4.5.3.1 Buttonhole Machine Types

The buttonhole sewing machines are categorised based on the following:

1. Stitch type such as lock stitch or chain stitch machines
2. Size and shape of the buttonhole
3. Method of creating buttonholes
   a. Button hole cut before stitch – The button hole is cut first and then finished with stitches. The stitches radiate from the centre to the outer edge of the buttonhole. It gives a neat appearance as the sewing thread covers raw edges of the hole effectively.
   b. Button hole cut after stitch – As the name indicates, the buttonhole edges are finished first and then slashed. It is created by a continuous line of sewing around the end and then the hole is cut subsequently.
4. Presence or absence of gimp

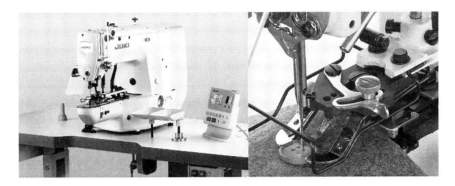

FIGURE 4.7
Button sewing machine. (Courtesy of Juki Corporation, Japan.)

4.5.4 Button Sewing Machine

A button sewing machine (Figure 4.7) is used to sew the button in the garment without damaging it. Various types of buttons like a button with two holes, four holes or shank could be sewn on this machine by making simple adjustments. The sewing action comprises a series of parallel stitches whose length is equal to the distance between the centres of the holes. The needle has only vertical movement but the button moves sideways by means of the button clamp for stitching. A hopper feed is a special attachment that automatically feeds the button to the clamp of the needle point of the machine. With this attachment, the button and needle are automatically positioned and the threads are clipped (Fredrerick 1999; Geršak 2000; Jana 2014a).

4.5.5 Feed of Arm Sewing Machine

It is used to stitch a tubular seam of narrow width on the edge of shirts and trousers (Figure 4.8). It is particularly utilised for sewing a lapped seam

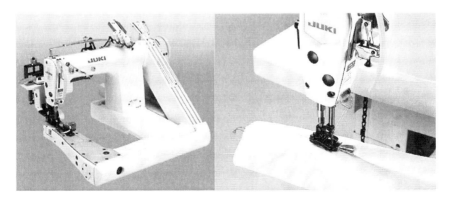

FIGURE 4.8
Feed of arm sewing machine. (Courtesy of Juki Corporation, Japan.)

FIGURE 4.9
Blind stitching machine. (Courtesy of Juki Corporation, Japan.)

which has to be closed such that the garment panels become a tube-like structure. These machines are common for sewing outside leg seams in jeans where the lap felled seam is used (Juki 1988; Geršak 2000; Jana 2014b).

4.5.6 Blind Stitch Machine

A blind stitch machine (Figure 4.9) is used for hem stitching in a knitted fabric since the hem stitch is too small in the right side of the garment and is invisible. In a few circumstances, the machine could be set to skip a stitch that is to pick up the fabric on alternate stitches only. But this type reduces the stability of the stitches (Juki 1988; Jana 2014a,b).

The sewing needle utilised for this machine is a curved one because it does not pierce through the fabric fully, but partially. Based on the application and type of fabric, the blind stitch can be grouped into two types; for fine fabric producing long and narrow stitches and for heavy fabric with short and wide stitches.

4.6 Stitch-Forming Mechanisms

The stitch-forming mechanisms are the mechanical components; with perfect synchronisation between the parts they form stitches. The various types of stitch-forming components are

- Thread control devices, which include tensioners and take-ups
- The needle
- The feed dogs
- Throat plate, tongues and chaining devices
- The presser foot
- The rotary sewing hook, loopers and spreaders
- Bobbin and the bobbin case

Proper selection of these components and precise synchronisation of them are crucial for proper formation of a stitch.

4.6.1 Thread Control Devices

It comprises thread guides, tension devices and take-up lever, which are essential to provide even and uniform thread movement. Thread guides direct the positioning and movement of the sewing thread. Faulty or damaged guides could lead to damage of sewing thread and may cause thread breakage and weaker seams. Tensioning devices control the tightness and flow of the sewing thread as it moves through the stitch-forming elements of a machine. It comprises a set of tension disks, a tension spring and a thumb nut that can be adjusted to tension on the thread (Juki 1988; Laing and Webster 1998; Clayton 2008; Jana 2014a,b). The kinds of tensioning device can be

1. Direct – It has two canvas discs, tension spring and tension screw to provide tension to the thread.
2. Indirect – These are cylindrical and conical in shape with a hook that is placed over the tension disc to provide extra tension.
3. Auxiliary – These are placed somewhere between the actual tension disc and the needle to give extra tension to the thread.

Higher thread tension could lead to thread breakage or seam puckering and lower thread tension leads to slack stitches. Thread tension must be tuned based on sewing thread, fabric construction and needle type and size (Brother Industries Limited 2000; Bheda 2002).

The take-up lever supplies adequate thread required to form each stitch. It provides additional thread to the needle to form the loop but takes it away to set the stitch. Types of take up levers are

1. Oscillating levers – These kinds of arrangements are used in a single needle machine and the gap will be 1″ for oscillating of the take up leaver
2. Rotating levers – This type of lever rotates to provide looser and tighter thread.

4.6.2 Sewing Needles

Needles hold and carry the thread through the fabric to form a stitch. Accurate formation of a stitch is dependent on a proper needle thread loop formation below the fabric that can be caught by a bobbin hook or looper (Schmetz 2001; Ukponmwan 2001).

4.6.3 Lower Stitch-Forming Devices

Lower stitch-forming components consist of a bobbin hook found on bobbin cases of lock stitch machines and loopers in chain stitch machines. A hook is a rotating device surrounding the bobbin case that picks up the loop formed by the needle thread to form a lock stitch. If it misses the loop, no stitch will be formed. This leads to skipped stitches. Hence, precision and timing are vital for stitch formation.

4.6.3.1 Loopers

It is metal piece having specific cyclic motion synchronised with the needle motion and feed dog to pick up the needle thread and aids to form stitches. It may hold lower threads to interlock with needle threads or with other looper threads. Spreaders work in combination with a looper to aid loop formation (Pareek 2012; Sewing Attachments 2015). They move the thread but do not carry the thread because of an absence of an eye in it. Two kinds of loopers based on profile and construction are

1. *Eye loopers* – Eye loopers are utilised mainly for class 400, class 600 and for all class 500 stitches other than class 501. They carry the bottom thread by means of an eye in it. The main functions of loopers are to grab the thread from the needle and to interlock the bobbin with the needle thread.
2. *Blind loopers* – Blind loopers do not have an eye and do the function of only grasping the thread from the needle. It is used for sewing machines without a bobbin and bobbin case arrangement. They are used mainly in class 100, 101, 102 and some class 500.

4.6.3.2 Stitch Tongues or Chaining Plates

These are pointed metal projection parts which are attached to throat plates. These are vital for the creation of three-dimensional stitches and their length and profile differ based on the requirements of the particular type of stitch and process. For example, thin pointed and longer stitching tongues are required for tight and close stitching like in a rolled edge, whereas a wider and flatter tongue is used for flat seaming.

4.6.3.3 Loop Spreader

Loop spreader aids the looper in creating the stitches. The blind loopers normally have two dull points, the point that grips the needle thread from the needle is the looper point and the other point that spreads the needle thread loop is a loop spreader.

4.6.3.4 Thread Finger

Thread finger is a metal link with an eye which may be static or dynamic. The static links direct the covering thread, whereas the dynamic links carry thread back and forth across the needle path. They are utilised for producing class 600 stitches and are generally synchronised with the needle.

4.6.4 Throat Plate

It encloses the area that embraces the bobbin. It has a hole for the sewing needle to pass through, a set of slots for the feed dog to move up and down and stitching guidelines. The needle hole may be a single hole (for straight stitches) or an oblong hole (for zigzag stitches).

4.6.5 Stitch Formation Sequence in Lock Stitch Machine

The first stage in making a stitch is the formation of a loop from the needle thread (Figure 4.10). If the needle thread loops are not formed properly, then it leads to several stitching problems like skipped stitches, breakage of sewing thread, loose stitches, etc.

Proper loop formation of the needle thread depends on the propensity of the needle thread to bulge away from the needle due to inertia and friction against the fabric as it is pulled in an upward direction subsequent to reaching the lowest point of its stroke (Clapp et al. 1992; Schmetz 2001; Gardiner 2003; Carr and Latham 2006; Carvalho et al. 2009). The most frequent circumstances under which the thread fails to form a loop are that the fabric is not detained tightly by the presser foot where the needle moves through, allowing the fabric to move up along with the needle as it rises as shown in Figure 4.11. This is normally called 'flagging' and in this situation, either no needle loop is formed at all or it is formed too late leading to skipped or broken stitches (Kunz 2004; Pareek 2012).

The second stage is receiving the top needle thread to go around the bobbin. This is accomplished using a bobbin hook coming across at the precise time and catching the top needle thread and drawing it around the bobbin. The timing sequence of lock stitch formation is shown in Figure 4.12.

> *Step 1*: After the sewing needle reached its lowest position, it starts to rise which causes the needle thread to bulge away from the needle to form a loop.

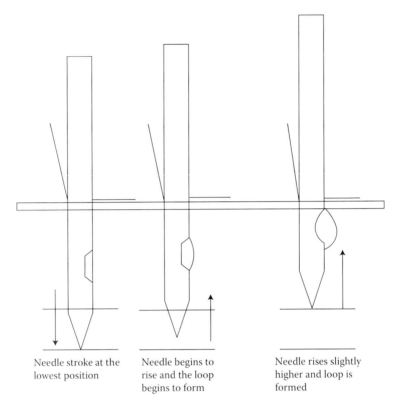

| Needle stroke at the lowest position | Needle begins to rise and the loop begins to form | Needle rises slightly higher and loop is formed |

FIGURE 4.10
Loop formation.

Step 2: The needle thread loop is then picked up by the point of the bobbin hook.

Step 3: As the needle keeps rising and the hook progresses in its rotation, the take-up lever provides extra needle thread so that it can be drawn down through the fabric to increase the size of the loop.

Step 4: On the first rotation of the sewing hook, it carries the needle thread around the bobbin case, the inside of the loop sliding over the face of the bobbin case while the outside passes around the back, to encompass the bobbin thread. As the take-up lever starts to rise, the needle thread loop is drawn up through the 'cast-off' opening of the bobbin hook before the revolution is complete.

Step 5: During the second revolution of the bobbin hook, the take-up lever completes its upward stroke, pulling the slack needle thread through the fabric to set the stitch. In the meantime, the feed dog moved the fabric along with it against the presser foot drawing the required length of under thread from the bobbin.

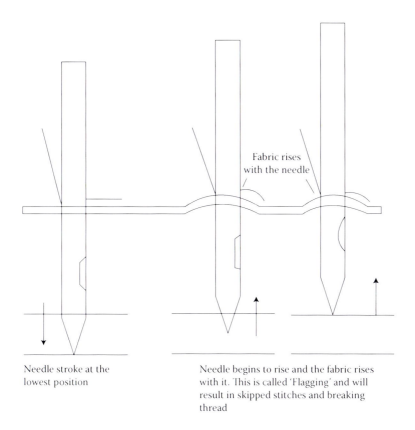

Needle stroke at the lowest position

Fabric rises with the needle

Needle begins to rise and the fabric rises with it. This is called 'Flagging' and will result in skipped stitches and breaking thread

FIGURE 4.11
Fabric flagging.

1   2   3   4   5

FIGURE 4.12
Timing sequence of stitch formation in lock stitch machine.

4.7 Embroidery Machine

In machine embroidery either a sewing or embroidery machine is utilised to make patterns on clothing materials. It is generally used in product branding, decorative purposes and corporate advertising. The types of machine embroidery are free-motion machine embroidery, link stitch embroidery and computerised machine embroidery (Geršak 2001; Jana 2014a,b).

4.7.1 Free-Motion Machine Embroidery

A basic type of zigzag sewing machine can be utilised to produce embroidery designs in the case of free-motion machine embroidery. Tightly banded fabric has to be moved beneath a needle to create a design in this kind of machine embroidery. In this case, the embroidery has to be developed manually by the operator using the machine's settings so that the tight stitches form a design or an image on a fabric.

These types of machines have only one needle, hence the operators have to stop and manually rethread for every colour in a multicolour design, which consumes lot of time. Any design created by this machine is very unique and cannot be accurately reproduced, unlike with computerised embroidery as this is a manual process rather than a digital production system (Geršak 2001; Jana 2014a,b).

4.7.2 Computerised Machine Embroidery

Modern embroidery machines (Figure 4.13) are mostly computer controlled and exclusively designed for embroidery. The embroidery machines normally comprise a frame that holds the framed area of fabric with tension below the sewing needle and automatically moves it to make a preprogrammed design which is saved in the machine.

Based on its capabilities, it requires various kinds of input from the user in a specified digital format for sewing the embroidery designs. In the case of multineedle industrial embroidery machines, threading has to be done before the running of the design and does not necessitate rethreading (Jacob 1998; Carvalho et al. 2012; Organ Needles 2014). The fundamental steps for producing the embroidery designs using a computerised embroidery machine are

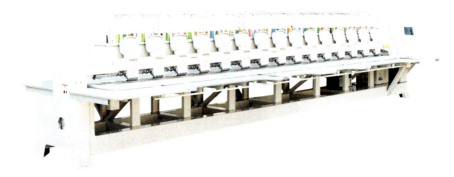

FIGURE 4.13
Computerised embroidery machine.

- Creating or obtaining a digitised design file
- Editing embroidery designs
- Loading final embroidery file into the machine
- Stabilising the fabric with adequate tension and position it on the machine
- Starting the embroidery process and monitor the process

4.7.2.1 Design Files

Digitised design files can be created on our own or purchased with embroidery software. Generally, embroidery file formats come under two categories, namely the source format, which is specific to the software and the machine format, which is particularly for a specific brand of embroidery machine. Embroidery machines commonly support one or more deign formats such as Tajima's .dst, Melco's .exp/.cnd and Barudan's .fdr based on the brand of the machine. Machine formats normally comprise primarily stitch data (offsets) and machine functions (trims, jumps, etc.) and editing of these files is very difficult and needs extensive manual work.

4.7.2.2 Editing Designs

After a design has been digitised, the editing of designs or combining it with other designs can be carried out using the embroidery software. Most embroidery software allows the user to supplement text rapidly and effortlessly.

4.7.2.3 Loading the Design

After completion of editing work, the final design file is loaded into the machine in the form of floppy disks, CDs or USB interface cables. The design format required by the machine will vary depending on the particular brand.

4.7.2.4 Stabilising the Fabric

During embroidery, wrinkling and other related issues can be avoided by stabilising the fabric before the embroidery process. The fabric stabilising method depends on the fabric characteristics, type of embroidery machine and the complexity of the design. For better stabilisation of fabric, generally additional fabric pieces known as 'interfacing' are placed on the bottom or top of the fabric or both sides.

4.7.2.5 Embroidering the Design

Finally, the machine is switched ON and the embroidery process is monitored continuously. Many designs necessitate more than one colour and may consist of extra processing for appliqués, foams and other kinds of special effects.

References

Abernathy, F. H. and J.T. Dunlop. 1999. *A Stitch in Time – Apparel Industry.* Blackwell Scientific Publications, Oxford, UK American Efird. 2014. Industrial sewing threads. http://wwwamefirdcom/products-brands/industrial-sewing-thread (accessed on January 12, 2015).

Anon. 1993. A system for made-to-measure garments by Telmat Informatique France. *Journal of SN International* 12(93):31–4.

Araujo, M., T.J. Little, A.M. Rocha, D. Vass and F.N. Ferreira. 1992. *Sewing Dynamics: Towards Intelligent Sewing Machines.* NATO ASI on Advancements and Applications of Mechatronics Design in Textile Engineering Side, Turkey.

Beazley, A. and T. Bond. 2003. *Computer-Aided Pattern Design and Product Development.* Blackwell Publishing, Oxford.

Bheda, R. 2002. *Managing Productivity of Apparel industry.* CBI Publishers and Distributors, New Delhi.

Brother Industries Limited. 2000. *Industrial Sewing Machine Handbook.* Brother Industries Limited.

Carr, H. and B. Latham. 2006. *The Technology of Clothing Manufacture.* Blackwell Science, Oxford.

Carvalho, H., A.M. Rocha and J.L. Monteiro. 2009. Measurement and analysis of needle penetration forces in industrial high-speed sewing machine. *International Journal of Clothing Science and Technology* 100(4):319–29.

Carvalho, H., L.F. Silva, A. Rocha and J. Monteiro. 2012. Automatic presser-foot force control for industrial sewing machines. *International Journal of Clothing Science and Technology* 24(1):20–7.

Chmielowiec, R. and D.W. Lloyd. 1995. The measurement of dynamic effects in commercial sewing machines. In: *Proceedings of the Third Asian Textile Conference,* Hong Kong, 2:814–28.

Chuter, A.J. 1995. *Introduction to Clothing Production Management.* Blackwell Scientific Publications, Oxford, UK.

Clapp, T.G., T.J. Little, T.M. Thiel and D.J. Vass. 1999. Sewing dynamics: Objective measurement of fabric/machine interaction. *International Journal of Clothing Science and Technology* 4(2/3):45–53.

Clayton, M. 2008. *Ultimate Sewing Bible – A Complete Reference with Step-by-Step Techniques.* Collins & Brown, London.

Coats. 2003. *The Technology of Thread and Seams.* J&P Coats Limited, Glasgow.

Coats. P.L.C. 2014. Thread numbering. http://wwwcoatsindustrialcom/en/information-hub/apparel-expertise/thread-numbering (accessed on March 22, 2015).

Cooklin, G., S.G Hayes and J. McLoughlin. 2006. *Introduction to Clothing Manufacture.* Blackwell Science, Oxford.

Fairhurst, C. 2008. *Advances in Apparel Production.* Woodhead Publication, Cambridge.

Gardiner, W. 2003. *Sewing Basics.* Sally Milner Publishing, Australia.

Geršak, J. 2000. The influence of sewing processing parameters on fabric feeding. *Proceedings of the 29th Textile Research Symposium,* Mt Fuji Shizuoka, 141–52.

Geršak, J. 2001. Directions of sewing technique and clothing engineering development. *Tekstil* 50(5):221–9.

Gersak, J. 2013. *Design of Clothing Manufacturing Processes: A Systematic Approach to Planning Scheduling and Control.* Woodhead Publishing, UK.

Glock, R.E. and G.I. Kunz. 2004. *Apparel Manufacturing – Sewn Product Analysis.* Prentice Hall, Englewood Cliffs, NJ.

Jacob, S. 1998. *Apparel Manufacturing Handbook—Analysis Principles and Practice.* Columbia Boblin Media Corp, New York, USA.

Jana, P. 2014a. *Automation in Sewing Room: Pocket Attaching in Shirt.* Apparel Resources, New Delhi.

Jana, P. 2014b. *Sewing Machine Resource Guide.* Apparel Resources, New Delhi.

Jana, P. and A.N. Khan. 2014. The sewability of lightweight fabrics using X-feed. *International Journal of Fashion Design Technology Education* 7(2):133–42.

Juki, C. 1988. *Basic Knowledge of Sewing.* Juki Corporation, Japan.

Laing, R.M. and J. Webster. 1998. *Stitches and Seams.* Textile Progress, The Textile Institute, Manchester.

Laing, R.M. and J. Webster. 1998. *Stitches and Seams.* Woodhead Publishing Limited, UK.

Organ Needles. 2014. http://organ-needlescom/english/product/downloadphp (accessed on March 21, 2015).

Pareek, V. 2012. *Stitches per inch.* http://wwwfibre2fashioncom/industry-article (accessed on January 18, 2015).

Sewing Attachments. http://wwwjanomecojp/e/pdf/home/AccessoryCatalog.pdf (accessed on March 14, 2015).

Shaeffer, C. 2000. *Sewing for the Apparel Industry.* Woodhead Publication, Cambridge.

The Schmetz. 2001. *The World of Sewing: Guide to Sewing Techniques.* Ferdinand Schmetz GmbH Herzogenrath.

Tyler, D.J. 1992. *Materials Management in Clothing Production.* Blackwell Scientific Publications, Oxford, UK.

Ukponmwan, J.O., K.N. Chatterjee and A. Mukhopadhyay. 2001. *Sewing Threads.* Textile Progress, The Textile Institute, Manchester.

5

Sewing Thread and Needles

5.1 Sewing Thread

Even a small sewing thread failure leads to losses on investments in material, equipment, garment engineering and labour. Sewing performance and seam quality could be influenced by sewing thread parameters, selection of proper thread and utilisation of thread. Sewing thread is a unique type of yarn, engineered and constructed to move through a needle and other components of a sewing machine swiftly. The fundamental task of a sewing thread is to produce aesthetic and performance in stitches and seams (Shaeffer 2000; Coats 2003).

5.1.1 Factors Influencing the Aesthetic Characteristics of Sewing Thread

Fineness, colour and lustre must be taken into account when selecting a sewing thread for decorative purposes. Other considerations influencing the aesthetics of sewing thread are

- Hue and shade
- Colour fastness
- Stitch selection
- Even stitch formation

5.1.2 Factors Affecting Performance of Sewing Thread

Sewing thread used in garments should be durable to withstand the abrasion during wear and needle heat that occurs while sewing, finishing of garments and during wear. Sewing thread performance could be assessed from its

- Seam strength
- Abrasion resistance
- Elasticity

- Chemical resistance
- Flammability
- Colour fastness

5.1.3 Basic Requirement of Sewing Thread

5.1.3.1 Sewability

Sewability is a capability of sewing thread to produce a seam with a minimum number of sewing thread breakages and the slightest damage to the thread and the fabric during the stitching process. The parameters that determine sewability of thread are

- No thread breakages in high-speed sewing
- Consistent stitch formation
- No skipped stitches
- Evenness of yarn
- Higher abrasion resistance
- Adequate smoothness of thread surface

5.1.3.2 Thread Performance in Seam

It is the capacity of a thread to produce the desired functional serviceability in a desired seam.

$$\text{Seam Efficiency Index (SEI)} = \frac{\text{Seam Tensile Strength} \times 100}{\text{Fabric Tensile strength}}$$

Proper choice of thread needs consideration of its performance during sewing as well as under conditions of wear and cleaning. As like other textile materials, sewing threads are composed of a fibre type, yarn construction and a finish, which all have an influence on appearance and the performance of the thread. Extensive ranges of thread sizes are available; the selection of the proper size depends on the fabric to be sewn as well as the needle size used in the machine (Blackwood and Chamberlain 1970; Shaeffer 2000).

5.1.4 Properties of Sewing Thread

Requirements of good quality sewing thread are given below:

- *Good tensile strength* to grip the seam firmly during wash and wear. The thread strength should be higher than the fabric so it will not rupture during wear.

- *Higher initial modulus* of thread guarantees the least thread deformation during shock loading in sewing. The sewing thread should be moderately stiff to form the loops for stitch formation.
- *Smooth surface and negligible faults* in thread provides minimum friction between the needle and sewing thread while sewing at high speed. It should be adequately lubricated to improve the abrasion resistance and its sewability.
- *Uniform thickness* provides even sewing thread, which ensures smooth passage of thread through the needle and the fabric.
- *Good elasticity* facilitates the thread to recover its original length instantly after the tension has been released. The elastic property of thread influences the strength and quality of a stitched seam.
- *Good colour fastness* of thread creates resistance to the various chemical agents the thread is exposed to during garment manufacturing and washing. The thread should be dyed evenly and uniformly.
- *Lower shrinkage* characteristics of the thread comparable to the fabric shrinkage to avoid shrinkage puckering on garments. Cotton threads usually undergo washing shrinkage puckering while synthetic threads suffer from the thermal shrinkage during ironing.
- *Better chemical resistance* is an enviable characteristic for sewing thread in garments that could undergo washing, bleaching or dry-cleaning during wear.
- *Higher abrasion resistance* provides a good sewing performance and makes the thread more durable.
- *Good colour fastness* could maintain the original colour of sewing thread without any fading while it is exposed to washing and sunlight.
- *Minimum metamerism* could be attained by the measurement of thread colour with colour matching cabinets. Metamerism is an intrinsic characteristic of a thread when the same thread colour appears to be dissimilar under diverse lighting conditions.

5.1.5 Classification of Sewing Thread

Sewing thread could be classified in numerous ways. Some common classifications are those based on:

- Substrate
- Construction
- Finish

5.1.5.1 Classification Based on Substrate

- *Natural:* The utilisation of sewing thread produced from natural fibres is very rare in industrial applications and cotton is the most frequently used natural sewing thread.
- *Synthetic:* The synthetic fibres have several advantageous characteristics compared to natural fibres such as high tenacity, better resistance to chemicals and higher abrasion resistance. Further, they are also not considerably influenced by rot, mildew, insects, bacteria and moisture (Shaeffer 2000; Coats 2003).

5.1.5.2 Classification Based on Thread Construction

5.1.5.2.1 Spun Threads (Figure 5.1)

It is produced by utilising natural as well as synthetic fibres. Spun polyester is the most frequently used sewing threads in garments. Spun threads have a hairy yarn surface, which provides better lubrication properties and softer hand. It gives exceptional sewing performance, however, it is lesser than the strength of continuous filaments (Coats 2003).

1. Cotton threads
   a. Soft cotton threads
   b. *Glazed cotton thread:* The glazed process gives the thread a hard finish that shields the thread from abrasion and improves ply security
   c. *Gassed thread:* Gassing process otherwise known as singeing process is used to remove the protruding hairs and produce a lustrous thread. It is produced by moving the cotton sewing thread over a flame at a higher speed to reduce the hairy fibres on the surface of thread

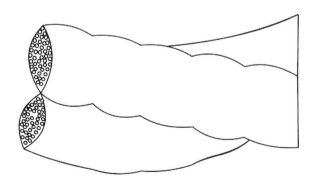

FIGURE 5.1
Spun sewing thread construction.

    d. *Mercerised cotton thread:* The cotton yarns are treated with caustic soda with 16%–18% concentration to improve the strength and lustre

2. Linen thread
3. Silk thread
4. Spun synthetic – fibre threads
5. Spun blended sewing threads

5.1.5.2.2 Core Spun Threads

Core spun thread (Figure 5.2) is a mixture of staple fibres and filaments. The most commonly used core spun sewing thread has a multiple-ply structure, with each ply comprising a core polyester filament wrapped by the cotton or polyester staple fibres. The strength of thread is provided by the filament and sewability by means of cotton or polyester fibre wrap (Coats 2003).

5.1.5.2.3 Continuous Filament Threads

It is produced by extruding the filaments from the synthetic polymer and is given a twist to improve the strength. The strength of these threads is stronger than spun threads for the same thread size (Coats 2003).

5.1.5.2.3.1 Monofilament Threads

Monofilament sewing thread (Figure 5.3) is produced from a single continuous fibre with a specific fineness. Although the monofilament sewing threads are stronger, more uniform and cheaper, they lack flexibility and are rough in feel. Because of this limitation, it is limited to sewing of hems, draperies and upholstered furniture.

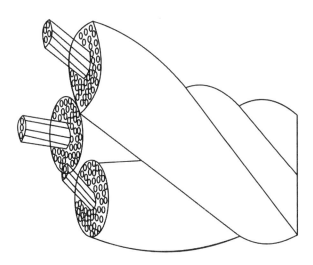

FIGURE 5.2
Construction of core spun sewing thread.

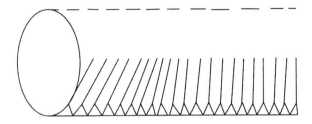

FIGURE 5.3
Monofilament.

5.1.5.2.3.2 Multifilament Threads   Multifilament sewing thread (Figure 5.4) is generally produced from nylon or polyester and is utilised where high strength is a principal requirement. It comprises two or more continuous filaments twisted together to give more strength. It is frequently used to sew leather garments, shoes and industrial products. Three kinds of multifilament sewing threads are lubricated, bonded and braided threads.

5.1.5.2.3.3 Textured Threads   The texturisation enhances texture to the continuous filament yarns by providing softness and bulk. They are then slightly twisted and heat set to make it permanent. The texturised sewing threads give exceptional seam coverage. Although these threads provide

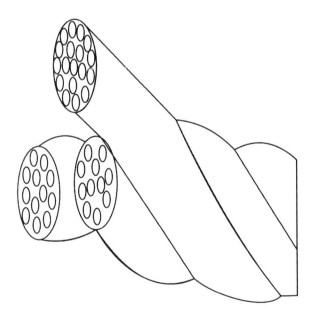

FIGURE 5.4
Multifilament.

more cover and high extensibility, they are more subject to snagging. The types of textured sewing threads are (i) false twist textured filament threads, (ii) air textured filament threads and (iii) air-jet intermingled filament threads

5.1.5.2.4 *Embroidery Threads*

- Mainly required for decorative purposes
- Colour and lustre are two main requirements for embroidery threads
- Mostly made from mercerised cotton, silk, viscose rayon and polyester fibre/filament yarns

5.1.5.2.5 *Technical Threads*

- Specifically developed for technical/industrial uses
- Perform satisfactorily in adverse climatic, industrial conditions and heavy duty applications
- Generally made from aramids, glass, ceramics, etc.

5.1.5.3 Classification Based on Thread Finish

Normally special finishes are provided to the sewing thread for two purposes:

1. To enhance the sewability of the thread – Certain finishes improve the thread strength, lubrication property and abrasion resistance.
2. To accomplish a specific functional requirement – Several types of finishes impart the special finishes such as fire retardant, water repellent, anti-fungal and anti-static finishes.

5.1.6 Twist of the Sewing Thread

It denotes the turns per unit length needed to keep the fibres or yarns together to provide the required strength as well as flexibility to the yarn/thread. Higher twist in thread leads to twist liveliness which could result in snarling, knots and loops which restricts the formation of a stitch. Sewing thread direction could be either 'S' for left twist or 'Z' for right twist (Figure 5.5). The majority of single needle lock stitch machines and other sewing machines are intended for 'Z' twist threads as 'S' twist in sewing threads tends to untwist during the formation of a stitch (Abernathy and Dunlap 1999; Fairhurst 2008; Jana 2014).

Normally single yarns of 'Z' twist are combined together and are twisted in an 'S' direction for sewing threads. Since the rotation of the bobbin hook could untwist some portion if it is twisted in a 'Z' direction during its normal

FIGURE 5.5
Direction of twist.

rotational direction, the doubling twist is normally kept in the 'S' direction (Alagha et al. 1996). For producing a sewing thread, single yarns are combined and twisted together to produce a ply thread. Further, the resultant threads are again combined and twisted together for producing a corded thread as shown in Figure 5.6.

5.1.7 Sewing Thread Size

Sewing thread size is the most significant factor in accomplishing the functional and aesthetic requirements of the finished garment. Sewing thread sizes could be expressed in direct or indirect numbering systems. Normally, metric count (Nm) is used to express synthetic, spun and core spun thread size

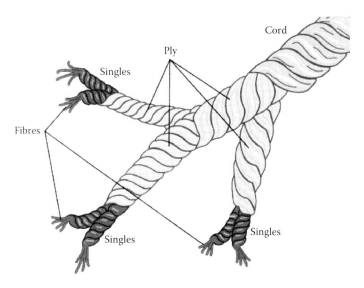

FIGURE 5.6
Construction of ply and cord threads.

while English count (Ne) is used to express the size of cotton thread. Filament sewing threads are generally expressed in denier or decitex (Mehta 1992).

In direct systems: Resultant thread size = Single yarn count/Number of plies

In indirect systems: Resultant thread size = Single yarn count × Number of plies

5.1.7.1 Ticket Numbering

Ticket numbering is a commercial sewing thread numbering method for expressing the sewing thread size. These are simply the manufacturer's reference values for the size of a given sewing thread. The metric system, English system and Denier systems are generally used to arrive at the ticket number of the finished sewing thread. A value of a ticket number in one type of sewing thread will not be the same as in another. For example, Ticket 40 cotton is not the same as Ticket 40 core spun thread. Generally, the higher the ticket number, the finer the thread and vice versa (Mehta 1992).

5.1.7.1.1 For Cotton Sewing Threads (IS: 1720-1978)

Normally, the ticket number is related to 3-ply construction as a base; the number indicates the single yarn count (Ne). For example, Ticket number 60 represents 3/60s or 2/2/80s or 3/2/120s of which resultant count is 20s.

$$\text{Cotton ticket number} = \frac{\text{Single yarn English count (Ne)} \times 3}{\text{Number of ply}}$$

5.1.7.1.2 For Synthetic Sewing Threads (IS: 9543-1980)

$$\text{Metric ticket number} = \frac{\text{Single yarn metric count (Nm)} \times 3}{\text{Number of ply}}$$

5.1.8 Sewing Thread Consumption

It is essential to know the consumption of sewing thread in a garment to

1. Estimate the number of cones needed.
2. Calculate the cost of sewing thread required to produce the finished product.

Thread consumption can be determined in numerous ways such as

- Determining the actual amount of thread consumption in a particular length of seam.

- Calculating the thread consumption by using stitch formulas.
- Calculating the thread consumption via thread consumption estimates.

5.1.8.1 Measurement of Actual Sewing Thread Consumption

For the determination of sewing thread consumption, the threads from the garment should be removed from a particular length of each different seam. Then, thread consumption could be determined by dividing the actual length of sewing thread after unravelling from the garment by the seam length.

Example:

Seam length = 220 cm
Stitch class 401 = Two-thread chain stitch
Seam length for which sewing thread is removed = 25 cm
Length of needle thread removed = 25.5 cm
Needle thread factor = 25.5/25 = 1.02
Length of bottom thread (looper thread) removed = 75.0 cm
Looper thread factor = 75.0/25 = 3.00
Total length of needle thread = 220 × 1.02 = 224.4 cm
Total length of bottom thread = 220 × 3.00 = 660 cm
Total sewing thread consumption = 224.4 + 660 = 884.4 cm
15% thread wastage can be added = 854 × 1.15 = 1017 cm

5.1.8.2 Determination of Thread Consumption Using Thread Consumption Ratios

A simple technique of calculation of thread consumption is to utilise the standard thread consumption ratios for a range of stitch classes as given in Table 5.1. By connecting these thread consumption ratios to the seam

TABLE 5.1

Thread Consumption Ratio for Various Stitch Types

Type of Stitch	Thread Ratio	No. of Needles	Needle Thread %	Looper Thread %
101 – Chainstitch	4.0	1	100	0
301 – Lockstitch	2.5	1	50	50
304 – Zigzag Lockstitch	7.0	1	50	50
401 – Two Thread Chain Stitch	5.5	1	25	75
406 – Three Thread Covering Stitch	18.0	2	30	70
503 – Two Thread Overedge Stitch	12.0	1	55	45
504 – Three Thread Overedge Stitch	14.0	1	20	80
602 – Four Thread Covering Stitch	25.0	2	20	80
605 – Five Thread Covering Stitch	28.0	3	30	70

Note: The thread ratios are attained by considering a stitch density of 7 stitches/cm.

length using each stitch type, the total thread consumption could be determined.

Example:

> Length of seam = 220 cm
> Stitch class 301 = Single thread lock stitch
> Thread ratio obtained from the table (thread length per cm of seam) = 2.5 cm
> Total sewing thread consumption = 220 × 2.5 = 550 cm
> Consumption of needle thread = 1210 × 0.25 = 275 cm
> Consumption of looper thread = 1210 × 0.75 = 275 cm
> By adding 15% wastage = 550 × 1.15 = 632.5 cm of thread per seam

5.1.9 Applications of Sewing Threads

The applications of different types of sewing threads in various types of fabrics are given in Table 5.2.

5.1.10 Sewing Thread Packages

Sewing threads are wound in several kinds of packages as shown in Figure 5.7. Small length spools are engaged in retail store distribution, whereas larger spools are utilised to a limited extent industrially (Jana 2014).

> *Spools:* It has relatively short length of thread and is wound as a parallel winding. It has a flange on both ends that obstructs with off winding on industrial sewing machines. Hence, these are made for home sewing use.

TABLE 5.2

Application of Types of Sewing Threads

Fabric	Thread
Delicate fabrics	Fine mercerised cotton and
Tulle, chiffon, organza	synthetic thread, silk thread
Lightweight fabrics	Mercerised cotton thread,
Satin, suiting, knits, deep pile fabric	synthetic thread, silk thread
Medium weight fabrics	Heavy duty mercerised cotton
Gabardine, denim coating, furnished fabric	Cotton thread, synthetic thread
Heavy weight fabrics	Mercerised cotton thread,
Dungaree, canvas fabric, upholstery	synthetic thread
Synthetic knit and stretch fabrics	Nylon, mercerised cotton,
Polyester double knit, jersey, velvet, nylon tricot	cotton, silk
Leather materials	Mercerised cotton, silk and
Suede, kidskin, capeskin, lambskin	synthetic thread

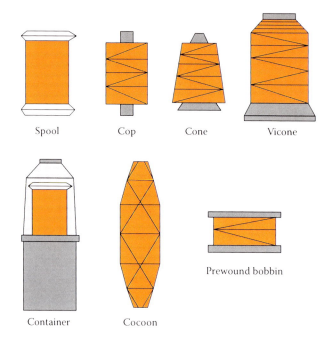

Spool   Cop   Cone   Vicone

Container   Cocoon

Prewound bobbin

FIGURE 5.7
Sewing thread packages.

Cops: Cops are utilised mainly in lockstitch machines where a range of colours are used. Sewing thread is cross-wound on the cop package to increase in off winding. The length of thread in a small cop ranges between 100 and 2000 m and 400 and 4000 m on a larger size cop.

Cones: Cones are tapered shapes made of paper or plastic material. They may contain fairly longer lengths of sewing thread about 1000–25,000 m with length of traverse ranging from 10 to 15 cm. It gives good off-winding performance for high-speed sewing machines. It is the most economical form of thread package where the consumption is high.

Vicones: It is a parallel tube having a flange at the bottom, which is designed to contain thread lengths of 1000–5000 m with length of traverse 6.5–9 cm.

Containers: Containers are intended to handle smooth and lively monofilament sewing threads that are complicated to control with the traditional thread packages.

Cocoons: Cocoons are centre-less sewing thread package forms created for insertion in multi-needle quilting machines and some kinds of embroidery machines.

Pre-wound bobbins: These are precision-wound packages intended to substitute metal bobbins in lockstitch machines. Normally, more sewing thread is available and the length is more consistent in these packages. It eliminates rewinding of bobbins, and hence improves the productivity and off winding is also improved because of precision winding.

5.2 Sewing Machine Needles

The manner in which fabric is pierced by the needle during stitching has a direct impact on the strength of the seam as well as garment appearance (Organ Needles 2015). The purposes of the sewing needles are to

- Make a hole in the fabric so that the sewing thread could pass through it to form a stitch without causing any damage to the fabric while doing so.
- To carry the needle thread through the fabric to form a loop. This is then taken up by the hook in a lockstitch machine or by means of the looper in chain stitch machines.
- Pass the needle thread through the loop created by the looper mechanism on a chain stitch machine.

5.2.1 Parts of a Needle

The different parts of a sewing needle are shown in Figure 5.8.

Shank: It is the top portion of the needle, which positions inside the needle bar. It could be designed as cylindrical or have a flat side, based on the method of holding it on to the needle bar. It is the principal support of the entire needle and is larger in diameter than the remaining part of the needle to give the strength.

Shoulder: It is the part in-between the shank and the blade, with the blade forming the longest portion of the needle up to the needle eye.

Blade: It undergoes an enormous amount of friction from the fabric through which the needle passes. In case of needles specifically designed for high-speed sewing, the shoulder is normally extended into the upper part of the blade to give a thicker cross section. This arrangement of reinforced blade strengthens the needle and produces the enlarged hole in the fabric while the needle is at its lowest

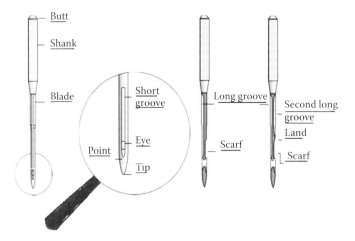

FIGURE 5.8
Parts of a sewing machine needle.

point, thus minimising the friction between it and the material. On the other hand, the blade could be designed as a tapered one, reducing its diameter gradually from shank to tip to minimise the friction.

Long groove: It gives a shielding channel for the sewing thread while it is carried down into the fabric for stitch formation thus reducing the abrasion and friction with the fabric.

Short groove: It is located on the reverse side of the long groove, that is, towards the hook or looper; it extends slightly above and below the needle eye. It assists in the formation of the needle thread loop.

Eye: It is the hole or opening in the sewing needle, lengthened through the blade along the long and short grooves on the needle. The profile of inside part of the eye at the top is vital in reducing sewing thread damage and in producing a good loop formation.

Scarf: The scarf otherwise known as clearance cut is a nook across the whole face of the needle immediately above the needle eye. Its objective is to facilitate closer setting of the bobbin hook or looper to the needle so that the needle thread loop could be entered more easily by the point of the hook or looper.

Point: It is tapering portion of the needle created to give a better penetration of the needle on various kinds of fabric. It should be properly selected to prevent damage of the fabric to be sewn.

Tip: It is the ultimate end of the point, which combines with the point in defining the penetration performance of the needle.

5.2.2 Special Needles

Several over-edge and safety stitch sewing machines utilise curved needles instead of straight needles. These needles are costly though the life of the needle is lesser compared to straight needles. However, the sewing machines utilising curved needles (Figure 5.9a) could achieve higher speeds than by using straight needles. Blind stitching machines also utilise needles that are curved, but the purpose here is to avoid penetration right through the fabric. Sewing machines (pick stitching machine) that imitate hand stitch (class 209) utilise a double-pointed sewing needle with an eye in the middle (Figure 5.9b), through which is threaded the short length of thread with which this machine sews (Gardiner 2003; Coats 2014).

5.2.3 Identification of Sewing Needle

Three parameters are generally used for the identification of sewing needles such as system, point and size.

5.2.3.1 System

It describes the elements of a needle to suit the sewing machine type. Based on the type of sewing machine and type of stitch, the needle is designed with variants in blade length, shank thickness, type of needle eye, etc. It is worthwhile to ensure with the sewing machine manufacturer for appropriateness of needle system to machine (Schmetz 2001; Gardiner 2003).

5.2.3.2 Point

A needle point is broadly categorised into two types:

1. Round point needle – set or cloth points
2. Cutting or leather point needle

5.2.3.2.1 Cutting Point Needles

Cutting point sewing needles have spiky tips like blades and a wide range of cross-sectional profiles such as rounded, triangular, square and

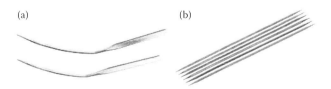
(a)   (b)

FIGURE 5.9
Special needle shapes. (a) Blind stitch needle. (b) Pick stitch needle.

lens exist. They are normally used to sew highly dense and non-fabric-based materials. Five universal kinds of cutting point sewing needles are shown in Figure 5.10, along with their profile of incision produced when used in a machine with the commonest threading direction (Organ Needles 2015).

The narrow wedge point needle: It cuts the material at right angles (90°) to the seam direction and permits to go for a high stitch density (SPI) while leaving adequate material between the needle holes to retain seam strength of material. On soft leather material, stitch densities as high as 12 per centimetre are achievable. It is the most frequently utilised cutting point needle for stitching uppers in the shoe industry (Clayton 2008).

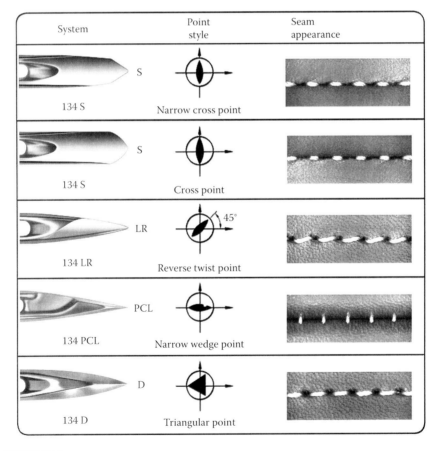

FIGURE 5.10
Cutting point needles.

The narrow reverse point needle: It produces cut that lies 45° to the seam direction, and produces a seam where the thread is turned to the left on the surface of the material.

The narrow cross point needle: It makes a cut along the line of the seam and necessitates a longer stitch length. Heavy decorative seams could be made where thicker sewing threads are used at lower stitch densities, that is, longer stitch length.

Numerous kinds of other point types exist for the variety of leathers, seams, sewing machines and strength and appearance requirements that arise. This involves triangular cross sections for multi-directional sewing (Brother Industries Limited 2000; American and Efrid 2013; Beckert 2014).

5.2.3.2.2 *Cloth Point Needles*

These kinds of needles are used for sewing textile materials instead of the leather/sheet materials as in the case of cutting point needles. The points have a round cross section contrasting to the various cutting profiles of the cutting point needles and the tip at the end of the point can vary in profile to suit the particular material being sewn (Carr and Latham 2006).

- The contour of the tip of the needle point which attains the deflection rather than penetration is a fine ball shape and the needle is called a *light ball point* needle which is utilised primarily for sewing knitted fabrics.
- The tip of the needle point which attains the penetration has the shape of a cone and is known as a *set point* needle which is utilised for sewing woven fabrics. Both ball and set point needles are available in a number of types, illustrated in Figure 5.11.

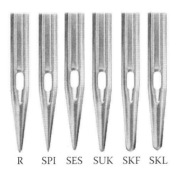

R   SPI   SES   SUK   SKF   SKL

FIGURE 5.11

Types of cloth point needles. R: set cloth point, SPI: slim set point, SES: light ball point, SUK: medium ball point, SKF: heavy ball point, SKL: special ball point.

Slim set point (SPI): It is generally used for sewing denser woven fabrics and aids in achieving a straighter stitch which could minimise seam pucker. Generally used for heavy woven fabrics, coated fabrics and topstitching of collars and cuffs.

Medium set point needle: It is the general purpose needle in no-problem sewing situations. It is commonly used for sewing a range of woven fabrics and in many circumstances could be used for knitted fabrics also.

Set cloth point (R): It is generally utilised for sewing standard fabrics with regular seams.

Acute set point: This kind of needle is used while sewing very dense fabrics like shirting fabric and interlining in collars and cuffs, where a straight line of stitching is required.

Heavy set point: These needles are used for sewing buttons as the button can be deflected to some extent into the correct position; thus, the needle can pass through the holes.

Light ball point (SES): It can be used for sewing lightweight knitted fabric and densely woven material.

Medium ball point (SUK): It is utilised for sewing denim fabrics of medium to coarser weight and knits.

Heavy ball point (SKF): It is utilised for sewing heavier woven elastic materials as well as coarser knits.

Special ball point (SKL): It could be utilised for sewing heavy knits and coarse elastics.

5.2.3.3 Needle Size

The needle size is normally expressed in two ways. One of the basic methods of representation is by a metric number (Nm). This system represents the diameter of the needle blade in hundredths of a millimetre measured just above the scarf area. For example, a needle size of Nm 100 is 1.0 millimetre in diameter as shown in Figure 5.12. Another standard needle sizing method

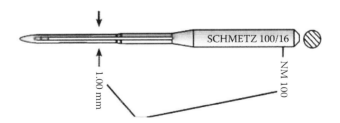

FIGURE 5.12
Metric needle sizing.

is the Singer system, otherwise called the American system, which uses a number that represents a size (Juki 1988; Beckert 2014).

Needles are offered in a wide range of sizes and the selection of needle size is based on the combination of fabric and sewing thread which is to be sewn. If the selected sewing needle is too small for the sewing thread size, the thread will not fit well into the long groove of the needle and will suffer from extreme abrasion. The use of too fine a needle while sewing heavy plies of fabric could lead to the deflection of the needle, which could influence the stitch loop pick up and cause slipped stitches or even needle breakage. Use of a larger sewing needle for the particular sewing thread resulted in poor control of the loop formation which could lead to slipped stitches (Grace and Ruth 2004; Cooklin 2006).

5.2.4 General Purpose Needles

The general purpose needles used in apparel manufacturing are given in Table 5.3.

5.2.5 Specialty Needles

The specialty needles used in garment manufacturing are given in Table 5.4.

5.2.6 Surface Finishing of Sewing Needles

Sewing needles are normally made from steel and during their final manufacturing stage they are polished, specifically in the needle eye area. They are then electroplated using chromium or nickel to provide resistance to mechanical wear, corrosion resistance and reduction of friction during sewing. One of the main requirements of the surface finishing of needles is it should not pick up any elements of synthetic fabric or thread which could melt due to excessive heat generation. By considering this aspect, the chromium-plated sewing needles are superior compared to nickel-plated needles.

TABLE 5.3

Application of General Purpose Needles

Needle	Description	Fabric	Sizes
Ball-point	It has a medium tip, rounded compared to universal needle	Knits	70/10 – 100/16
Sharp	It has a slim shaft and sharper needle point	Fine woven fabrics	60/8 – 90/14
Universal	Needle point is marginally rounded; however, it is sharp enough to pierce woven fabrics	Woven and knitted fabrics	60/8 – 120/19

TABLE 5.4

Application of Specialty Needles

Needle	Description	Fabric	Sizes
Denim	It has a thicker and stronger shaft and a sharp needle point	Denim and heavy woven fabrics	70/10 – 110/18
Leather	It has a wedge-shaped needle point	Leather and nonwoven fabrics	80/12 – 110/18
Machine embroidery	It has a larger eye and a special scarf to protect the sewing thread	For embroidery	70/10 – 90/14
Metallic	It has a larger eye than the embroidery needle, and a sharp point to avoid thread breakage	For metallic threads	80/12
Quilting	It has a tapered as well as sharp needle point	Machine quilting	75/11 and 90/14
Spring needle	It has a shaft surrounded by the wire coil, which acts as a presser foot	Quilting	–
Stretch needle	It has a deep scarf to avoid skipped stitches	Lightweight knitted fabrics	75/11 and 90/14
Topstitch needle	It has an extra-large eye and a large groove	Topstitching	80/12, 90/14 and 100/16
Twin needle	It is constructed with two needles attached to a single shaft	Double topstitching	1.6/70 – 4.0/100
Triple needle	It is constructed with three needles attached to a single shaft	Triple topstitching	2.5/80 and 3.0/80
Hemstitch needle	It has a widened shaft and produces a decorative hole in tightly woven fabrics	Decorative stitching	100/16 and 120/19

References

Abernathy, F.H. and J.T. Dunlop. 1999. *A Stitch in Time – Apparel Industry.* Blackwell Scientific Publications, Oxford, UK.

Alagha, M.J., J. Amirbayat and I. Porat. 1996. A study of positive needle thread feed during chainstitch sewing. *Journal of Textile Institute* 87:389–95.

American and Efrid. 2013. Industrial sewing thread. http://www.amefird.com/prodcts-brands/industrial-sewing-thread (accessed on January 12, 2015).

Beckert, G. 2014. Groz Beckert needles. http://www.groz-beckert.com. (accessed on October 12, 2014).

Blackwood, W.J. and N.H. Chamberlain. 1970. *The Strength of Seams in Knitted Fabric.* Technical Report No.22, Clothing Institute.

Brother Industries Limited. 2000. *Industrial Sewing Machine Handbook.* Brother Industries Limited, USA.

Carr, H. and B. Latham. 2006. *The Technology of Clothing Manufacture.* Blackwell Science, Oxford.

Clayton, M. 2008. *Ultimate Sewing Bible – A Complete Reference with Step-by-Step Techniques.* Collins & Brown, London.

Coats. 2003. *The Technology of Thread and Seams*. J&P Coats Limited, Glasgow.

Coats, P.L.C. 2014. Thread numbering. http://wwwcoatsindustrialcom/en/information-hub/apparel-expertise/thread-numbering (accessed on March 22, 2015).

Cooklin, G., S.G. Hayes and J. McLoughlin. 2006. *Introduction to Clothing Manufacture* Second ed. Blackwell Science, Oxford.

Fairhurst, C. 2008. *Advances in Apparel Production*. The Textile Institute, Woodhead Publication, Cambridge.

Gardiner, W. 2003. *Sewing Basics*. Sally Milner Publishing, Australia.

Grace, K. and G. Ruth. 2004. *Apparel Manufacturing: Sewn Products Analysis*. Prentice Hall, Englewood Cliffs, NJ.

Jana, P. 2014. *Sewing Machine Resource Guide*. Apparel Resources, New Delhi.

Jana, P. and A.N. Khan. 2014. The sewability of lightweight fabrics using X-feed. *International Journal of Fashion Design and Technology Education* 7(2):133–42.

Juki, C. 1988. *Basic Knowledge of Sewing*, Juki Corporation, Japan.

Mehta, P.V. 1992. *An Introduction to Quality Control for Apparel Industry*. CRC Press, Boca Raton, FL.

Organ Needles. http://organ-needlescom/english/product/downloadphp (accessed on November 12, 2015).

Shaeffer, C. 2000. *Sewing for the Apparel Industry*. Woodhead Publication, Cambridge.

The Schmetz. 2001. *The World of Sewing: Guide to Sewing Techniques*. Ferdinand Schmetz GmbH Herzogenrath.

6

Seams and Stitches

6.1 Seams

In garment assembling process, two or more plies of fabric or other materials are detained by rows of stitches known as a seam. Seams are generally categorised based on the seam type as superimposed seam, lapped seam, bound seam, flat seam and based on their location in the completed garment such as centre back seam, inseam and side seam. Seams finishing can be carried out with a range of methods to prevent unraveling of fabric raw edges as well as to neaten seam edges on the inside of garments. The type of seam and sewing thread used will vary with each application (Blackwood and Chamberlain 1970; Chuter 1995).

6.1.1 Classification of Seam

Seams are classified based on the type/number of fabric components used. Eight classes of seams are defined by ISO 4916:1991. Conventionally, the seams were classified as flat, superimposed, lapped or bound seam and stitching was defined as edge finishing or ornamental. The ranges of seams are given below together with their descriptions under the above and the new system of seam classification (Chuter 1995; Shaeffer 2000; Coats 2003).

Class 1: Superimposed seam

Class 2: Lapped seam

Class 3: Bound seam

Class 4: Flat seam

Class 5: Decorative/ornamental stitching

Class 6: Edge finishing/neatening

Class 7: Attaching of separate items

Class 8: Single ply construction

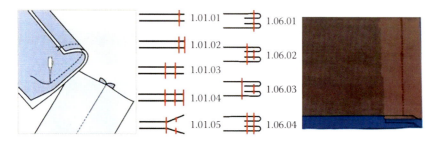

FIGURE 6.1
Superimposed seam.

6.1.1.1 Class 1: Superimposed Seam (SS)

The arrangement of fabric layers for superimposed seams is shown in Figure 6.1. In this kind of seam, normally two or more fabric panels are superimposed over one another and seaming was done near an edge, with one or several rows of stitches. The various kinds of seams within this class are shown in Figure 6.1.

A simple SS can be produced using 301 or 401 class of stitches and can also be sewn with other classes of stitches. It is used in seams of jeans, in side seams of skirts, dress slacks, finishing belt ends, ends of waist bands on jeans, collars or cuffs and attaching elastic to waistline (Chuter 1995; Shaeffer 2000; Coats 2003).

6.1.1.2 Class 2: Lapped Seam (LS)

In this type of seam, two or more fabric plies are lapped with the raw edges, flat or folded and attached with one or several rows of stitches as shown in Figure 6.2. The lap felled seam is the mostly used seam in this class, which involves one stitching operation. It is a strong seam with fabric edges, generally used to safeguard jeans fabric from fraying. The 401 chain stitch class is commonly used for lap felled seams in jeans because of its strong

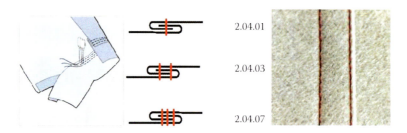

FIGURE 6.2
Lapped seam.

construction. Another class of lapped seam, French seam, comprises a two-stitching process with a superseding folding operation. It is a flat, folded seam with only one row of stitching noticeable on the face side of fabric and involves a minimum of two components and can have different variations comprising a number of rows of stitching. It is commonly used for rain wear, and edge stitching front facings on jackets and dresses (Chuter 1995; Shaeffer 2000; Coats 2003).

6.1.1.3 Class 3: Bound Seam (BS)

In this kind of seam, the binding strip is folded on the edge of the base fabric plies and is stitched at the edges along with the fabric plies with one or several rows of stitches (Figure 6.3). This makes a neat fabric edge on a seam exposed to view. A 401 chain stitch or 301 lock stitch class of stitches is normally used for seaming bound seams and it is utilised for finishing sleeve hems, necklines, finishing seams on unlined jackets and coats, finishing raw edges, continuing the motif design of lace, etc. (Chuter 1995; Shaeffer 2000; Coats 2003).

6.1.1.4 Class 4: Flat Seam (FS)

This seam is also called a butt seam as the edges of the fabrics do not overlay one another, they will be butted together. In this seam type, two fabric edges in flat or folded conditions are brought together and oversewn with stitches as shown in Figure 6.4. The main purpose of this seam is to provide a joint without any extra thickness of fabric at the seam, as needed in underwear or foundation garments. The bottom threads (looper thread) should be softer as well as stronger and the cover thread should be decorative as well as stronger. The flat seam is normally sewn with a zigzag lock stitch, chain stitch or covering stitch. This kind of seam will comprise two components and could be seen on knitted garments where seams are required to be free from bulk (Chuter 1995; Shaeffer 2000; Coats 2003).

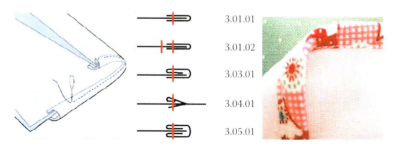

FIGURE 6.3
Bound seam.

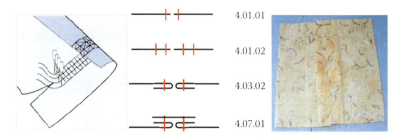

FIGURE 6.4
Flat seam.

6.1.1.5 Class 5: Decorative/Ornamental Stitching

The decorative or ornamental stitch (Figure 6.5) is a series of stitches down a straight or curved line or following an ornamental design on a single fabric ply. A more intricate kind of stitch involves various systems of piping, producing an elevated line along the fabric surface. The stitching in a single fabric ply resulted in decorative effects on the fabric surface like pin tucks.

6.1.1.6 Class 6: Edge Finishing/Neatening

Edge neatening stitch (Figure 6.6) could be seen where the edge of a single fabric ply is folded or covered with a stitch. The simplest of this process is known as 'Serging' where the raw edge of the fabric is secured by overedge stitching to prevent fraying of edges as well as edge neatening. This seam class involves seams whereby the edges are neatened by means of stitches and could be utilised in circumstances where the raw edge of fabric needs finishing.

6.1.1.7 Class 7: Edge Stitched Seam

This kind of seam involves seams that need the inclusion of another component at the edge of a fabric ply, for example, elastic braid inserted onto the

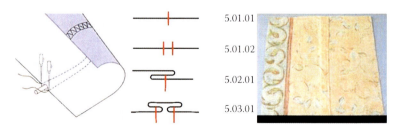

FIGURE 6.5
Decorative stitching.

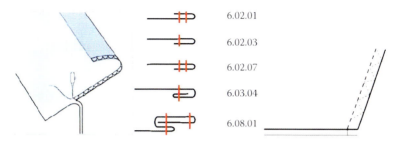

FIGURE 6.6
Edge neatening.

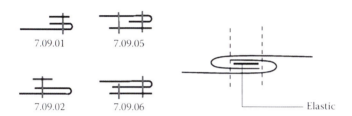

FIGURE 6.7
Edge-stitched seam.

edge of ladies briefs as shown in Figure 6.7. This kind of seam requires two components.

6.1.1.8 Class 8: Enclosed Seam

In this kind of seam class, only one piece of strip of fabric is turned on both edges. The general application of enclosed seams could be found in belt loops or belts for which a folder attachment can be done on the machine as shown in Figure 6.8.

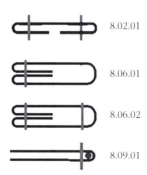

FIGURE 6.8
Enclosed seam.

6.1.2 Numerical Expressions of Seams

Each seam is recognised by a numerical description consisting of five digits.

- The first digit of seam expression represents the seam class (1–8).
- The second and third digits represent the counting numbers (0–99) to specify the differences in the position of the needle penetrations.
- For better communication of seam type, the description of the type of stitch has to be mentioned after the designation of the stitched seam.

6.1.3 Seam Quality

Though the type of stitches selected for a seam depends on the functional as well as aesthetic requirements, the following factors have to be taken into account for seam quality.

- *Seam size*: Expressed by seam length, seam width and depth.
- *Seam slippage strength*: It is the force required to draw out a 0.25″ of the opposing sets of yarns which are perpendicular to the seam line.
- *Seam strength*: It is the force required to break the seam either by breaking the sewing thread or by breaking the sewn material (Alagha 1996; Amann 2015).

6.2 Stitches

A series of repetitive stitches of one pattern is termed a stitch. BS 3870 (1991) categorises the several types which are available into six stitch classes which cover the demands of joining fabrics together, neatening raw edges, or providing decoration (Carvalho et al. 2009; ASTM D 6193-11; ASTM D 7722-11). The stitch could be formed in any of the three methods given below:

- *Interlooping*: It is created by passing the loop of one thread through the loop of another sewing thread.
- *Intralooping*: It is created by passing the loop of one thread through the loop of the same thread.
- *Interlacing*: One thread passes over another thread.

The six stitch classes included in the British Standard are as follows:

1. Class 100: Chain stitch
2. Class 200: Hand stitch

3. Class 300: Lock stitch
4. Class 400: Multi-thread chain stitch
5. Class 500: Over-edge chain stitch
6. Class 600: Covering chain stitch

6.2.1 Class 100: Chain Stitches

These kinds of stitches are formed from one or several needle threads, and are described by intra looping. One or several loops of needle thread are passed through the fabric and secured by intra looping with a subsequent loop after they are passed through the fabric (Figure 6.9). As each loop is reliant on the subsequent one, these kinds of stitches are very insecure and unravel very easily (Carvalho et al. 2012). The front and rear side of the class 100 stitches in the fabric is shown in Figure 6.10.

The class 101 stitch is the simplest one in this class produced from a single sewing thread. Since this stitch is insecure, it could be easily removed, and it is used for 'basting' operation in tailor-made garments. This kind of seam is normally not preferred for seaming operation as it is highly insecure but is widely used in multi-needle machines.

In the assembling of a garment it is vital to start and finish at a fabric edge, and that edge could be a small piece of temporary fabric secured to the garment edge by the stitches. After that, the fabric edges could be attached securely through the ends of the chain stitching to avoid the running back of stitches (Carvalho et al. 2012).

6.2.2 Class 200: Hand Stitches

The class 200 stitch types are categorised as hand stitches. These stitches are described by a single sewing thread and the stitch is held by a single line of

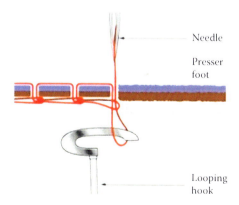

FIGURE 6.9
Intralooping of thread.

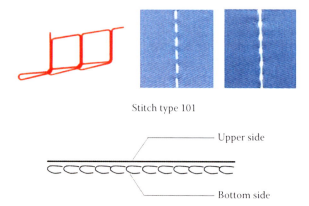

Stitch type 101

FIGURE 6.10
Front and rear side of single-thread chain stitched fabric.

FIGURE 6.11
Front and rear side of hand-stitched fabric.

thread passing through in and out of the fabric. Hand stitching is used at the high-priced garment production as the customer expects it at that price, and it may be the only way to a perfect finish. The front and rear side of the hand-stitched fabric is shown in Figure 6.11.

The sewing machines have been developed to replicate hand stitching (stitch type 209), which is used around the outer edges of tailored jackets. The machine is called a pick stitching machine. A double-pointed, centre-eyed needle sews short lengths of thread in a simulation of hand-sewing. The pick stitching machine could be set to sew a longer stitch on the top than at the bottom or vice versa (Chmielowiec 1995; Chuter 1995).

6.2.3 Class 300: Lock Stitches

The Class 300 stitches are formed using two or more sets of sewing threads, and are characterised by interlacing of the two or more threads. Loops formed by one group of threads are passed through the fabric and are held by the second group of thread. The top thread is called the needle thread and the bottom thread is known as the bobbin thread. The interlacing of thread in this class makes them secure and difficult to unravel. Straight lock

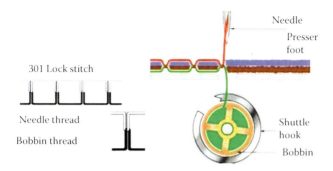

FIGURE 6.12
Interlooping of thread in lock stitch.

stitch, 301 (Figure 6.12), with a single needle thread and a bobbin thread, is still the most common stitch used in the apparel industry (Chuter 1995; Shaeffer 2000).

Lock stitch has adequate strength for most purposes, provided appropriate sewing thread is used, with sufficient stretch, when it is correctly balanced. It has a similar appearance on both sides of the fabric (Clapp et al. 1992). The front and rear side of the lock stitched fabric is shown in Figure 6.13.

The zigzag version of stitch (Class 304) is generally utilised for joining trimmings like lace and elastic where a wider row of stitching is required. The main disadvantage of the lock stitch is that it uses a small bobbin comprising only a limited length to give the lower thread. Hence, it will exhaust

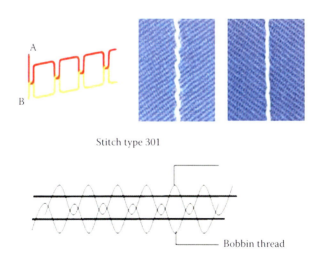

FIGURE 6.13
Front and rear side of lock-stitched fabric.

quickly and changing of bobbins is time consuming in production. The two main disadvantages of lock stitch machines are

1. Multi-needle stitching with many closely spaced needles is not viable due to space required for the bobbin. So the maximum number of needles generally used on lock stitch machines is two.
2. The limited stretch of lock stitch because of interlacing of threads which is unsuitable for edge neatening.

6.2.4 Class 400: Multi-Thread Chain Stitches

The class 400 stitches are created using two or more sets of sewing threads, and are characterised by interlooping of two sets of threads known as needle thread and looper thread. Loops formed in one set of sewing threads are passed through the fabric and are held by interlooping and interlacing with loops formed by another set of threads. The simplest version of this class of stitch, 401, is shown in Figure 6.14.

The chain stitch has the appearance of lock stitch in the front side of the fabric but has a double chain effect created by a looper thread in the backside of the fabric (Brother Industries Limited 2000; Carr and Latham 2006). A two-thread chain stitch is stronger than a similar lock stitch and since no threads are interlocked with each other within the fabric, there is less probability to cause the type of seam pucker that occurs when tightly woven fabrics are distorted by the sewing thread (Clapp et al. 1992; Coats 2014).

The great advantage of this class of stitch is that both the needle and looper threads are run from large packages (cone) on top of the machine; therefore, there are no issues with running out of bobbins like with a lock stitch machine. It is often used on long seams in garments like trousers. The

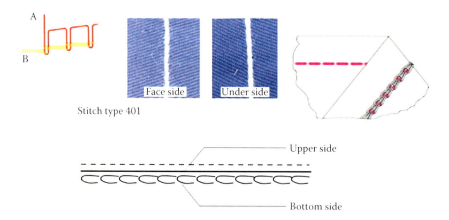

FIGURE 6.14
Front and rear side of multithread chain stitched fabric.

maximum sewing speed in lock stitch machine would be around 6,000 spm while in chain stitch machine 8,000 spm could be achieved (Coats 2014).

6.2.5 Class 500: Over-Edge Chain Stitches

These types of stitches are created using one or more sets of sewing threads, and have characteristic features that loops formed by at least one set of threads goes around the raw fabric edge. These stitches are generally called 'overlocking stitches'. The most regularly used stitch types in this class has one or two needle threads and one or two looper threads and they form a narrow group of stitching line along the fabric edge with threads intersecting at the edge and preventing the fabric from fraying as shown in Figure 6.15.

All classes of stitches in this category have high elasticity, they do not unravel easily, and a trimming knife on the machine makes sure there are neat edges prior to sewing. Stitch class 503 is formed with one needle and one looper thread, and is less versatile, mainly used for edge neatening. Stitch class 504 is created from one needle thread and two looper threads and is utilised for edge neatening and, in the case of knitted fabrics, for joining seams. A combination of 401 and 503 stitch class is sewn simultaneously on one machine, where a joined and neatened seam is required that does not need to be pressed open and is generally called a safety stitch (Coats 2014).

Overlock stitches are categorised by a number of ways and the most common way of classification is based on the number of sewing threads used in a stitch such as 1, 2, 3, 4 or 5 threads overlock stitches. Each of these stitch classes has a distinctive application and benefits as given below.

- 1-thread overlock stitches are used for 'butt-seaming'.
- 2- and 3-thread overlock stitches known as 'merrowing' are utilised for seaming and edge neatening on woven and knitted garments.
- 4-thread formations known as 'mock safety stitches' provide extra strength while retaining flexibility.
- 5-thread formations, which employ two needle threads known as safety stitches, create stronger seams which are used for apparel manufacturing.

Stitch type 503

FIGURE 6.15
Front and rear side of over-edge chain stitched fabric.

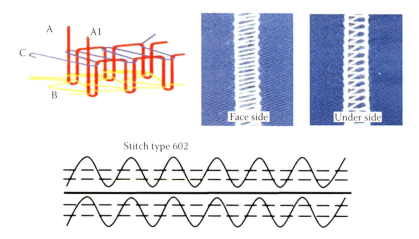

Stitch type 602

FIGURE 6.16
Front and rear side of covering chain stitched fabric.

6.2.6 Class 600: Covering Chain Stitches

The stitches in this class are made by utilising three sets of sewing threads, namely, needle (A), looper (B) and spreader (C) thread as shown in Figure 6.16. The loops formed by the needle threads are passed through the loops of the spreader threads, which are already case on the fabric surface and then passed through the fabric where they are outer looped with loops formed by looper threads on the rear side of the fabric (Hurt and Tyler 1976; Jana 2014; Jana and Khan 2014).

Stitches in this class are the most complicated of all types and could have up to a total of nine threads including four needle threads creating a broad, flat joining of elastic, braid of binding to the edges of garments like briefs with the possibility for a decorative top cover as well as the functional bottom cover over the raw edge of the garment fabric (Coats 2003, 2014).

The complicated type of stitch in this class is known as a flat lock (606), which can be utilised to join fabrics that are butted together. Two trimming knives in the machine ensure that neat fabric edges butt together and four needles and nine threads provide a smooth join with good extensibility. It is commonly used on knitted underwear fabrics to provide a seam with low bulk that can be worn comfortably against the skin (Laing and Webster 1998; Clayton 2008).

6.3 Seam Quality Issues

The major seam quality problems associated with the garment manufacturing process are discussed in Table 6.1.

TABLE 6.1

Seam Quality Issues

Sl. No	Defect	Causes	Remedial Measures
1	Skipped stitches 301 Lock stitch – skipped stitches 401 Skipped stitch	• Bobbin hook or looper does not enter thread loops at the correct time • Thread loop failure caused by incorrect needle size/style for the particular thread size/type • Thread loop failure because thread control mechanism is incorrectly set, thereby leading to thread loop starvation • Fabric flagging due to improper presser foot control or larger throat plate hole • Deflection of needle • Improper loop formation of thread	• Check the machine clearances and the timings • Check if the needle is inserted and aligned correctly • Change the needle size/style • Restart and check loop formation • The presser foot pressure should be checked and readjusted if necessary • Change the throat plate to match the needle size • Readjust tensions
2	Staggered stitches	• Needle vibration or deflection • Incorrect or blunt needle point • Incorrect needle-to-thread size relationship • Improper fabric control and bouncing of presser foot	• Needle size can be increased or tapered needle can be tried • Either sewing needle or sewing thread can be changed • Can go for positive sewing feed mechanism
3	Variable stitch density	• Improper control of fabric feed	• Increase the foot pressure • Can go for positive sewing feed mechanism *(Continued)*

TABLE 6.1 (Continued)
Seam Quality Issues

Sl. No	Defect	Causes	Remedial Measures
4	Seam grin 	• When two fabric panels are opened at opposite sides of one another to the seam, a gap is revealed between the two fabric panels	• Increase stitching tensions • Stitches per inch (SPI) can be increased
5	Seam slippage 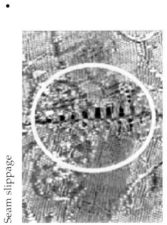	• This happens when the yarns in the fabric are pulled out of the seam and are more frequent in fabrics made from continuous filament yarns	• French seam type can be tried • Seam width can be increased • Stitch density should be optimised

(Continued)

TABLE 6.1 (Continued)

Seam Quality Issues

Sl. No	Defect	Causes	Remedial Measures
6	Needle thread breakage	• Thread gets trapped at the thread guide • Snarling of thread before tension disc • Excessive needle thread tension • Irregularities or damages in needle guard, throat plate, bobbin case and needle eye • Excessive needle heat, groove or eye blocked with melted fabric • Overheating of hook • Quality of needle thread is inferior	• Use a foam pad or a similar device to prevent the package from tilting • Ensure that the rethreading is done correctly • The needle thread tension can be reduced and the condition of the disc tensioner should be checked • Replace the needle with one of better quality • Change to a correctly finished thread of a better quality
7	Bobbin/looper thread breakage	• Bottom thread not wound properly on the bobbin • Bottom thread tension is very high • Damages in bobbin case, looper eyelet • Improper fitting of bobbin case	• Adjust the alignment of the bobbin winder • Adjust the bottom thread tension • Polish the edges and the correct surfaces of bobbin hook and case
8	Thread fusing	• Incorrect sewing thread • Poorly finished woven fabric • Damaged needle or overheating of needle	• Use better quality sewing thread • Change to more suitable needles. Apply needle coolants
9	Imbalanced/variable stitching	• Incorrect sewing tensions • Incorrect threading • Needle thread getting snagged on bobbin case or positioning finger • Variable tension caused by poor thread lubrication	• Check for snarling, adjust the thread tension • Rethread the machine • Polish the bobbin case surface • Reset the positioning finger and the opening finger, if fitted • Switch to superior quality threads from coats

6.3.1 Seam Puckering

Seam puckering is defined as seam gathering during sewing or after sewing, or laundering of garments, leading to an undesirable seam appearance to the garment. It is more frequent on woven fabrics especially on tightly woven fabric than knitted fabric (Mathews 1986; Mehta 1992; Amann 2015). Seam puckering is typically caused by the following conditions:

- Inherent puckering (structural jamming)
- Tension puckering
- Machine puckering
- Shrinkage puckering

6.3.1.1 Seam Puckering Due to Structural Jamming

Seam puckering is more common on tightly woven fabrics (high thread density) since the yarns are aligned in very tight layers that could not move easily to compensate for the sewing thread as it is introduced into the seam. In case of densely woven fabrics, there could be insufficient space to accommodate a sewing thread without displacement of yarns (Dorkin and Chamberlain 1961; Needles 2014; Amann 2015). Hence, stitching along a straight line will distort and push the adjacent yarns in the fabric, which cause the seam to pucker and is commonly known as 'structural jamming' or 'inherent pucker' as shown in Figure 6.17.

To check whether the defect is due to structural jamming, sewing threads should be cut between nearby needle penetrations through the seam line and have to be observed whether the seam puckering exists in the fabric. If the seam puckering is noticed even after the cutting of sewing threads, then yarn displacement is the possible cause as shown in Figure 6.18.

Remedial Measures for Structural Jamming

- Use of finer sewing thread which will retain sufficient seam strength.
- Use of finer needles that will not lead to sewing problems.

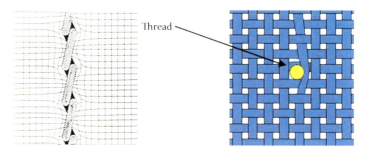

Thread

FIGURE 6.17
Structural jamming.

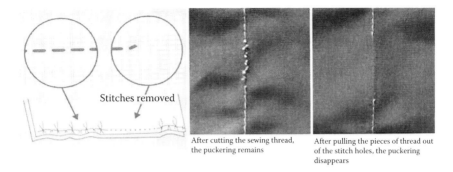

Stitches removed

After cutting the sewing thread, the puckering remains

After pulling the pieces of thread out of the stitch holes, the puckering disappears

FIGURE 6.18
Identification of seam puckering due to structural jamming.

- SPI (stitches per inch) should be reduced, hence less yarns are exiled from the stitch line.

6.3.1.2 Tension Puckering

If a sewing thread tension is higher in the seam, it will be in stretched condition during the stitching process and it will try to relax after sewing. This leads to seam puckering instantly as the seam is coming out from beneath the presser foot. This incident also happens after the garment is laundered causing the seam to pucker. Excessive sewing thread tension will not only lead to seam puckering but also cause other problems such as skipped stitches and sewing thread breakage (Tyler 1992; Ruth and Kunz 2004 Fairhurst 2008).

To check whether the puckering is due to structural jamming or thread tension, the top and bottom threads of all stitches along a seam have to be cut for a few centimetres, without displacement of yarns in the fabric (Jacob 1988; Tyler 1992; Cooklin 2006). If the seam pucker is disappeared over this length, then it was caused by sewing thread tension and subsequent recovery as shown in Figure 6.19.

Remedial Measures for Tension Puckering

- Optimise needle thread and bobbin thread tensions.
- Synchronisation of timing of feeding has to be set correctly as incorrect feed timing can lead to the need to apply excessive tension to the needle thread, in order to create a properly balanced stitch.
- Positioning finger should be set correctly to permit the sewing thread to go through the bobbin hook easily in case of lock stitch machines.
- Stich balance should be adjusted on chain stitch machines in a manner that the needle loops on the bottom side of the seam lay over at least halfway to the next needle penetration when the looper thread is unravelled out of the seam (Sumathi 2002).

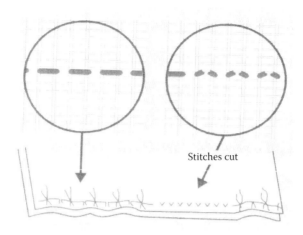

FIGURE 6.19
Identification of tension puckering.

6.3.1.3 Feed Puckering

It occurs when different fabric plies are fed at variable rates than one another. This leads to a gathering effect in the over-fed ply. Ply mismatching of fabric plies as shown in Figure 6.20 occurs

1. When the presser foot holds back on the upper fabric ply as the bottom fabric ply is being fed at a greater rate by the feed dog
2. When the operator grips the bottom fabric ply but shoves the top fabric ply to the seam line hence the fabric edges will come out evenly

To identify the feed puckering, two perpendicular cuts across a sewn seam have to be done where the puckered condition is the maximum. Then, the

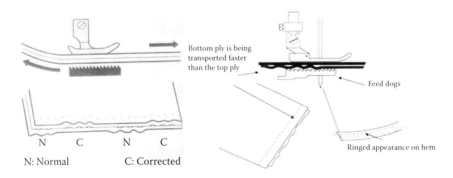

FIGURE 6.20
Feed puckering.

sewing thread has to be removed from the seam and ensure whether two fabric plies are of equal length. If one fabric is longer than the other, then the puckering is being caused by the uneven feeding of fabric (Abernathy and Dunlop 1999; Schmetz 2001).

Remedial Measures for Feed Puckering

- The presser foot pressure exerted on the fabric should be less to keep up uniform feeding. The clamping of fabric by the presser foot should be ensured at the front as well as the back of the needle.
- Setting of feed dogs with respect to their height as well as back feeding should be ensured. The selection of feed dog with reference to the number of teeth per inch and number of rows of teeth should be done. The feed dog with 20–24 TPI (teeth per inch) for lightweight fabrics, 14–18 TPI for medium weight and 8–12 TPI for heavy weight fabrics are normally preferred.
- The presser foot and needle plate should have comparatively small needle holes with respect to the needle size being used.
- Sewing machines equipped with more positive feeding mechanisms are advisable.

6.3.1.4 Shrinkage Puckering

Shrinkage puckering could happen when one fabric panel in the seam shrinks differently compared to the other fabric panel as shown in Figure 6.21. Typical components include the base fabric, interlining, zipper tapes, stay tapes and the thread. All these components should have minimum shrinkage to produce a pucker-free seam (Ukponmwan 2001).

- *Wash pucker*: If the sewing thread shrinks during the washing process, it pulls the fabric with it causing puckering and is more prominent with the use of cotton sewing threads.

FIGURE 6.21
Shrinkage puckering on fabric.

- *Ironing pucker*: It occurs while using synthetic sewing threads in the garment. The application of heat changes the molecular structure of the fibres in the thread, which results in shrinkage leading to puckering.

To identify the shrinkage puckering, two perpendicular lines at a distance of 10″ across a seam should be marked using an indelible ink pen that normally shows extreme seam puckering after laundering. Two perpendicular lines against the seam line which has been marked before should be connected with a line running parallel to the seam (Gardiner 2003; Pfaff Industrial 2009). The length of the seam should be verified after the garment is subjected to finishing and pressing cycles. The gap between the two marked lines will be less than 10″ if there is seam shrinkage. To minimise this puckering, the sewing thread having low shrinkage characteristics has to be selected (Hurt and Tyler 1975; Juki 1988).

References

Abernathy, F.H. and J.T. Dunlop. 1999. *A Stitch in Time – Apparel Industry*. Blackwell Scientific Publications, Oxford, UK.

Alagha, M.J., J. Amirbayat and I. Porat. 1996. A study of positive needle thread feed during chain stitch sewing. *Journal of Textile Institute* 87:389–95.

Amann Group. Preventing Seam Pucker in Service & Technik – Information for the Sewing Industry. http://wwwamanncom/fileadmin/download/naehfaden/b_nahtkraeuseln_ENpdf (accessed on February 12, 2015).

ASTM D 6193-11. Standard Practice for Stitches and Seams.

ASTM D 7722-11. Standard Terminology Relating to Industrial Textile Stitches and Seams.

Blackwood, W.J. and N.H. Chamberlain. 1970. *The Strength of Seams in Knitted Fabric*. Technical Report No 22, Clothing Institute.

Brother Industries Limited. 2000. *Industrial Sewing Machine Handbook*. Brother Industries Limited, USA.

BS 3870-1. 1991. Stitches and Seams Classification and Terminology of Stitch Types.

BS 3870-2. 1991. Stitches and Seams Classification and Terminology of Seam Types.

Carr, H. and B. Latham. 2006. *The Technology of Clothing Manufacture*. Blackwell Science, Oxford.

Carvalho, H., A.M. Rocha and J.L. Monteiro. 2009. *Measurement and Analysis of Needle Penetration Forces in Industrial High-Speed Sewing Machine*. Taylor and Francis, Boca Raton, FL, 100(4):319–29.

Carvalho, H., L.F. Silva, A. Rocha and Monteiro, J. 2012. Automatic presser-foot force control for industrial sewing machines. *International Journal of Clothing and Science and Technology* 24(1):20–7.

Chmielowiec, R. and D.W. Lloyd. 1995. The measurement of dynamic effects in commercial sewing machines. In: *Proceedings of the Third Asian Textile Conference*, Hong Kong, 2:814–28.

Chuter, A.J. 1995. *Introduction to Clothing Production Management*. Blackwell Scientific Publications, Oxford, UK.

Clapp, T.G., T.J. Little., T.M. Thiel and D.J. Vass. 1992. Sewing dynamics: Objective measurement of fabric/machine interaction. *International Journal of Clothing and Science and Technology* 4(2/3):45–53.

Clayton, M. 2008. *Ultimate Sewing Bible – A Complete Reference with Step-by-Step Techniques*. Collins & Brown, London.

Coats. 2003. *The Technology of Thread and Seams*. J&P Coats Limited, Glasgow.

Coats, P.L.C. 2014. Thread numbering. http://wwwcoatsindustrialcom/en/information-hub/apparel-expertise/thread-numbering (accessed on March 22, 2015).

Cooklin, G., S.G. Hayes and J. McLoughlin. 2006. *Introduction to Clothing Manufacture*. Blackwell Science, Oxford.

Gardiner, W. 2003. *Sewing Basics*. Sally Milner Publishing, Australia.

Dorkin, C. and N. Chamberlain. 1961. *Seam Pucker its Cause and Prevention*. Technological Report No 10.

Fairhurst, C. 2008. *Advances in Apparel Production*. The Textile Institute, Woodhead Publication, Cambridge.

Hurt, F.N. and D.J. Tyler. 1975. *Seam Damage in the Sewing of Knitted Fabrics II – Material Variables*. HATRA Research Report No 36.

Hurt, F.N. and D.J. Tyler. 1976. *Seam Damage in the Sewing of Knitted Fabrics III – The Mechanism of Damage*. HATRA Research Report No 39.

Jacob, S. 1988. *Apparel Manufacturing Handbook – Analysis, Principles and Practice*. Columbia Boblin Media Corp, New York, USA.

Jana, P. 2014. *Sewing Machine Resource Guide*. Apparel Resources, New Delhi.

Jana, P. and A.N. Khan. 2014. The sewability of lightweight fabrics using X-feed. *International Journal of Fashion Design and Technology Education* 7(2):133–42.

Juki, C. 1988. *Basic Knowledge of Sewing*. Juki Corporation, Japan.

Laing, R.M. and J. Webster. 1998. *Stitches and Seams*. Woodhead Publishing Limited, UK.

Mathews, M. 1986. *Practical Clothing Construction – Part 1 & 2*. Cosmic Press, Chennai.

Mehta, P.V. 1992. *An Introduction to Quality Control for Apparel Industry*. CRC Press, Boca Raton, FL.

Organ Needles. 2014. http://organ-needlescom/english/product/downloadphp (accessed on March 21, 2015).

Pfaff Industrial. 2009. What are the causes of seam pucker? http://wwwpfaffindustrialcom/pfaff/en/service/faqs/generalsewing/seampucker/view?searcherm1/4Speed responsive (accessed on April 2, 2015).

Ruth, E.G. and G.I. Kunz. 2004. *Apparel Manufacturing – Sewn Product Analysis*. Prentice Hall, Englewood Cliffs, NJ.

Shaeffer, C. 2000. *Sewing for the Apparel Industry*. Woodhead Publication, Cambridge.

Sumathi, G.J. 2002. *Elements of Fashion and Apparel Designing*. New Age International Publication, New Delhi, India.

The Schmetz. 2001. *The World of Sewing: Guide to Sewing Techniques*. Ferdinand Schmetz GmbH Herzogenrath.

Tyler, D.J. 1992. *Materials Management in Clothing Production*. Blackwell Scientific Publications, Oxford, UK.

Ukponmwan, J.O., K.N Chatterjee and A. Mukhopadhyay. 2001. *Sewing Threads*. Textile Progress, The Textile Institute, Manchester.

7

Sewing Machine Feed Mechanisms and Attachments

The sewing machine feeding systems are used for handling fabrics in a controlled manner during stitching for continuous sewing. For producing an accurate straight line of stitches, the fabric must be moved through the stitch-forming area of the machine precisely and accurately. The feeding mechanism comprises three components, namely, presser foot, throat plate and feed dog. Each of these components has many variations in shapes and sizes. Application of these appropriate components depends on the type of sewing machine, number of needles used, types of attachments used and types of operations (Araujo et al. 1992; Chmielowiec 1995; Carvalho et al. 2009).

7.1 Elements of Feeding Mechanism

7.1.1 Presser Foot

It is the upper component of the feeding system that grips the fabric during the feeding action and stitch formation. The presser foot, which is fitted on to the presser bar, controls the quantity of pressure applied on the fabric panel as it is fed through the sewing machine. The extent of pressure to be exerted on the fabric can be varied based on the stitching speed and fabric construction and weight. Higher sewing speeds may require more pressure to control the movement of the fabric (Araujo et al. 1992; Chuter 1995).

A basic presser foot could be assembled as a single unit or hinged to facilitate movement over bulky seams. The universal type is the flat presser foot, which comprises a shank that attaches to the presser bar and the shoe that rests on the fabric surface is shown in Figure 7.1.

Variations in the basic presser foot take place in the shoe component, which involves the sole, heel and toe. The sole is the flat area that has direct contact with the fabric, which could be made smooth, toothed and so on. The toe is the front portion of the shoe that is accountable for guiding, holding and positioning the unsewn fabric. The heel is the back part of the shoe that is mainly responsible for holding fabric and retaining its established position for the feeding and stitching action to take place (Chuter 1995; Amann Group 2010a,b).

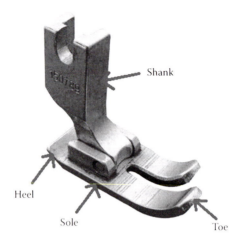

FIGURE 7.1
Parts of presser foot.

Some of the modifications of presser foot are offset soles for stitching along raised edges of fabric, short toes for sewing curves, long toes for long straight seams, channelled soles for fitting over bulky lapped seams, solid narrow feet for sewing close to raised edges and so on (Shaeffer 2000).

7.1.2 Feed Dog

The main element in a feeding system is the feed dog, which transports the fabric by a preset length between succeeding stitches. A set of feed dogs, which look like short, thin metal bars, are crosscut with grooves, move to and fro in the throat plate slot which is marginally bigger than the feed dog as shown in Figure 7.2. The typical four motions performed by the feed dogs such as forward, then downward, then backward and then upward aid to transport the fabric for continuous stitching. The feed dogs are in contact with the fabric panels on the forward movement and are pulled down below the main plate on the backward movement by the sewing machine's mechanism (Geršak 2001; Carr and Latham 2006).

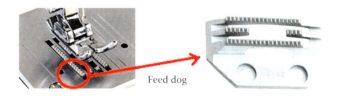

FIGURE 7.2
Feed dog.

7.1.3 Throat Plate

They are removable metal plates secured to an adapter plate directly under the needle. Throat plates keep the fabric panel as the needle penetrates to form the stitch. Throat plates have openings for needles to pass through and for the feed dog to come up. The selection of the throat plate with respect to its shape and size is based on the type of sewing machine and its specific function in the stitch-forming process. Throat plates are interchangeable and must be compatible with other stitch forming and fabric-carrying parts in order for stitches to be formed properly (Geršak 2000, 2013; Fairhurst 2008).

7.2 Types of Feed Mechanism

The feed mechanism on a sewing machine could be categorised based on its application and end-use as

- Manual feed
- Drop feed
- Differential feed
- Needle feed
- Compound feed
- Unison feed
- Puller, roller feed
- Cup feed

7.2.1 Manual Feed

It is also known as free motion or darning feed. In this kind of feeding mechanism, the operator moves work under the needle. The sewing machine has an upright motion presser foot which grips the fabric prior to the entrance of the needle into the fabric, and releases to permit the worker to handle the fabric between each stitch (Clapp et al. 1999). This feeding system is commonly utilised for darning, embroidery, freehand quilting, etc.

7.2.2 Drop Feed

The drop feed mechanism (Figure 7.3) uses a feed dog below the throat plate that raises up through the plate, grips the fabric counter to the presser foot to transport the fabric by one stitch, and then drops below the plate to come back to its original position.

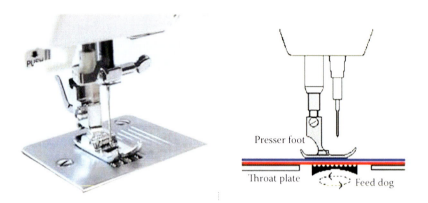

FIGURE 7.3
Drop feed mechanism.

7.2.2.1 Problems in Drop Feed

- While sewing with two plies of fabric, the lower ply moves forward by means of a feed dog positively but not the upper fabric plies. Therefore, two fabric plies are moving at different speed, that is, the lower ply at a faster speed than the upper ply. This is known as differential feed pucker or feed pucker.
- If the pitch of the feed dog teeth and the stitch are the same, then there is more possibility of fabric damage as the teeth of the feed dog and the fabric have repetitive contact at the same area.

7.2.3 Differential Feed

Differential feed (Figure 7.4) utilises a two-piece feed dog located beneath the throat plate that rises up and grips the fabric against the foot and then advances the fabric.

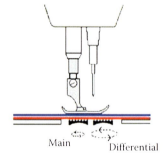

FIGURE 7.4
Differential feed mechanism.

The front (main) and rear feed dog could be fixed to move at the same or different speed/distances. When the rear feed dog is moving at a faster rate compared to the front, the fabric will be stretched. In contrary, when the front feed dog is moving faster than the rear feed dog, the fabric is gathered (shirring) (Laing and Webster 1998; Geršak 2001).

7.2.4 Top Feed Mechanism

In a top feed mechanism, the presser foot is made in two different sections. One section of the presser foot holds the fabric panel during the stitch formation by the needle and another presser foot has length on the lower side and wakes in a manner that the top ply is moved along positively when the needle is in and out of action on the fabric. A combination of adjustable feed and differential bottom feed can cause gathering of the top ply or bottom ply (Laing and Webster 1998).

7.2.4.1 Vibrating Presser Foot

In the case of a vibrating presser foot, the forward and backward motions of the presser foot are not driven; however, they are spring loaded as shown in Figure 7.5. The presser foot has teeth that aids in the movement of fabric along with the feed dog. It has a vibrating motion forward with the feeding process and backward with the return stroke. Generally it is constructed with a lifting motion during its return stroke to enable the presser foot to clear the fabric and to lower comparatively straight down onto an uneven section of the fabric without interference. It is commonly known as a walking foot or top feed (Laing and Webster 1998).

FIGURE 7.5
Vibrating presser foot.

7.2.4.2 *Alternating Presser Foot*

It has a couple of presser feet that alternately press against the fabric (Figure 7.6). When one foot is aiding in moving the fabric along, the other foot is raised to clear the fabric. These actions will take place alternatively. Out of two presser feet, one is normally a vibrating presser foot whereas another presser foot is a rising and descending one. The vibrating foot will facilitate in fabric feeding and the rising and descending foot will grip the fabric down between feeding motions (Laing and Webster 1998).

7.2.5 Needle Feed

Needle feed (Figure 7.7) utilises a feed dog beneath the throat plate that rises up through the plate, presses the fabric counter to the foot, in combination with the sewing needle, which is lowered through the fabric and then both the sewing needle and the feed dog move the fabric by one stitch. Then they separate and return to their respective original positions for the next stitch formation.

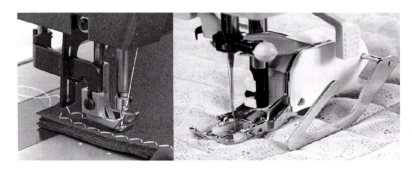

FIGURE 7.6
Alternating presser foot. (Adapted from http://textileapex.blogspot.in.)

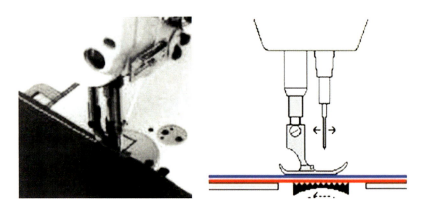

FIGURE 7.7
Needle feed mechanism.

Needles come into the fabric and stay in the fabric when moving the fabric perpendicular to the needle's normal direction, thus feeding the fabric along with the feed dog. It prevents the upper, middle and lower layers of fabric panels from slipping apart. It does not require any pressure from the top surface of the fabric during feeding, which could be useful for stitching delicate fabrics where the impression of the feed dog will be marked on the fabrics. It is commonly used in combination with drop feed and/or with upper feed. It is predominantly utilised in bulky sewing circumstances such as quilting fabric and for sewing heavy materials like leather, carpet, etc. The three main kinds of needle feed mechanisms are upper pivot needle feed, central pivot needle feed and parallel drive needle feed. The pivoting needle feed devices move the needle at a definite angle to assist the feeding of fabric; however, parallel drives simply move the needle back and forth (Clayton 2008).

7.2.5.1 Upper Pivot Needle Feed

The needle bar, which holds the sewing needle, is detained in a frame, and its movement is pivoted from the frame far from the needle. Hence, the sewing needle goes into the fabric at a leading angle with respect to the centreline of the needle and will exit the fabric at a trailing angle. This would appear to disturb the fabric and the sewing process, but practically it does not (Clayton 2008).

7.2.5.2 Central Pivot Needle Feed

In this system, the needle bar, which holds the needle and its movement, is pivoted at a point which is nearer to the middle of the frame. The sewing needle enters the fabric at a larger leading angle from the axis of the needle than the upper pivot system and exits the fabric at an equally larger trailing angle. There is less impetus of the needle bar in motion than the upper pivot system, and higher stitching speeds could be achieved.

7.2.5.3 Parallel Drive Needle Feed

The needle bar, which secures the needle, is permanently parallel relative to its earlier and successive movements. It remains perpendicular to the fabric during entry as well as exit from the fabric. This kind of needle feed is appropriate for sewing higher grams per square metre (GSM) fabrics.

7.2.6 Compound Feed

It is a combination of a drop feed mechanism and a needle feed mechanism. Feeding of fabric happens while the needle is in the fabric by means of combined motion of needle bar and feed dog. Compound feed (Figure 7.8) utilises a feed dog beneath the throat plate that raises up and presses the fabric

FIGURE 7.8
Compound feed mechanism.

against the presser foot in combination with a needle, which is still in the fabric, moves the fabric together by a one stitch. Then the needle is out of the fabric and moves to its respective position to form the next stitch with one step advance. This kind of feed mechanism is useful in bulky sewing circumstances like quilting the fabric, wadding and for slapping fabrics. In this feed mechanism, the change of stitch length warrants setting of both needle and feed dog.

7.2.7 Unison Feed

Unison feed as shown in Figure 7.9 is the conjunction of a needle feed and a compound feed mechanism. As the needle penetrates the fabric, the top (presser foot) as well as the bottom feed dogs compress the fabric, and all three components (feed dog, presser foot and needle) move the fabric by one stitch, then all are released from the fabric as the presser foot drops to hold the fabric, and all return for the next stitch (Geršak 2001).

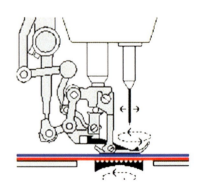

FIGURE 7.9
Unison feed mechanism.

The word unison feed is utilised in two different manners. One is its application to any of two or more feed systems working in combination. A second application is to depict the uncommon feed system of a vibrating presser foot, along with needle feed, and a drop feed, working in combination, but from a one-piece frame. This is the only feed mechanism where it is impossible for the upper and lower feed mechanisms to become out of synchronisation. All other kind of feed mechanisms are synchronised by linkage or electronic controls system (Glock and Kunz 2004).

7.2.8 Puller Feed

A puller feed is a method for providing a positive control of all fabric plies as they depart basic feeding mechanisms such as drop feed in the sewing machine. In this kind of feed mechanism, feeding is normally carried out by feed rolls as shown in Figure 7.10. The fabric passes between an upper roller and a sewing bed, or a lower roller and a presser foot. These feed rollers provide a dragging motion on the fabric behind the foot. The top roller is normally driven by the machine and the lower roller moves due to the pressure of the top roller. The surface speed of the puller roller is slightly higher than the speed of the feed dog to presser ply shifting roping. It is useful in multi-needle machines particularly for attaching the waist band (Tyler 1992; Bheda 2002).

7.2.9 Wheel Feed

The wheel feed mechanism shown in Figure 7.11 uses a roller that moves the fabric one stitch at a time, in a ratcheting motion. In this kind of feeding, the foot has small rollers to enable easy movement of fabric. Wheel feed is more suitable under circumstances where the fabric to be sewn would be damaged by the tooth of the feed dog such as products like vinyl plastic and some leather products.

FIGURE 7.10
Puller feed mechanism.

FIGURE 7.11
Wheel feed mechanism.

FIGURE 7.12
Cup feed system.

7.2.10 Cup Feed

A cup feed system as shown in Figure 7.12 utilises one or two cup-shaped wheels that squeeze the edge of the fabric, allowing the sewing needle to sew across the edge of the material. It is generally called a fur machine, as it is perfect for sewing the narrow strips together to make a fur coat. In this kind of feed mechanism, the sewing needle works in the horizontal path and feeding can be done by moving the fabric between the two rotating discs or by moving the fabric between a disc and a presser surface (Glock and Kunz 2004).

7.3 Special Attachments to Sewing Machines

Sewing machine attachments make sewing machines easier and provide a variety of decorative sewing possibilities. The majority of the attachments are normally secured to the presser bar instead of the foot. A few sewing attachments have hooked ends that rest on the needle clamp (Solinger 1998; Clayton 2008). The following lists the classes and types of sewing machine attachments:

1. Position attachments
2. Guide attachments
3. Preparation and finishing attachments

7.3.1 Position Attachments

7.3.1.1 Hemmers

Hemmers (Figure 7.13) construct hems from 3/16″ to 7/8″ wide, right on the sewing machine. Machine hemming with the hemmer attachments could save plenty of time compared to hand turning and basting. The hemming portion is automatically turned by the hemmer, and simultaneously the line of stitching is guided close to the edge of the hem. Hems are normally done at various widths, which can be made with the hemmers, suitable for the common requirements (Solinger 1998; Clayton 2008; Fairhurst 2008).

7.3.1.2 Ruffler

The ruffler attachment has the capacity of doing gathering or pleated frills as shown in Figure 7.14. It is normally utilised for making children's clothes and curtains. The means of utilising the ruffler attachment varies with different sewing machines.

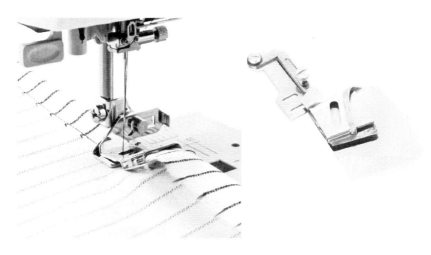

FIGURE 7.13
Hemmers.

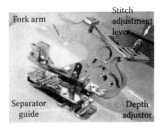

FIGURE 7.14
Rufflers.

7.3.1.3 Binder

It is a valuable attachment in a sewing machine. Though fine bindings can be created by hand, the binding using the attachment on the machine saves a lot of time and is precise and accurate as shown in Figure 7.15. It is commonly utilised for applying readymade bias binding to a straight or curved edge and is a useful attachment for trimming dresses, etc. The binder attachment has a small funnel-like portion for folding and guiding the binding over the edge of the fabric before it reaches the sewing needle. This attachment could be used for sewing straight, zigzag as well as decorative stitches (Abernathy and Dunlap 1990).

The quilt binder set (Figure 7.16) is used to make 1/2" binding, which can handle quilt fabric and other thicker and heavy weight fabrics. The two upper screws can be adjusted to line up the binding's top and bottom folds with the needle for precision. The quilt binder set makes straight line binding for large quilts and also adds an eye-catching accent to any small items such as placemats and hot pads.

The tape binder (Figure 7.17) is utilised for covering or finishing the raw edges of stretch fabrics. The tape binder folds the fabric to use as tape around the raw edge during sewing and binds any raw edge with ease (Fredrerick and Dunlop 1999).

7.3.1.4 Tucker

This attachment is used for creating uniform tucks from 1/8" to 1" width. Finest pin tucks having 3/4" width could be created easily without any basting. Delicate twin-needle pin tucks are a breeze with the grooves on the base of the presser foot as shown in Figure 7.18. The pin tucking foot is used in conjunction with a 2-mm twin sewing needle to make multiple rows of pin tucks. The grooves on the base of the pin tuck foot make it easy to stitch multiple rows parallel and uniformly spaced from each other (Ukponmwan 2001; Stitch attachments 2015).

7.3.1.5 Gathering Foot

The gathering foot attachment (Figure 7.19) is used for making soft gathers swiftly particularly in lightweight fabrics. The gather size in the garment

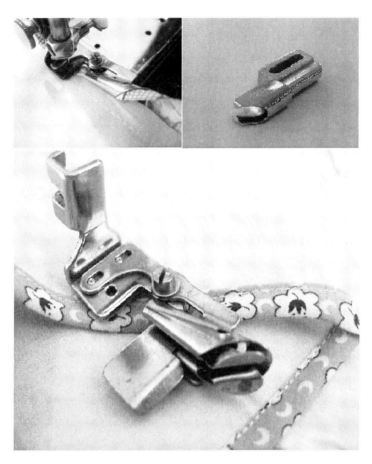

FIGURE 7.15
Binder.

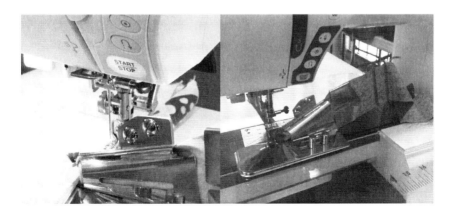

FIGURE 7.16
Quilt binder.

FIGURE 7.17
Tape binder.

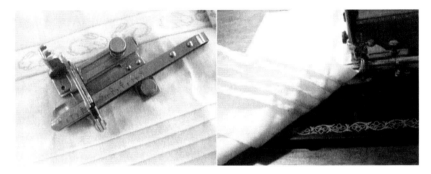

FIGURE 7.18
Pin tuck.

FIGURE 7.19
Gathering foot.

depends upon the fabric GSM, tension of the needle thread and stitches per inch. The base of the gathering foot is raised up at the back of the sewing needle and has a thicker bar section in front of the sewing needle for gathering and ruffling of fabric simultaneously.

7.3.2 Guide Attachments

7.3.2.1 *Zipper Foot*

The zipper foot could be set to stitch on both sides of the zipper (Figure 7.20). The edge of the foot directs the zipper to make sure placement is straight. Normally ready-to-wear garments will commonly have an invisible zipper fitted onto them (Schmetz 2001; Stitchworld 2012). Invisible (concealed) zippers are appropriate for all garments made from fine silk jersey through to suit weight wools and tweeds, which can be secured on the garments using a concealed zipper foot (Figure 7.21). The grooves underneath the concealed zipper

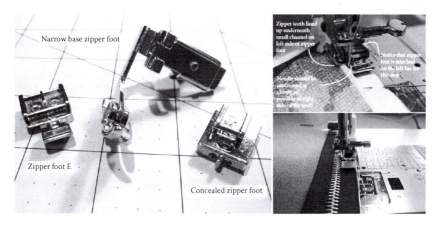

FIGURE 7.20
Zipper foot.

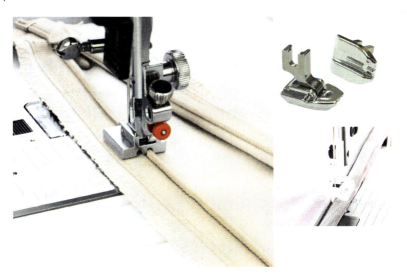

FIGURE 7.21
Concealed zipper foot.

FIGURE 7.22
Adjustable zipper foot.

foot contain the zipper teeth and hold them in place during stitching. The main criterion is to get the needle as close to the zipper as possible, which this foot achieves by slightly unrolling the zipper just before the needle. An adjustable zipper foot (Figure 7.22) can also be set to sew on each side of the zipper by regulating the location of the foot and tightening the screw (Gardiner 2003).

7.3.2.2 Cording Foot

The 3-way cording foot (Figure 7.23) will grip three fine cords or threads. Since it is attached to the presser foot, the requisite design can be easily followed and the cords are perfectly placed. A range of functional or decorative stitches could be sewn over the cords to put them onto base fabrics.

7.3.2.3 Circular Attachment

The circular attachment (Figure 7.24) is the most suitable one for sewing of circles using straight, zigzag or decorative stitches. Circles up to 26 cm in

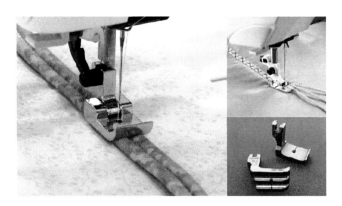

FIGURE 7.23
Three-way cording foot.

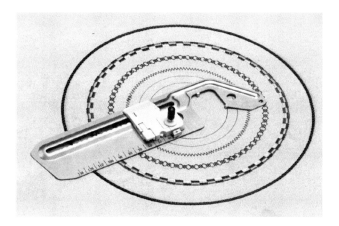

FIGURE 7.24
Circular attachment.

diameter can be stitched perfectly using this attachment, which is vital for craft and decorative work.

7.3.2.4 Button Sewing Foot

The two bars in the button sewing foot (Figure 7.25) are fixed to the shank of the presser foot to give additional firmness and it has a rubber sleeve for better gripping of the button during sewing.

7.3.2.5 Buttonhole Stabiliser Plate

With the buttonhole foot (Figure 7.26), which is secured to the buttonhole stabiliser plate, the machine feeds a range of fabrics and uneven layers smoothly instead of causing the needle to stick in position.

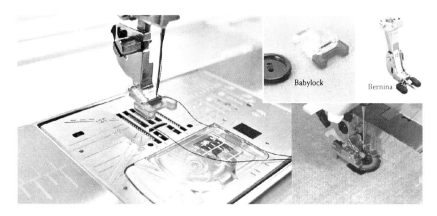

FIGURE 7.25
Button sewing foot.

FIGURE 7.26
Buttonhole stabiliser plate.

7.3.2.6 Buttonhole Foot

The buttonhole foot (Figure 7.27) is used for sewing buttonholes on the fabric. Two kinds of buttonhole foot are available, namely, transparent buttonhole foot and sliding buttonhole foot. The transparent buttonhole foot is used for stitching buttonholes on tight zones like cuffs and a sliding-type foot is used for stitching buttonholes on areas where more freedom of movement is essential (Gardiner 2003).

7.3.3 Preparation and Finishing Attachments

1. *Pinking:* It is a common finishing operation on garments. A power pinker is normally used for this purpose or pinking mechanism could be attached to the sewing machine. The two major actions carried out by the pinkers are chopping and cutting.

2. *Pressing attachments:* It is used for finishing garments after the fabric is sewed. For example, on a belt loop attachment process where a flat iron or rotary press pressing device is attached to the machine head.

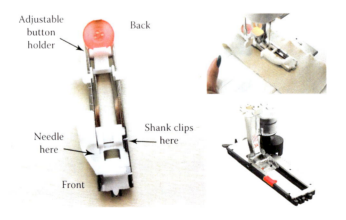

FIGURE 7.27
Buttonhole foot.

3. *Thread cutters:* These are extensively required alternatives that mini-mise production time and get rid of manual thread clipping. On a few machines, sewing threads are cut beneath the throat plate, and a wiper pulls the residue portion of cut thread out of the way in prepa-ration for the next process. Most of the 400, 500 and 600 class stitch machines have chain cutters and latch back devices built-in since the chain stitch formed by these kinds of machines should not be broken by a hand-tearing action.

4. *Chain cutters:* The chain cutters cut the chains in such a way that the stitch is secured against unravelling. Stitches produced on these machines cannot be cut as close as like in lock stitch machines, and some remnant thread remains.

5. *Tape cutters:* It could be used with the application of shoulder rein-forcements, neck bindings, elastic, lace and so on. As stitching is fin-ished, a photocell sensor finds the fabric end or piece and connects the cutter automatically. Tape may be cut at the beginning and end of the garment piece.

6. *Needle and stitch devices:* On several sewing machines, options are designed especially for assisting in the construction of the per-fect line of stitches such as needle positioners and stitch pattern regulators.

References

Abernathy, F.H.A. and J.T. Dunlop. 1999. *A Stitch in Time – Apparel Industry.* Blackwell Scientific Publications, Oxford, UK.

Amann Group. 2010a. Determining Your Sewing Thread Requirements. http://www.amann.com/en/download/industrial-sewing-threads.html (accessed on October 11, 2015).

Amann Group. 2010b. Preventing Seam Pucker in Service and Technik – Information for the Sewing Industry. http://www.amann.com./fileadmin/download/naehfaden/b_nahtkraeuseln_EN.pdf (accessed on October 11, 2015).

Araujo, M, T.J. Little., A.M. Rocha, D. Vass and F.N. Ferreira. 1992. *Sewing Dynamics: Towards Intelligent Sewing Machines.* NATO ASI on Advancements and Applications of Mechatronics Design in Textile Engineering Side, Turkey.

Bheda, R. 2002. *Managing Productivity of Apparel Industry.* CBI Publishers and Distributors, New Delhi.

Carr, H. and B. Latham. 2006. *The Technology of Clothing Manufacture.* Blackwell Science, Oxford.

Carvalho, H., A.M. Rocha and J.L. Monteiro. 2009. *Measurement and Analysis of Needle Penetration Forces in Industrial High-speed Sewing Machine.* Taylor and Francis, Boca Raton, FL, 100(4):319–29.

Chmielowiec, R. and D.W. Lloyd. 1995. The measurement of dynamic effects in commercial sewing machines. In: *Proceedings of the Third Asian Textile Conference,* Hong Kong, 814–28.

Chuter, A.J. 1995. *Introduction to Clothing Production Management.* Blackwell Scientific Publications, Oxford, UK.

Clapp, T.G., T.J. Little, T.M. Thiel and D.J. Vass. 1999. Sewing dynamics: Objective measurement of fabric/machine interaction. *International Journal of Clothing Science Technology* 4(2/3):45–53.

Clayton, M. 2008. *Ultimate Sewing Bible – A Complete Reference with Step-by-Step Techniques.* Collins & Brown, London.

Fairhurst, C. 2008. *Advances in Apparel Production.* The Textile Institute, Woodhead Publication, Cambridge.

Gardiner, W. 2003. *Sewing Basics.* Sally Milner Publishing, Australia.

Geršak, J. 2000. The influence of sewing processing parameters on fabric feeding. In: *Proceedings of the 29th Textile Research Symposium,* Mt Fuji, Shizuoka, 141–52.

Geršak, J. 2001. Directions of sewing technique and clothing engineering development. *Tekstil* 50(5):221–9.

Geršak, J. 2013. *Design of Clothing Manufacturing Processes: A Systematic Approach to Planning Scheduling and Control.* Woodhead Publishing, Cambridge.

Glock, R.E. and G.I. Kunz. 2004. *Apparel Manufacturing – Sewn Product Analysis.* Prentice Hall, Englewood Cliffs, NJ.

Laing, R.M. and J. Webster. 1998. *Stitches and Seams.* Textile Progress, The Textile Institute, Manchester.

Schmetz. 2001. *The World of Sewing: Guide to Sewing Techniques.* Ferdinand Schmetz GmbH Herzogenrath.

Shaeffer, C. 2000. *Sewing for the Apparel Industry.* Woodhead Publication, Cambridge.

Solinger, J. 1998. *Apparel Manufacturing Handbook: Analysis Principles and Practice.* Columbia Boblin Media Corp, New York, USA.

Stitch attachments. http://www.janome.co.jp/e/pdf/home/AccessoryCatalog.pdf (accessed on March 14, 2015).

Stitchworld. 2012. X-feed from typical: The ultimate feed system. *Stitch World* 23–4.

Tyler, D.J. 1992. *Materials Management in Clothing Production.* Blackwell Scientific Publications, Oxford, UK.

Ukponmwan, J.O., K.N. Chatterjee and A. Mukhopadhyay. 2001. *Sewing Threads.* Textile Progress, The Textile Institute, Manchester.

8

Fusing, Pressing and Packaging

8.1 Fusing

Each apparel manufacturer persistently attempts to manufacture garments with instant sales appeal. Nevertheless, one of the vital materials, fusible interlining, which is utilised for nearly almost every component of outerwear, has no sales appeal because it is imperceptible to the consumer. The method of fusing interlining to the garments started in Europe in about 1950 and in Japan in about 1960. Today, about 80% of all garments necessitate the use of interlining (Sang-Song 2001; Shim 2013).

Fusible interlining is the process where the wrong side of the fashion garment panel is fused with a thermoplastic resin and can be bonded with another strip of fabric by the proper application of pressure and heat at a specific temperature and time. The fusible interlinings improve the appearance of finished garments through

1. Stabilisation and control of crucial regions of the garment.
2. Strengthening of particular design features.
3. Without much change in the draping quality of the top cloth.
4. Maintaining the crisp and fresh look of the base fabric.

8.1.1 Purpose of Interlining

1. To make sewing easier and to increase throughput
   a. Since the speed of sewing machines is very high, the material must be in perfect structure and shape before sewing. Therefore, the operator efficiency could improve. Suppose if interlining is fused onto the material. It keeps its shape, therefore saving time and labour (Sang-Song 2001; Shim 2013).
2. Maintaining shape and improving appeal of the garment
   a. The interlining fabric improves the garment appearance while preserving the form of the garment.

3. Making a functional, easy to wear product
   a. By the use of a permanent press method, the sewing of garments becomes easy and a good quality product is made, which is easy to care and easy to wear. The main objective of pressing is to enhance the look and durability of the garment shape.

8.1.2 Requirements of Fusing

- The laminate formed by the fusing process should demonstrate the aesthetic properties necessary for the finished garment.
- The bonding strength between the base fabric and interlining fabric of the laminate should be satisfactory to bear up handling during further processes in the manufacturing sequence.
- Fusing should take place without either strike-through or strike-back taking place.
- The fusing process should not cause thermal shrinkage in the main fashion fabric after fusing.

8.1.3 Fusing Process

The elements of the fusing process are temperature and pressure, applied over a particular period of time on a fusing machine. By increasing the temperature at the 'glue line', the resin changes its state from a dry solid to a viscous fluid. By applying adequate pressure, the molten state of resin will adhere to the fibres in the main fashion fabric as well as fibres in the interlining. During the cooling process, the resin resolidifies and forms a durable bond between the two fabric panels. The heat has to go through the fabric to activate the resin, and this necessitates a certain holding time, which differs based on the construction of the fabric and the type of resin (Sang-Song 2001; Shim 2013). Apart from the outer fabric panel of the garment, the factors that decide the characteristics of the fused laminate are

- Base fabric of the interlining
- Type of fusible resin
- Pattern of application of resin to the base cloth

8.1.3.1 Base Fabric

The base fabric, otherwise called a substrate, is an interlining fabric on which the thermoplastic resin is coated. They are manufactured in a range of woven, knitted and nonwoven forms from natural or synthetic fibers. Irrespective of the interlining fabric construction and fibres used, the base fabric influences the following properties of the finished garment:

- Handle and bulk
- Shape retention
- Shrinkage control
- Crease recovery
- Appearance in wear
- Appearance after dry cleaning or washing
- Durability

8.1.3.2 Resins

The resins are applied to the base fabric for bonding. While the resins are exposed to pressure and heat for a specific period of time, it becomes the bonding or adhesive agent between the interlining and the top fabric. During the continuous application of pressure and heat, the molten resin could penetrate into the top fabric and while cooling the solidification of molten resin forms a bond between the interlining fabric and the top cloth (Abernathy and Dunlop 1999). The resins have to meet the following conditions:

- *Upper-limit temperature* – The resin should be converted into a viscous state at this temperature and should not damage the top fabric. While this temperature varies based on the consumption of the top fabric, it hardly ever exceeds 175°C.
- *Lower-limit temperature* – This is the minimum temperature at which the resin starts to become a viscous state. For most fusible interlinings, this is about 110°C, and is slightly lower for fusible interlining used for leather materials.
- *Cleanability* – The adhesive characteristics of the resin have to be durable enough to sustain repeated dry cleaning or washing operations during the life of the garment.
- *Handle* – The resin should contribute to the requisite handle and drape of the top cloth and should not act as a stiffening agent on the final garment.

The generally used adhesives in the fusing process are shown in Table 8.1.

8.1.3.3 Coating Systems

Coating is the method in which the thermoplastic resins are applied to the substrate material. Generally used coating methods are

TABLE 8.1

Types of Resins and Their Specifications Used in Fusing Process

Resin	Specification	Application	Fusing Temperature	Special Care
Polyamide	Used for a wide range applications	Used for all garments that are dry cleanable	120–160°C	Suitable for dry-cleaning but not suitable for high temperature applications
Poly vinyl chloride	For imparting soft finish	Specially suitable for higher GSM fabrics	130–160°C	Suitable for dry-cleaning as well as high temperature applications
High density polyethylene	It is resistant to all kinds of washing and cleaning	Suitable for top fusing of shirts	150–180°C	Suitable for dry-cleaning as well as high temperature applications
Low density polyethylene	It can be fused by iron	Provides only temporary fusing	130–160°C	Not suitable for dry-cleaning as the resin remelts and not appropriate for high temperature applications
Ethylene vinyl acid copolymer (EVA)	It can be fused by iron	Temporary fusing	120–150°C	Not suitable for dry-cleaning as the resin remelts and not appropriate for high temperature applications
Polyester	Mostly used when polyester-based fabrics are used	For polyester garments	130–160°C	Suitable for dry-cleaning as well as high temperature applications while used with polyester fabrics

1. *Scatter coating* – This method utilises electronically controlled scattering heads to set down the resin onto the moving fabric.
2. *Dry-dot-printing* – The resin is printed onto the fabric through an engraved roller having microgrooves to retain the resin.
3. *Paste coated* – The net-like structure is formed by heating the resin and then it is laminated on the fabric by applying heat and pressure.

8.1.4 Fusing Machinery and Equipment

In spite of what kind of fusible equipment are used, the process of fusing is influenced by four main factors such as time, pressure, temperature and cooling and these have to be precisely combined to attain the desired results.

1. *Temperature* – The temperature of the resin should be optimum as high a temperature leads to too viscous a resin and too low temperature does not allow the resin to penetrate into the top fabric.

2. *Time* – The time period which is a very decisive factor during the fusing process is while the top fabric and interlining material are kept under the influence of heat and pressure in the heating region of the fusing machine. This time of fusing for a particular fusible is based on

   a. Melting temperature of the resin used

   b. Weight of the interlining

   c. The nature and construction of top fabric being used, that is, thick or thin, dense or open

3. *Pressure* – When the resin becomes viscous after heating at a particular temperature, then pressure is applied to the top fabric and fusible assembly to make sure that

   a. Full contact is made between the fusible and the top fabric

   b. Heat transfer is at the optimal level

   c. There is more even dispersion and diffusion of the adhesive thermoplastic resin into the surface and internal structure of fibres of the top fabric

4. *Cooling* – Forced cooling is generally used. Therefore, the fused assemblies can be handled instantly after the fusing process. Cooling of fused fabrics could be done by means of several methods such as use of compressed air circulation, water-cooled plates and vacuum. Rapidly cooling the fused assemblies to 30–35°C makes a higher level of productivity.

The equipment used for fusing is classified as specialised fusing presses, hand irons and steam presses (Gutauskas et al. 2000).

8.1.4.1 Specialised Fusing Presses

8.1.4.1.1 Flat-Bed Fusing Press

These are specialised fusing equipment prevailing in a range of types from bigger, floor standing machines to smaller table type models. Mainly this type of fusing press comprises a top padding buck and bottom normal bucks along with the provision of heating elements in one or both the bucks as shown in Figure 8.1. The bottom buck is normally stationary and the top buck could be raised or lowered to open as well as close the trays, which travel horizontally to feed the components for fusing and remove it after the job is completed (Gutauskas and Masteikaite 1997; Gutauskas et al. 2000).

There are numerous kinds of specialised flat-bed pressing machines, which could have a magazine type or carousel action, which automatically

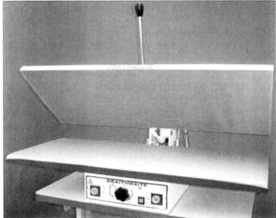

FIGURE 8.1
Flat-bed fusing press.

transports the fusing components from the loading point to the fusing zone in a machine and after the process is completed it returns the fused components to the operator for unloading. Flat-bed fusing machines are further classified as vertical action and scissor action closer, in which a vertical action closing type machine is considered more desirable owing to its capability to exert even and uniform pressure throughout the fabric area (Gutauskas and Masteikaite 1997). The advantages of flat-bed machines are

- Simple, less complex and thus easy to operate.
- Size of the machine is smaller and comparatively low cost.
- As the fabric is constantly held under pressure throughout the fusing process, the chances of fabric shrinkage are minimised.

8.1.4.1.2 Continuous Fusing Press

These machines normally have a motorised conveyor belt arrangement for transportation of fusing components through all the processes. The two conveyor belts systems in common use are

1. *End-to-end feed* – The garment panels are transferred from the loading area, through the fusing and cooling section, finally to the delivery side located at the other end of the machine.
2. *Return feed* – In this belt system, the fused components are returned to the same location where it is loaded. The upper belt transports the garment panels to the fusing processes whereas the lower belt returns the fused panels to its starting position.

The construction and features of a continuous fusing press machine are presented in Figures 8.2 and 8.3. The head could be offered in three methods:

* With a direct heating system, the endless conveyor belt transports the fusing components (top cloth and interlining fabric) to the direct contact with a heated drum or plate surface.

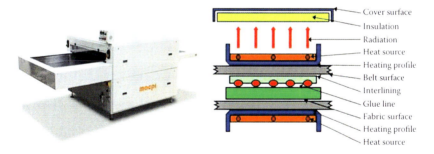

FIGURE 8.2
Continuous fusing press. (Reproduced by permission of Macpi Ltd, Palazzolo sull'Oglio, Italy.)

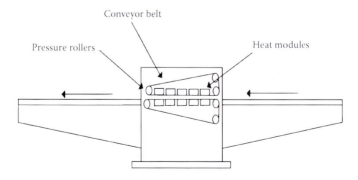

FIGURE 8.3
Features of continuous fusing press. (Reproduced by permission of Veit Corporation, Landsberg, Germany.)

- With an indirect heating system, the fusing components are transported into a heated chamber.
- With a low temperature gradient heating system, the fusing components are carried through a preheating zone.

Heating Mechanism:
The heating systems normally used for continuous fusing presses are

1. *Heating plates* – It comprises two heating surfaces, one above and one below the conveyor belt with separate temperature control.
2. *Cylinder heating* – The cylinder consists of two distinct portions, the inner cylinder, which is a stationary unit where the heating components are fixed and the outer cylinder, which rotates around the inner cylinder.

Pressure Mechanism:
Pressure is applied constantly and uniformly right through the process once fed into the machine. Real pressure is applied at the exit point where drums exert more pressure on heated fusible components.

Time Mechanism:
Fusing time depends on the conveyor belt, that is, as the conveyor moves at a faster speed, then fusing time will be less and vice versa. The residential period in the heating zone of a fusing machine could be altered by means of a speed controller depending upon the material to be fused.

8.1.4.1.3 High Frequency Fusing

Normally in fusing presses, heat is provided by electric heating elements. This confines the thicknesses of fabrics to be fused. If many layers of fabric and interlining could be stacked up and fused concurrently, productivity might be increased by means of high frequency energy. This machine contains an electric frequency generator. The high frequency fusing press is shown in Figure 8.4. The particular kind of polymer present in the thermoplastic adhesive resin system absorbs alternating waves generated by the high frequency generator, which causes the resin to heat up faster than the interlining fabric and the base garment fabric. This leads to better bonding at the glue line without the need of excessive heat (Gutauskas and Masteikaite 1997; Gutauskas et al. 2000).

8.1.4.2 Hand Iron

Hand irons are an appropriate method of fusing where the interlinings could be fused at comparatively low temperatures and pressures and in a relatively short time period. But only small components can be fused in this system (Cooklin 2006; Fairhurst 2008).

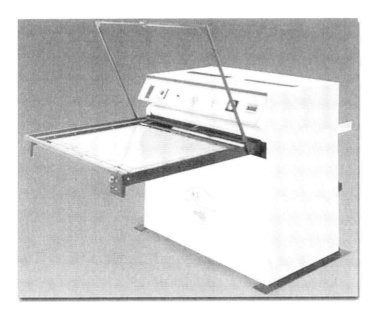

FIGURE 8.4
High frequency fusing press.

8.1.4.3 Steam Press

Normal steam pressing equipment is generally not preferred for fusing although some fusibles are made from these machines. The main limitations of pressing machines concerning fusing are

- Incapability to reach the heat levels requisite by most of the thermoplastic resins.
- Shape and size of bucks limit size of components that can be fused.
- Majority of the pressing machines do not have inbuilt programme control options and hence complete operation should be manually controlled.
- If the resin is activated using a heated steam, a similar process could happen on final pressing during garment production which could lead to serious problems.

8.1.5 Methods of Fusing

1. *Reverse fusing* – In this technique, the outer fabric is placed on top of the fusibles.
2. *Sandwich fusing* – This method could be carried out only when the heat is applied from both top and bottom of the fusibles like in a horizontal continuous press machine.

3. *Double fusing* – In this method, two types of interlining are fused to the outer fabric in a single process like fusing of shirt collars and men's jacket fronts.

8.1.6 Control of Fusing Quality

Fusible interlinings are accurate products and it is important that they are fused on the correct equipment under strict control of parameters. Factors affecting fusing quality are

- Temperature
- Time
- Pressure
- Peel strength
- Dry-clean/wash

8.2 Pressing

It is the process of application of heat, pressure and moisture to shape or crease garments or garment components into the geometric forms proposed by the designer. The pressing process influences the final garment appearance and hence the garment appeal. Finishing and pressing machines contour the semifinished garment panels as well as finished garments by bringing down the fibres in the fabric to an elastic state and then deforming and setting them (Gutauskas 2000; Fairhurst 2008).

8.2.1 Purpose of Pressing

- To flatten out the undesirable wrinkles, creases and crush marks.
- To make creases where the garment design needs it.
- To mould the garment to the silhouette of the body.
- To prepare garments for further sewing.
- To refinish the garment after completion of the production process.

8.2.2 Classification of Pressing

The basic processes that are involved in pressing can be divided into two groups:

1. *Under pressing* – It is the pressing operation performed on garment components as they are made up.

2. *Top pressing/Final pressing* – This refers to the finishing operation, which a garment undergoes after being completely assembled.

Both groups involve a huge number of individual processes, their extent determined by the cloth, quality and design of the garment (Gutauskas 2000).

8.2.3 Categories of Pressing

Based on the above factors, garments are divided into various classes according to the amount and kind of pressing required.

1. Garments, which require no pressing – Foundation garments, stretch swimwear, bras, briefs and underwear.
2. Garments requiring minimal pressing – Single ply garments such as slips, nightgowns, knitted synthetics and T-shirts.
3. Garments requiring the use of an iron in under pressing and final pressing – For the opening of seams and creasing of edges and for pressing garments with gathers and fullness and in situations where style change is frequent.
4. Garments requiring extensive under pressing and final pressing – Men's jackets, trousers and waistcoats, women's tailored jackets, skirts, top coats. Style change in these garments is infrequent.
5. Garments requiring pleating or 'permanent press' finishing.

8.2.4 Basic Components of Pressing

The main elements of the pressing process are heat, pressure and moisture, which deform fibres, yarns and fabrics to accomplish the required effect.

1. *Heat* – It is necessary to soften the fibres, stabilise and set the fabric in the desired shape. Temperature must be selected based on the fibres, yarns and fabrics.
2. *Steam (Moisture)* – It is fastest way of transmitting the heat onto the fabric. Steam and heat are essential to ease the fabric from tension and make the fabric with adequate flexibility so that it can be moulded to get the required contour.
3. *Pressure* – It is applied to change the form and increase the durability of the moulding. Pressure could be applied by means of a mechanical device or steam.
4. *Drying* – Subsequent to the steam and pressure application on the fabric, the garment panel or finished garment must be dried and cooled; thus, the fabric can return to its regular moisture content and steady condition. This could be done by removing the surplus water from the fabric by means of a vacuum action which cools it at the same time.

5. *Time* – The time period for which the garment is exposed to steam, pressure and drying depends on the type of fabric being pressed and there will be an optimal time period for each component.

8.2.5 Classification of Pressing Equipment

The mechanically operated pressing machine was invented in 1905 and developed continuously thereafter. The pressing machines are classified in three major categories based on how the machines are pressed (Carr and Latham 2006).

1. *Solid Pressure Equipment (Pressing Equipment)*
   a. Pressing irons
   b. Buck presses
   c. Mangle presses
   d. Block presses
   e. Form presses
   f. Pleating presses
   g. Creasing machines: Edge folders
2. *Moisture Pressure Equipment (Steaming and Wetting)*
   a. Wetting tanks: London shrinkers and auxiliary equipment
   b. Sponging machines
   c. Decaters
   d. Steam guns and jets
   e. Steam chambers
   f. Autoclaves
3. *Heat Energy Equipment (Heating and Baking)*
   a. Thermoelectric machines
   b. Hot plates
   c. Casting equipment
   d. Dry heat ovens

8.2.6 Types of Pressing Equipment

Solid surface pressing equipment uses a firm surface to apply pressure while steam and heat mould the fabric, garment or garment parts. Pressure may be applied through a rolling action, gliding action or compression (Holme 1999).

8.2.6.1 Hand Irons

Conventionally, the hand irons were heated on coal stoves and are steadily replaced by gas heating. Steam was created by lightly wetting the area to be pressed or by pressing through a dampened piece of linen. The most frequent but least productive pressing equipment is the hand iron, which is appropriate for executing a large number of jobs in processing with the aid of moisture and heat (Holme 1999).

More sophisticated and more productive but less prevalent are ironing presses, which allow a significant degree of mechanisation and automation of operation with the aid of pressure, moisture and heat. Fabrics are steamed to eliminate a lustrous appearance, which happens while utilising the pressing machines (Chuter 1995; Shaeffer 2000). The two basic kinds of irons used today are

1. *Dry iron*: These are lightweight irons weighing about 1.4 kg with 70°C–240°C heating range along with electronic temperature controls (Figure 8.5). This kind of iron is made in various shapes and is mostly used for smoothing of finishing operations where steam is unnecessary.
2. *Electric steam irons*: These are generally used as a kind of hand iron which could carry out numerous operations, specifically concerned with under pressing. Normally, the electric iron has a heating element, and steam is passed from a central or independent boiler into the steam chamber at the base of the iron (Figure 8.6).

Normally, hand irons are available in different shapes and weights.

- Narrow hand irons are used for seam opening on sleeves and trouser legs. The wrinkle marks on the garment are evaded by the narrow sole construction of the steam iron as well as curved and narrow ironing bucks. Teflon-coated soles should be used for ironing fabrics that are sensitive to lustre
- Wide ones for flat shapes
- Pointed shape

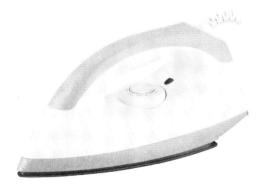

FIGURE 8.5
Dry iron.

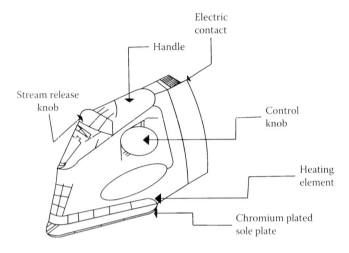

FIGURE 8.6
Electric steam iron.

8.2.6.1.1 *Iron Table*

The significant factor for the proper selection of an ironing station is the air flow through the garment to cool it and set it. Some types of ironing tables are listed below:

- Jacket seam ironing station
- Trouser seam ironing station standard
- Sleeve seam ironing station
- Dress board ironing tables
- All purpose table
- Flat top ironing tables
- Blouse and shirt ironing station
- Trouser leg ironing station
- Hip-bow ironing station
- Concave ironing station
- Convex ironing station
- Curtain ironing table

The stability of the covering of the ironing table is based on several factors such as hydrolysis resistance, pressure resistance and heat resistance of the used materials. All layers of fabrics in an ironing table together are accountable for the even distribution of steam. The covering starts at the metal surface of the ironing table.

1. *Rough wire mesh* – Besides the steam spreading it also improves the vacuum suction.
2. *Lower padding* – It should be durable as well as heat resistant.
3. *Lower layer* – Polyester wire screen mesh which distributes the steam.
4. *Intermediate padding* – This provides the softness of the covering.
5. *Upper layer* – It is a polyester wire mesh with inlet.
6. *Final top cover* – It must be less heat resistant than other lower layers.

8.2.6.2 Steam Presses

A steam press comprises a static buck and a head of harmonising shape, which closes onto it, therefore sandwiching the garment to be pressed. It can be used for either under pressing or top pressing.

8.2.6.2.1 Under Pressing

Basically this type of equipment can be general purpose or a special pressing machine.

1. *General*: This usually consists of a rectangular table with a built-in pressing area. It is equipped with a hand iron, which is connected to either a central steam supply or independent steam and vacuum supply system. The operator normally operates in a standing position controlled by foot pedals.
2. *Special purpose*: Edge pressing before top stitching or simultaneously pressing the shoulders and fusing the shoulder pads into position. Some of the special purpose steam pressing machines are shown in Figure 8.7.

8.2.6.2.2 Top Pressing

Numerous variety of pressing machines are available for top pressing and finishing, which are capable of several or all the operations necessary to fully top press a garment. The three kinds of pressing action of machines available are scissor action, cassette and carousel.

1. *Scissors press*: In this type of system, a lower buck is stationary and has a movable top buck system which can be lowered and raised using a lever type action as shown in Figure 8.8. A stationary bottom buck along with a top buck has a steam supply and the bottom buck is connected to a vacuum supply for drying purposes. Conventional pressing machines were operated through foot pedals and levers, but modern pressing machines are operated by compressed air.

   The typical pressing cycle includes (i) loading of a garment panel on the bottom buck, (ii) application of steam from the bottom buck, (iii) closing of the head or top buck, (iv) application of steam from

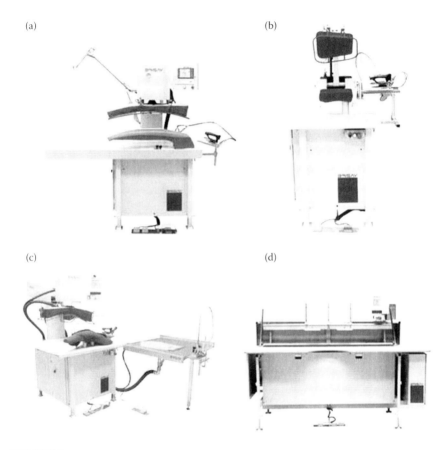

FIGURE 8.7
Special steam pressing machines. (a) Sleeve seam pressing machine, (b) shoulder seam-pressing machine, (c) elbow hem under-pressing machine and (d) jacket hem pressing machine. (Reproduced by permission of Veit Corporation, Landsberg, Germany.)

the head, (v) after the steaming period is completed, releasing of the head, (vi) application of vacuum to remove remaining moisture and drying the garment and (vii) placing the garment on a hanger. The duration of steam, pressure and vacuum vary based on the type of garment. Modern steam presses have vertical head movement action instead of a scissors action, which provides better control and a uniform distribution of pressure over the entire surface of the buck (Jevsnik and Gersak 1998).

2. *Cassette*: These kinds of equipment work as pairs such that while one pair is in a pressing action, another pair is loaded and ready for the action. In this kind of machine, the lower buck moves in the horizontal path from the front loading station to the back of the machine where the vertically moving top buck is located.

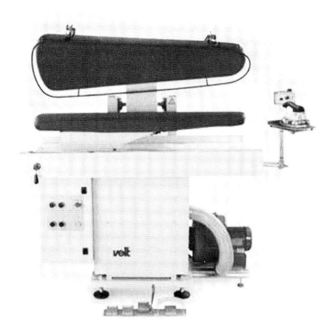

FIGURE 8.8
Universal steam press. (Reproduced by permission of Veit Corporation, Landsberg, Germany.)

3. *Carousel machines*: These kinds of machines are found in three different buck configurations. These machines comprise vertically mounted top bucks and bottom bucks mounted on a lower plate, which rotates through 180° or 120° (Figure 8.9). The swivelling action of the plate brought about the loaded bottom bucks into position of the top bucks and at the same time returning the pressed garments to the operator's position.

Pressing machines are available (dolly and tunnel finishers) that press the garments in a single operation, and are generally used for unconstructed garments. This is principally a mechanised tailor's dummy comprising a shaped inflatable nylon bag through which air and steam are blown. The garment is positioned onto the form and steam is forced through it. The setting and drying of the garment has been done by means of pressurised hot air, which inflates the nylon body (Jevsnik and Gersak 1998; Jones 2013).

8.2.6.3 Steam Finisher

This equipment is known as a form press or a 'dolly' press. It has a compressed air system, frame for a steam distribution system and a pressing form made

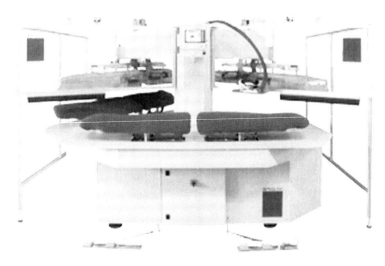

FIGURE 8.9
Carousel pressing machine. (Reproduced by permission of Veit Corporation, Landsberg, Germany.)

of a canvas bag in the suitable silhouette of the garment to be pressed. The pant steam finisher and universal steam finisher are shown in Figure 8.10.

8.2.6.4 Tunnel Finisher

This equipment comprises a conveyor fed unit through which the garments are moved while being steamed and dried. A smaller capacity form of this

FIGURE 8.10
Steam air finisher. (Reproduced by permission of Veit Corporation, Landsberg, Germany.)

equipment is known as a cabinet tunnel, which could process separate batches of 4 or 5 garments at a time automatically. The production capacity of cabinet tunnel machines are about 10% of that of a larger version of a tunnel finisher and are typically used by small garment industries (Jevsnik and Gersak 1998; Jones 2013). The construction of a tunnel finisher with its salient features is shown in Figures 8.11 and 8.12.

1. Conveyor and hook systems
2. Air curtain entrance avoids condensation from forming, eliminating moisture drops on garments
3. The cotton care unit is used to moisten the garment
4. Roller unit
5. Exhaust steam is recycled and reclaimed
6. Preconditioning module

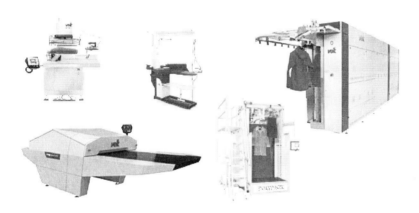

FIGURE 8.11
Steam tunnel finisher. (Reproduced by permission of Veit Corporation, Landsberg, Germany.)

FIGURE 8.12
Features of steam tunnel finisher. (Reproduced by permission of Veit Corporation, Landsberg, Germany.)

7. Variable steam quality while processing natural fibres (moist steam) or synthetics (dry steam)
8. Condensation traps
9. Self-regulating valves
10. Lower steam shut off position for normal-sized garments to increase energy efficiency
11. Reduction of condensation by means of improved steam tube configuration
12. Airlock provided
13. Extended air circulation module with blowers
14. New BUS control
15. Blower system combines high performance with energy efficiency
16. Steam doll effect is achieved by airflow from the bottom, thus inflating the garment
17. Hook system

The main objective of the steam is to relax natural fibres. With the garments on hangers, the tension due to gravity removes the wrinkles, and the turbulence of air blowing gives additional energy to remove wrinkles in woven fabrics (Jevsnik and Gersak 1998; Jones 2013).

8.2.6.5 Press Cladding

Bucks of steam presses and the ironing tables used with hand irons are normally covered with silicone foam. This is covered on the outer side normally by a top cover of polyester woven fabric. The heads of the steam presses could be covered with several layers of materials like a layer made of metal gauze for uniform steam distribution, a layer of synthetic felt to shield the next layer, the main layer of cotton knitted padding, and a last layer of outer cover as on the buck (Kim et al. 1998).

8.2.6.6 Creasing Machines

This unique kind of small press performs an exceptionally useful function. Creasing machines fold over the fabric and press the edges of fabric panels such as in pockets or cuffs to make them ready for easier sewing. For example, a patch pocket, which already has a done top hem seam, is pressed ready for the operator to sew it to the garment (Figure 8.13).

The operator positions the garment component panel over a specifically shaped die and blades control the fabric to make the creases around it and applies pressure during the pressing cycle. The means of pressing may be heat alone coming from the element in the machine (Kim et al. 1998; Glock and Kunz 2004).

FIGURE 8.13
Pocket creasing machine. (Reproduced by permission of Veit Corporation, Landsberg, Germany.)

8.2.6.7 Pleating

It is a special type of pressing used to produce a range of creases in a garment. The pleats may be smaller which are made by means of machine pleating, or it can be larger which are produced by hand pleating. Crystal pleating, hand pleating, box pleats and fan-shaped pleats are some of the examples shown in Figure 8.14.

There are two principle types of machine pleating known as rotary machine pleating and blade machine pleating. In the case of a rotary machine pleating (Figure 8.15) the rollers are built-in with paired dies similar to gears. The tiny pleats such as crystal pleats and accordion pleats are created using this machine (Mathews 1986; Laing and Webster 1998; Clayton 2008).

In a blade machine pleating (Figure 8.16), the pleats are created by the thrust action of the blades. Then the pleats are fixed or set by means of application of pressure and heat as it passes through a pair of rollers (Frings 1999; Wei and Yang 1999).

8.2.6.8 Block or Die Pressing

In die pressing, the fabric is kept over a fixed die prior to the application of steam, heat and pressure. This is normally used for shaping and moulding of hat and gloves during the manufacturing process. Another variety of an

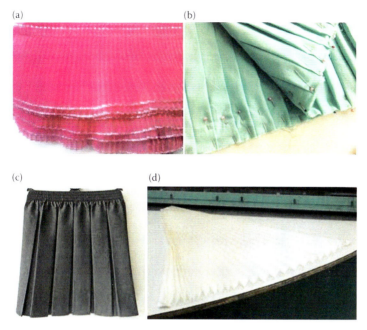

FIGURE 8.14
Types of pleats. (a) Crystal pleat, (b) hand pleats, (c) box pleats and (d) fan-shaped pleats.

FIGURE 8.15
Rotary pleating machine.

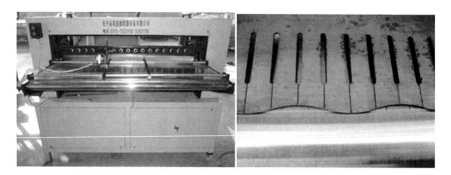

FIGURE 8.16
Blade pleating machine.

automated die pressing machine is utilised for combined folding and creasing of patch pockets as well as pocket flaps, in which the operator has to keep the components to be pressed over a die and engages the machine for folding and creasing (Mathews 1986; Clayton 2008).

8.2.6.9 Permanent Press

The permanent press method normally results in reduction of fabric strength. This method was developed for producing better crease recovery of cellulosic fabrics. The process involves processing the fabrics during its manufacture with a resin. A permanent press fabric is processed after the resin treatment and is then made into garments. The method is commonly used for trousers to introduce the creases at the seams and hems and down the front and back. The garments are then passed through an oven to cure the resin in the fabric (Mathews 1986; Mehta 1992; Clayton 2008).

8.3 Garment Packaging

This is the final process in the production of garments, which prepares the finished merchandise for delivery to the customer. These operations come under the materials handling methods and are no less important than other systems used in the factory. After completing the entire manufacturing task, apparel is required to be packed. After packing, it is placed in cartons as per instructions and then it is stored in a store section before it is delivered to the respective buyer (Solinger 1988; Glock and Kunz 2004).

Packaging refers to the container that carries a product. Two basic objectives of packaging are preventing any damage to the product during transportation and enhancing the features of the product to the consumer for a sale of it. Packaging has two major functions:

- Distribution
- Merchandising

The main purpose of distribution packaging is packaging the garment in a way that it allows the garment manufacturers to transport the garment at a minimum cost and in the shortest time to the retailer or purchaser, without deteriorating the quality of the product. The merchandising function deals with showcasing the garment product in a way that it stimulates consumer desire for purchasing the particular product (Solinger 1988; Sumathi 2002; Glock and Kunz 2004).

8.3.1 Types of Package Forms

The basic types of package forms used in apparel and allied products are

- Bags
- Boxes
- Cartons
- Cases
- Crates
- Twine
- Wrappers

8.3.2 Types of Packing Materials

The simple packaging materials used in garment and related items are paper, plastic, film, wood, nails, staples, cords, gum tape and metal bands.

1. Wood cases and crates are generally used as packing materials for bulk exports or rugged shipments where shipment handling is higher.
2. Paper and plastic film packaging materials are used in the garment and related industries. The paper types such as kraft, crepe, tissue, paper foil, paper board and waterproof are typically used as packing materials. Plastic films have a major advantage over paper because of clarity in range.

8.3.3 Quality Specifications for Packing Materials

Quality specifications for packaging paper and film are similar to that of fabric. The basic quality factors in paper and films are

1. Properties
   a. Clarity
   b. Thickness

    c. Width and length

    d. Weight

    e. Yield

2. Characteristics

    a. Tensile strength

    b. Elongation

    c. Bursting and tearing strength

    d. Flammability

    e. Porosity

    f. Air/moisture permeability

    g. Sunlight transference

    h. Resistance to odours

    i. Dimensional stability to heat and sunlight

8.3.4 Package Design

A product's packaging mix is the outcome of numerous requirements that decide the way the packaging achieves the distribution and merchandising functions. A package must promote or sell the product, protect the product, aid the consumer to utilise the product, offer reusable options of the package to the consumer and has to satisfy legal requirements. The main two criteria for package design are functional and sales requirements (Sumathi 2002).

8.3.4.1 Functional Requirements

Package design for a specific product should fulfil five groups of functional criteria such as in-store, in-home, production, distribution and safety and legal. In-home requirements normally usually state that packaging should be convenient to use and store and reinforce consumer's expectations of the product. For in-store conditions, packaging must draw the attention of the buyer, identify the product and differentiate it from the competition, and tempt the customers to purchase the product. The package should be designed such that the retailer could easily store the product, keep the stock on the floor, and it must be simple to process at a check-out counter (Tyler 2008).

Production demands influence primarily the cost of a package. During packaging of products, their distribution and safety are vital. If an undesirable segment of the products is damaged during transportation, distribution, or storage, then the package has failed. The last group of functional packaging requirements relates to laws and legislation. Several federal laws have been created to safeguard the consumers from parody and unsafe products. The major significant class of laws that influence packaging is labelling (Sumathi 2002; Tyler 2008).

8.3.4.2 Sales Requirements

Apart from functional aspects, product packaging should be designed in such a way that it appeals to customers. The four most important merchandising requirements of package design are its apparent size, impression of quality, attractiveness of a package and finally readability of the brand name.

Apparent size involves designing of packages to look as good as possible without misrepresenting the actual product contents. This is accomplished by using larger package sizes and displaying the brand name in the visible portion of the package. The obtrusiveness and aesthetics of the package design determines the attention drawing power of the package. Based on the type of product and manufacturer's policy, the package could be made to emerge as attractive, exciting, soft, intriguing, or to evoke some other emotion. Normally bright colours, prominent carton displays and other elements can acquire positive attention of consumers (Solinger 1988; Sumathi 2002; Glock and Kunz 2004).

A quality impression is a vital sales requirement for packaging because the products that are perceived to be of low quality are normally believed to be a low value, regardless of cost. Readability is the fourth sales requirement for successful design of a package (Gardiner 2003). This parameter is of extreme significant for products such as food items which are kept next to numerous competing products and brands. Among other important parameters, logos and letters of the product should be sufficiently large enough to read and printed in the same style as that used in complementary print and advertisement.

8.3.5 Types of Garment Packing in Finishing Section

The flowchart shown in Figure 8.17 gives the idea of selection of garment packing methods to ensure merchandise is floor ready.

The most commonly used types of garment packing are given below.

8.3.5.1 Stand-Up Pack

This type of packing is commonly used for shirts and hence termed as 'shirt packing'. For this type of packing, the garments have to be pressed prior to packing and are packed with additional packing materials like tissue paper, back support, pins or clips, inner collar patty, outer patty, etc. (Solinger 1988; Sumathi 2002; Glock and Kunz 2004). The stand-up garment package and the accessories used are shown in Figure 8.18.

The advantages of the stand-up pack are

- It is an attractive pack so it enhances the appeal of the garments to the customer.
- It is a safer pack as it has inner and outer cartons, therefore the packed garments can be handled easily.
- On account of its better presentation, it can increase the sales of a product.

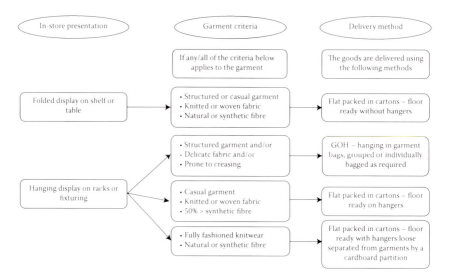

FIGURE 8.17
Flow chart for determining packing methods.

The disadvantages of the stand-up pack are

- It is costlier.
- It needs many packing materials.
- It involves a lot of effort as well as time.
- Unpacking of this kind of package needs more time and once unpacked it is tough to repack.
- In case it is crushed by any source, creases and wrinkles are formed on the garments and thus the pressed condition is disturbed.

FIGURE 8.18
Stand-up pack and accessories.

FIGURE 8.19
Flat pack garment

8.3.5.2 Flat Pack

In this packing method, the garments are pressed and folded well as like in a stand-up pack, however with less additional packing materials. It is generally normally used for ladies' garments and has a flat surface (Figure 8.19). The size of the folding is based on the garment style and specifications of the buyer. The common sizes of flat pack are 8″ × 10″ and 10″ × 12″.

The merits and demerits of flat pack are

- It is less expensive than the stand-up pack as it requires less material.
- It is less attractive than the stand-up pack.
- For shirts it does not present the beauty of the collar portion very well.
- The disadvantages are the same as that of the stand-up pack.

8.3.5.3 Hanger Pack

It is a simple garment packing method where the garments are secured in a poly bag with a hanger after pressing (Figure 8.20). Here polybag is the only material used. This type of packing can be used for all types of garments especially for blazers, coats, pants, etc.

FIGURE 8.20
Hanger pack garments in display.

The merits and demerits of a hanger pack are

- Because of its simplicity it reduces the cost of packing and materials.
- All the components/panels of the garments could be seen easily without removing the bag.
- The time for packing and unpacking is less.
- Material handling is not easy.

8.3.5.4 Deadman Pack

This kind of packing is used for shirts. Here, the sleeves are folded in front of the pack and pinned with each other. Next, the garments are folded in the center. As it resembles the appearance of dead body, it is called a 'deadman pack'. It is a simple packing method using only pins or clips and polybags. The merits and demerits of this pack are

- The costs of packing materials and packing are less compared with other methods due to its simplicity.
- The packing and unpacking time is less.
- Garments can be examined in the packed condition.
- This type of packing enables easy handling of garments.
- This type of packing is not suitable for shirts because it does not show the collar and the collar point as in the stand-up pack; hence, it is less attractive.

8.3.6 Types of Carton Packing

After garment packaging, the process of cartoning is carried out based on the size of the apparel and its color. Most used packing types are given below.

1. Solid colour solid size pack
2. Solid colour assorted size pack
3. Assorted colour solid size pack
4. Assorted colour assorted size pack

Information provided in carton boxes is given below:

- Carton box number
- Order number
- Style, colour
- Number of pieces in each colour and style

TABLE 8.2

Packing Instructions for Different Types of Fibres and Fabrics

	Blouses	Blazers and Jackets	Coats	Foundations	Jacket & Pant	Skirts & Dresses	Pants	Robes & Sleepwear	Swimwear	Kids Wear	Special Occasion & Bridal	Accessories	All Other
Velvets	H	H	H	F	H	H	H	F	F	H	H	F	F
Wool	F	H	H	F	H	H	F	F	F	F	H	F	F
Structured with shoulder pads, boning or piping	H	H	H	F	H	H	H	F	F	H	H	F	F
Leather & suede	H	H	H	H	H	H	H	H	F	H	H	F	F
Linen	F	H	H	F	H	F	F	F	F	F	H	F	F
Cotton	F	F	F	F	F	F	F	F	F	F	F	F	F
Denim/twill/ rayon	F	F	F	F	F	F	F	F	F	F	F	F	F
Ornamentation, beads or sequins	H	H	H	F	H	H	H	F	F	F	H	F	F
Fully pleated (seam to seam)	H	H	H	F	H	H	H	H	F	H	H	F	F
Jersey, knits and sweaters	F	F	F	F	F	F	F	F	F	F	F	F	F
100% Silk	F	H	H	F	H	H	F	F	F	F	H	F	F

Note: F = Flat packing, H = Hanger pack.

- Total number of pieces
- From address and To address
- Contact number
- Net weight of the carton box
- Dimension of the carton box

8.3.7 Requirements of Packing

The plastic bags are most commonly used for garment packing either at the completion of production or when they arrive at the finished goods stores. Apparel such as shirts and underwear is usually bagged and boxed immediately after final inspection and enters the stores in prepacked form. Other hanging garments like jackets, dresses and skirts are usually bagged when they enter the stores. A carton package made of quite strong corrugated material is normally preferred while transporting the boxed or hanging garments in bulk form. The packed garment boxes are sealed by contact adhesive paper tape or bound with a plastic tape (Solinger 1988; Ukponmwan et al. 2001; Sumathi 2002; Glock and Kunz 2004). The recommended packing methods for different types of fibre and fabrics are given in Table 8.2.

References

Abernathy, F.H. and J.T. Dunlop. 1999. *A Stitch in Time – Apparel Industry.* Blackwell Scientific Publications, Oxford, UK.

Carr, H. and B. Latham. 2006. *The Technology of Clothing Manufacture.* Blackwell Science, Oxford.

Chuter, A.J. 1995. *Introduction to Clothing Production Management.* Blackwell Scientific Publications, Oxford, UK.

Clayton, M. 2008. *Ultimate Sewing Bible – A Complete Reference with Step-by-Step Techniques.* Collins & Brown, London.

Cooklin, G. 2006. *Introduction to Clothing Manufacture.* Blackwell, UK.

Fairhurst, C. 2008. *Advances in Apparel Production.* The Textile Institute, Woodhead Publication, Cambridge.

Frings, G.S. 1999. *Fashion Concept to Consumer.* Prentice Hall, Englewood Cliffs, NJ.

Gardiner, W. 2003. *Sewing Basics.* Sally Milner Publishing, Australia.

Glock, R.E. and G.I. Kunz. 2004. *Apparel Manufacturing – Sewn Product Analysis.* Prentice Hall, Englewood Cliffs, NJ.

Gutauskas, M. and V. Masteikaite. 1997. Mechanical stability of fused textile systems. *International Journal Clothing Science and Technology* 9(5):360–6.

Gutauskas, M., V. Masteikaite and L. Kolomejec. 2000. Estimation of fused textile systems shrinkage. *International Journal Clothing Science and Technology* 12(1):63–72.

Holme, I. 1999. Adhesion to textile fibers and fabrics. *International Journal Adhesives* 19:455–63.

Jevsnik, S. and J. Jelka Gersak. 1998. Objective evaluation and prediction of properties of a fused panel. *International Journal Clothing Science and Technology* 10(3/4):252–62.

Jones, I. 2013. The use of heat sealing hot air and hot wedge to join textile materials (Chapter 11). In: *Joining Textiles: Principles and Applications*, Jones, I. and Stylios, G.K. Eds. Woodhead Publishing, UK.

Kim, S.J., K.H. Kim, D.H. Lee and G.H. Bae. 1998. Suitability of non-woven fusible interlining to the thin worsted fabrics. *International Journal Clothing Science and Technology* 10(3/4):273–82.

Laing, R.M. and J. Webster. 1998. *Stitches and Seams*. Textile Institute Publications, UK.

Mathews, M. 1986. *Practical Clothing Construction* – Parts 1 and 2. Cosmic Press, Chennai.

Mehta, P.V. 1992. *An Introduction to Quality Control for Apparel Industry*. CRC Press, Boca Raton, FL.

Sang-Song, L. 2001. Optimal combinations of face and fusible interlining fabrics. *International Journal Clothing Science and Technology* 13(5):322–38.

Shaeffer, C. 2000. *Sewing for the Apparel Industry*. Woodhead Publication, Cambridge.

Shim, E. 2013. *Bonding Requirements in Coating and Laminating of Textiles In: Joining Textiles: Principles and Applications*. Woodhead Publishing Limited, Cambridge.

Solinger, J. 1988. *Apparel Manufacturing Hand Book – Analysis Principles and Practice*. Columbia Boblin Media Corp, New York, USA.

Sumathi, G.J. 2002. *Elements of Fashion and Apparel Designing*. New Age International Publication, New Delhi, India.

Tyler, J. D. 2008. *Carr and Latham's Technology of Clothing Manufacture*. Blackwell, UK.

Ukponmwan, J.O, K.N. Chatterjee and A. Mukhopadhyay. 2001. *Sewing Threads*. Textile Progress, The Textile Institute, Manchester.

Wei, W. and C.Q. Yang. 1999. Predicting the performance of durable press finished cotton fabric with infrared spectroscopy. *Textile Research Journal* 69:145–51.

9

Fullness and Yokes

9.1 Gathers

The fullness can be distributed effectively over a given area by means of gathering, which enhances the garment appearance. Gathers can be noticed at the waistline, neckline, yoke line and upper and lower edge of the sleeve in the case of children's and ladies garment. In general, gathering requires twice the amount of fabric as that of the waist circumference measurement (Chuter 1995). The different ways in which gathering can be done are mentioned below.

9.1.1 Gathering by Hand

Gathering is achieved by fastening a thread followed by securing with two rows of tack stitches. The ends of the thread are drawn until the section measures the desired length and the thread is wound around over a bell pin. Fabrics of any weight can be gathered using the hand and the size of the folds can be easily changed by varying the width of the stitches, which cannot be performed in a machine. Figure 9.1 shows gathering effects made by hand. Two hand sewing needles are threaded – the threads are doubled and knotted securely at the ends. Then two rows of running hand stitches, which are 1/4″ from the fabric edge and 1/4″ apart from each other, are made (Shaeffer 2000).

9.1.2 Gathering by Machine

The gathering effect made by machine is illustrated in Figure 9.2. In this case, the machine is adjusted to produce long stitches and the upper tension is reduced slightly. With such a configuration, two rows of machine stitches, each 1/4″ apart, are made. Both the bobbin threads are pulled together and stitched to evenly distribute the fullness. For gathering large sections of fabric, the specialised attachments in sewing machines such as gathering foot or gathering rufflers could be used. Gathering by machine is much faster for lightweight fabrics like voile, organza, netting, cotton lawn, etc. Usually, two

FIGURE 9.1
Gathering by hand.

FIGURE 9.2
Gathering by machine.

rows of machine stitching are done down the length to gather 1/4″ from the edge and 1/4″ apart (Abernathy and Dunlop 1999; Fairhurst 2008).

9.1.3 Gathering Using Elastic

A narrow band of elastic is stretched and stitched onto the portion of the garment that needs to be gathered. After stitching, the relaxation of the stretched elastic causes the gathering effect. This is the fastest technique and is suitable for any weight fabric. The length of the strip of elastic is the same as the final gathering length and one end of the elastic is anchored to the fabric by stitching and backstitching a couple of times (Abernathy and Dunlop 1999; Fairhurst 2008). The broken zigzag stitch is preferred for gathering as shown in Figure 9.3.

FIGURE 9.3
Gathering using elastic.

9.1.4 Methods of Controlling Gathers

The fullness of gathers can be controlled by the following methods:

- Two rows of machine basting (with a slightly loose upper tension) can be used. One row is made exactly on the line of stitching and another row of stitches is made above it.
- By providing three rows of machine basting (with a slightly loose upper tension), one row is made exactly on line of stitching, and the further two rows of stitches are made above and below the first row. The stitches should not damage the fabric when removed. Figure 9.4 illustrates the controlling of gathers using heavy duty threads.

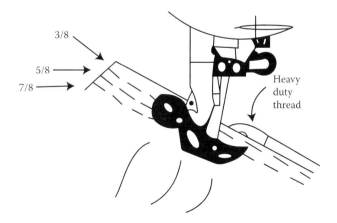

FIGURE 9.4
Controlling of gathers using heavy duty threads.

9.1.4.1 Process of Gathering

The sequence of steps to produce gathering effects in garments is explained in detail below:

- At first, the desired method of controlling fullness should be selected.
- The pattern markings are matched with the seam lines of garment pieces by pinning along the line of stitching on the side of the fullness.
- The bobbin thread or cord is drawn between each set of markings for fullness.
- The fullness is distributed evenly using a fingernail or pin.
- The long threads are secured in position by wrapping them around a bell pin. When sufficient gathering has been achieved, additional pins (right at the stitching line) as per the requirement can be added.
- Machine basting is done on the seam line with the gathered side up. The pins are removed one by one as the basting progresses.
- The even distribution of the gathers during sewing must be ensured.
- Finally, stitching (permanent) is made over the top of the machine basting.

9.2 Pleats

Pleats are provided at the waistline of dresses and especially in skirts with the intention of evenly distributing the fullness around the entire area of the fabric. The construction of pleats is like tucks, but the pleats are wider than tucks. In garments, the box pleats, knife pleats or inverted pleats are commonly used. The pleat design can be varied to control the garment fullness. These can be pressed or unpressed, soft or crisp (Carr and Latham 2006).

9.2.1 Knife Pleat

Knife pleats can be used as an alternative for gathers. These pleats generally have a width of about 1/2–2″ and are turned towards the same direction. Figure 9.5 illustrates a skirt with knife pleats. (Kunz and Glock 2004; Clayton 2008).

9.2.2 Box Pleats

A box pleat is formed when two consecutive knife pleats are folded in opposite directions – one to the left and one to the right (Kunz and Glock 2004;

FIGURE 9.5
Knife pleat skirt.

FIGURE 9.6
Box pleat skirt.

Clayton 2008). This is used in frocks and skirt waistline. A skirt with box pleats is shown in Figure 9.6.

9.2.3 Inverted Pleat

Inverted pleats can be obtained by reversing the box pleat. An inverted pleat is made while two knife pleats are twisted nearer to each other in a manner

FIGURE 9.7
Inverted pleat skirt.

that the folds meet in the centre on the face side of the garment (Kunz and Glock 2004; Clayton 2008). These kinds of pleats are utilised commonly in uniforms and skirts. Figure 9.7 shows an inverted pleat skirt.

9.2.4 Accordion Pleat

Accordion pleats (Figure 9.8) are a series of very narrow and straight pleats of equal width. These folds have a striking resemblance to the bellows of an accordion, hence the name. The width of the pleats ranges from 3 to 13 mm. These pleats are close to each other and have a uniform depth from the waist to hem (Mathews 1986).

9.2.5 Sunray Pleat

Sunray pleats (Figure 9.9), originating from the waist and gradually spread out. These pleats are capable of being set in synthetic fabrics or blends with a higher proportion of synthetic fibres (over 50%) in order to sustain the pleats. Circular skirts and wedding dresses make use of sunray pleats (Mathews 1986; Clayton 2008).

FIGURE 9.8
Accordion pleat skirts.

9.2.6 Kick Pleat

Any of the above discussed pleats like knife pleat, box pleat or inverted box pleat could be used to construct kick pleats in skirts. After pleating, a top stitch is made near the fold and extended to the desired length (should not be made until the hem edge) and then decorated as required. Kick pleats are usually made in pencil skirts. Figure 9.10 illustrates a kick pleat skirt (Mathews 1986; Clayton 2008).

9.2.7 Cartridge Pleat

Cartridge pleats (Figure 9.11) are basically round pleats utilised in door and window curtains. These pleats are used for curtains as they provide good drape and fullness. Fabrics with a firm and heavy construction are well suited for cartridge pleats (Clayton 2008).

9.2.8 Pinch Pleat

Pinch pleats (Figure 9.12) are another type of pleat used in curtains and draperies. These pleats are constituted by stitches from the top which are extended down for a part length. Markings are made for the rest of the length as in the case of tucks. In this case, three small pleats of equal width are grouped together and then basted along the top and front edge.

FIGURE 9.9
Sunray pleat wedding dress.

Following this, they are pressed and machined together across the bottom ends (Mathews 1986).

9.3 Flounces

Flounces are nothing but a piece or strip of decorative material that is usually gathered or pleated and attached by one edge onto a garment. These are

FIGURE 9.10
Kick pleat skirt.

generally observed in women's fashion dresses. Such fabrics are characterised by a wave-like appearance by sewing the fabric strip on one edge only and allowing it to hang freely along the other edge. Flounces are shown in Figure 9.13. Flounces can be made in cuffs, collars, blouses, or the hemline or neckline of women's clothing. Depending on the method of making and style, a variety of flounces are available. Some of them are neck flounces that can be used in the form of a bow, flounce collars and curved band flounces (Bheda 2002; Nayak and Padhye 2015).

The use of a flounce in fashion actually dates back to the 1920s and it is typically a decorative fashion technique. A flounce imparts uniqueness to the product. Today, several fashion designers use flounces to give a retro look to their clothes (Mathews 1986; Clayton 2008).

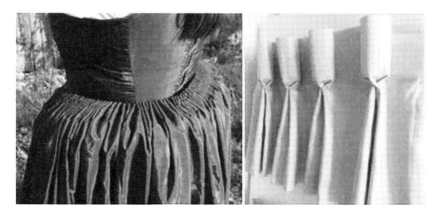

FIGURE 9.11
Cartridge pleat skirt and curtain.

FIGURE 9.12
Pinch pleat curtain.

FIGURE 9.13
Ladies top with flounces.

9.4 Tucks

Tucks are stitched folds of fabric mainly used to decorate garments. Sometimes, released tucks may be used for shaping the fabric to the body (Solinger 1988). The common types of tucks are illustrated in Figure 9.14.

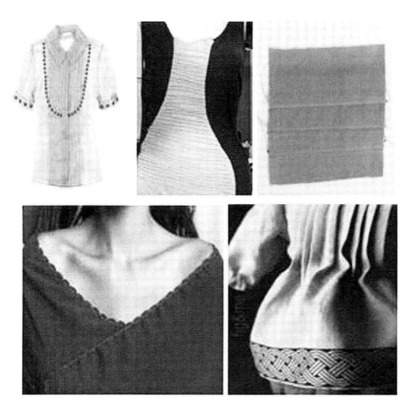

FIGURE 9.14
Pin, spaced, blind, shell and released tucks.

- *Pin tucks:* These are minor ones.
- *Spaced tucks:* These tucks have gaps or spaces between the stitching.
- *Blind tucks:* The stitching of one tuck overlaps the previous tuck.
- *Shell tucks:* These can be stitched using a sewing machine or by hand. Blind stitches are used to produce a uniform design.
- *Released tucks:* These tucks are partially stitched and are not stitched along the complete tuck length.

9.5 Darts

Darts help in shaping the fabric to fit the body and thus provide comfort to the wearer. They provide fullness to natural body curves. Darts are very rarely used for decorative purposes like providing a design line. The fitting, marking, stitching and pressing of darts should be done accurately (Tyler 2008). The different types of darts are discussed below:

9.5.1 Straight Dart

It is a straight line of stitching from the point to the seam line (Figure 9.15). This can be noticed in the underarm of the front bodice, back skirt, shoulder, elbow and back neckline.

9.5.2 Curved Outward Dart

The stitch line curves outward along the path from the point to the seam line (Figure 9.16). This gives a snugger fit to the garment. This is sometimes used on a bodice front to make a mid-body fit snug.

9.5.3 Curved Inward Dart

The stitch line curves inward from the point to the seam line. This facilitates a better fit along the body curve (Figure 9.17). It is frequently used in pant and skirt fronts.

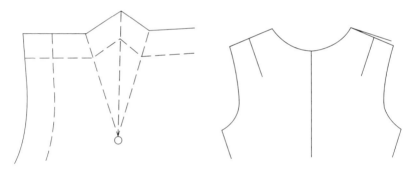

FIGURE 9.15
Straight dart.

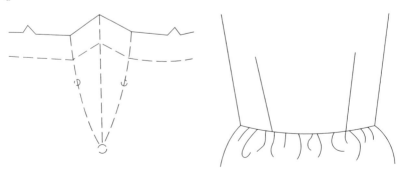

FIGURE 9.16
Curved outward dart.

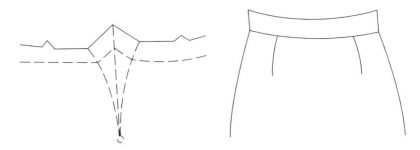

FIGURE 9.17
Curved inward dart.

9.5.4 Neckline Dart

This is usually a solid line marking on the back neckline indicating a straight dart of 1/8″ (Figure 9.18).

9.5.5 Double Pointed Dart

This dart is unique as it tapers in a straight line from the middle to both the ends (Figure 9.19) and is clipped at the widest part. It is usually made from the waistline (widest point). It finds application in princess and A-line dresses, over blouses and jackets.

9.5.6 Dart in Interfacing

In this case, a slash is made on the fold line. Then the cut ends are lapped along the line of stitching and zigzagged to keep in place (Figure 9.20).

9.6 Yoke

It is a fashioned pattern piece used in garments, which generally fits around the shoulder and neck region. The yokes offer support for slacker parts of the

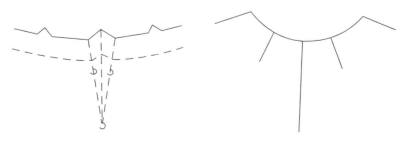

FIGURE 9.18
Neckline dart.

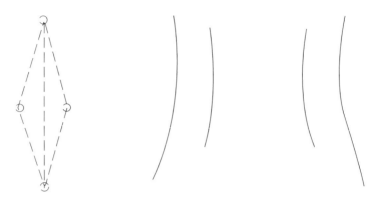

FIGURE 9.19
Double pointed dart.

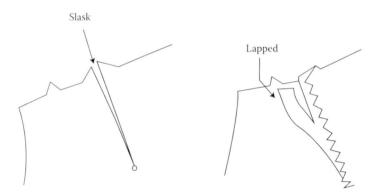

FIGURE 9.20
Darts in interfacing.

garment like in a gathered skirt or the body of a shirt and thereby aid in regulating the fullness of the garments. They are effectively horizontal panels near the shoulders or waist, which are often used for shaping because dart values can sometimes be absorbed into this seam line. Hence, the yokes are responsible for a trim and smooth upper area of the waistline in garments. Yokes are less often designed for decoration of garments (Nayak and Padhye 2015). Figure 9.21 illustrates the different types of suitable yokes for ladies and men's garments.

9.6.1 Selection of Yoke Design

The important factors influencing the yoke design are given below.

9.6.1.1 Design of the Fabric

The form of the yoke should go along with the fabric design. For fabric designs such as large checks or stripes, yokes with round or curved shapes

(a)   (b)   (c)

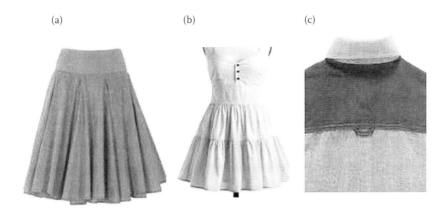

FIGURE 9.21
Types of yokes: (a) Hip yoke, (b) midriff yoke, (c) shoulder yoke.

are not well suited. Instead, straight line yokes are more appropriate. Floral designs or curved line fabric will go in harmony with round or curved yokes (Mehta 1992).

9.6.1.2 Design of the Garment

It also influences the design of the yoke. It is preferable to design yokes that are similar in shape to the design details of the dress like collar, cuff, pockets, etc. but with slight variations to avoid a monotonous display.

9.6.1.3 Purpose and Use of the Garment

Yokes find best utility in school uniforms and home-wear apparel. For party wear garments, innovative and fancy yoke designs like scalloped, asymmetrical shapes, etc. with contrasting material and decorative edging can be implemented.

9.6.1.4 Sex and Age of the Wearer

Round and curved yokes go well with girls while straight yokes are more suited to boys. Simple yoke designs without decorations are preferred by the older generation.

9.6.1.5 Figure and Personality of the Wearer

The physical stature of a person also decides the nature of the yoke. In the case of a short, plump figure, deep narrow yokes with vertical decorations should be adopted. This causes a vertical eye movement giving an impression of added height and reduced width, thus making the person look taller and slimmer. Conversely, horizontal lines in yokes make a person look shorter and fatter and are suitable for thin figures. Also, a yoke that is wide at the shoulder and pointed toward the waistline gives an idea of a narrow waist and wide shoulder.

9.6.2 Creating Variety in Yoke Design

9.6.2.1 Variety in Shape and Size

A yoke can be designed with a number of shapes like square, round, straight line, scalloped, triangular and asymmetrical. The width and depth of the yoke can be changed to produce desired effects. The yoke with a panel has a part of the yoke stretching out to the full length of the garment. Whereas in a partial yoke, the yoke may extend into the sleeve or it may extend for a certain part of the garment (Solinger 1998). Figures 9.22 and 9.23 show the yoke with a panel and without a panel for a frock.

9.6.2.2 Variety in Material and Grain

For garments with light shades, yokes with contrasting colours are used and vice versa. Likewise, yokes with prints can be attached to plain garments

FIGURE 9.22
Yoke with panel.

FIGURE 9.23
Yoke without panel.

or vice versa to provide a good appearance. With respect to grain, the yoke is cut in an inclined direction to the lengthwise grain while the garment is along the lengthwise grain.

9.6.2.3 Designing Seam Line of Yoke

The yoke attachment to the main panel of the garment can be carried out in a decorative manner by inserting ruffles, lace, faggoting, decorative stitches or top stitches with contrasting coloured threads.

9.6.2.4 Decoration within the Yoke

Additionally, the yoke can be beaded, quilted, embroidered, shirred, smocked, tucked or pleated to enhance the appearance.

9.6.2.5 Introducing the Yoke at Different Positions

Basically, the yoke can be introduced in three positions: at the top of the garment (shoulder yoke), above the waistline (midriff yoke) or below the waist line (hip yoke).

9.6.2.6 Designing Yokes Which Release Fullness in Various Forms

The fullness in the body of the garment can be released in the form of gathers, pleats and tucks originating from the edge of the yoke.

9.6.3 Preparing Patterns of Different Types of Yokes

9.6.3.1 Yoke without Fullness

This type of yoke comes in a wide range of shapes and sizes. The pattern for such yokes can be prepared by two methods. For a curved or 'V' shaped yoke (Figure 9.24), the yoke line from the shoulder to the center front is constructed in the front bodice as desired. In the case of a straight line yoke, a line is drawn from the armhole to the centre front of the bodice pattern and both sections are labelled.

9.6.3.2 Yoke with Fullness

These yokes involve decoration of the fabric with any fullness (tucks, pleats, gathers, shirring and embroidery). The fullness must be completed prior to the attachment of the yoke pattern. The required amount of fabric is cut and desired types of tucks are stitched according to the design (Figure 9.25). Now, the paper pattern is placed over the tucked fabric and the yoke is cut with the required seam allowance (Tyler 2008).

9.6.4 Attaching Yokes

A plain seam or lapped seam can be used to append the yoke to the lower section. The skirt is gathered such that the width of the gathered skirt is the same as the width of the yoke. The yoke is now placed over the skirt right side facing up and, subsequently, the notches are matched. Pinning

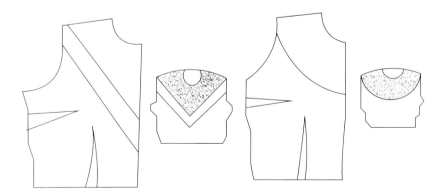

FIGURE 9.24
Yoke without fullness.

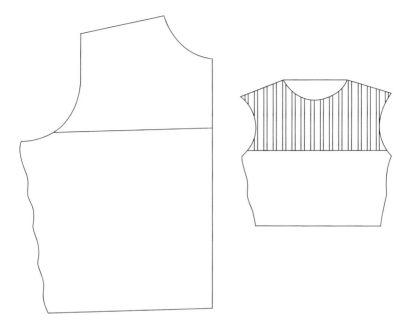

FIGURE 9.25
Yoke with fullness.

and tacking are done in order to distribute gathers evenly. After attaching the yoke, the tacking should be removed. The yoke is turned up on the right side and pressed. The straight and decorative yokes are shown in Figures 9.26 and 9.27, respectively.

In yokes consisting of both curved and straight lines as in a skirt with panel, the seam runs almost at right angles to the corner. The lower segment

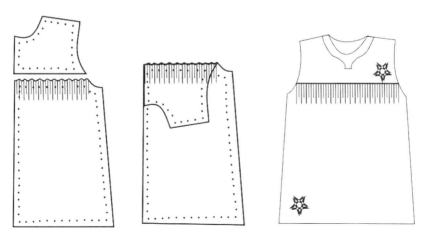

FIGURE 9.26
Straight line yoke.

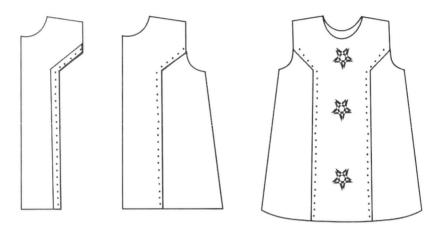

FIGURE 9.27
Decorative yoke.

of the garment is gathered and stitched initially. In the yoke, stitches are made nearer to the seam line and the corners are reinforced. Now, the seam allowance is folded to the back side and tacked with small stitches close to the fold. To make a flat seam, cuff the notches into the seam allowance (Solinger 1998).

By placing the yoke over the top of the lower section, the seam lines are matched. Tacking followed by top stitching close to the folded edge of the yoke is done. If necessary, a tucked seam effect can be obtained by doing the top stitching away from the folded edge of the yoke.

Apart from skirts, yokes frequently appear in men's shirts, trousers and coats. Depending on the form of the shoulder line, the yokes on shirts, tops or coats can be cut in two variants. In the first type, the yoke is cut as a single pattern piece, thus the back and front pieces could be merged along the shoulder line. The other variant includes the back and front yokes as two separate pieces, especially for drop shoulders as two separate pieces are required to retain the curve of the shoulder line. Alternatively, the yoke can be cut in one piece and darts can be provided along the shoulder line to impart the necessary curved shaping (Mathews 1986). Moreover, on trousers and skirts, the yoke region facilitates the dart value to be absorbed into a single panel.

References

Abernathy, F. H. and J.T. Dunlop. 1999. *A Stitch in Time – Apparel Industry.* Blackwell Scientific Publications, Oxford, UK.

Carr, H. and B. Latham. 2006. *The Technology of Clothing Manufacture*. Blackwell Science, Oxford.

Chuter, A.J. 1995. *Introduction to Clothing Production Management*. Blackwell Scientific Publications, Oxford, UK.

Clayton, M. 2008. *Ultimate Sewing Bible – A Complete Reference with Step-by-Step Techniques*. Collins & Brown, London.

Fairhurst, C. 2008. *Advances in Apparel Production*. The Textile Institute, Woodhead Publication, Cambridge.

Kunz, G. and R. Glock. 2004. *Apparel Manufacturing: Sewn Products Analysis*. Prentice Hall, Englewood Cliffs, NJ.

Mathews, M. 1986. *Practical Clothing Construction – Part 1 and 2*. Cosmic Press, Chennai.

Mehta, P.V. 1992. *An Introduction to Quality Control for Apparel Industry*. CRC Press, Boca Raton, FL.

Nayak, R. and R. Padhye. 2015. *Garment Manufacturing Technology*. Woodhead Publication, Cambridge.

Shaeffer, C. 2000. *Sewing for the Apparel Industry*. Woodhead Publication, Cambridge.

Solinger, J. 1998. *Apparel Manufacturing Handbook-Analysis Principles and Practice*. Columbia Boblin Media Corp, New York, USA.

Tyler, D.J. 2008. *Carr and Latham's Technology of Clothing Manufacture*. Blackwell, UK.

10

Collars

10.1 Introduction

Collars contribute to the style and ultimate look of garments. The season's fashion trend decides the collar design, style and shapes. Collars can have square or pointed corners or adjusted edges. Collars could be in one piece or cut in two pieces or as a portion of the variety of the pieces of garments. Collars come in different shapes and styles ranging from simple collars such as the stand collar to more complex collars such as the shirt collar but the majority of collars have the same basic construction (Chuter 1995; Shaeffer 2000). A portion of the collars lay level, some fold near the neckline and others stand up. Irrespective of the style, a neckline should case the wearer's face, draping elegantly around the neck area, free of pulls, ripples or wrinkles. The parts of a collar are highlighted in the Figure 10.1.

10.1.1 Construction of Collars

The neckline components are normally cut as per the pattern guide sheet. The under neckline is trimmed by about 1/8" less than the upper neckline at the middle edges, close to the neck area as shown in Figure 10.2.

The centre back or centre front, shoulder line and notches should be marked carefully in the neckline panel. Likewise, the centre front, centre back and notches where the neckline edges must be found are also marked in the bodice neckline (Fairhurst 2008).

Then, the interfacing (Figure 10.3) of suitable weight is assembled onto the upper neckline. The interfacing can be of fusible or sew-in type. The interfacing is attached to the upper neckline with seam allowance indicated through the right half of the completed neckline (Fredrerick 1999). Sew-in interfacing is set up when the neckline is stay sewed and then the interfacing close to the stitches should be trimmed. This is followed by trimming of interfacing at the corners to lessen the thickness.

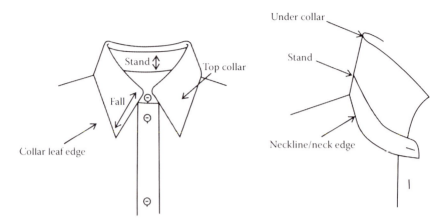

FIGURE 10.1
Parts of the collars.

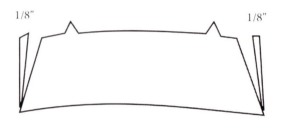

FIGURE 10.2
Under collar.

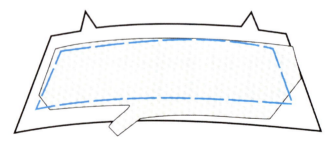

FIGURE 10.3
Interfacing.

Fusible interfacing must be attached before the neckline is stay sewed. Before attachment, the fusible interfacing is trimmed to 1/8″ for crease recompense (Figure 10.4). The fusible interfacing should also be trimmed at the corners to reduce the thickness.

The appropriate side of the upper collar is pinned to the corresponding side of the under collar to match all the cut edges. For a pointed collar, the outer

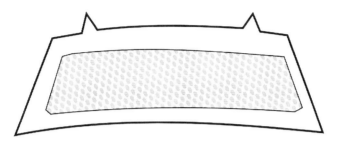

FIGURE 10.4
Fusible interfacing.

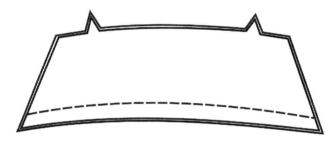

FIGURE 10.5
Outer edge sewing.

edge is sewn first (Figure 10.5) followed by grading and under stitching the seam (Figure 10.6). Folding of ends, matching cut edges, sewing and under stitching have to be carried out as much of the seam as possible.

In the case of a rounded collar, the outer edge is sewn from the centre to the neckline edge. This avoids distortion of the collar's shape. Then the

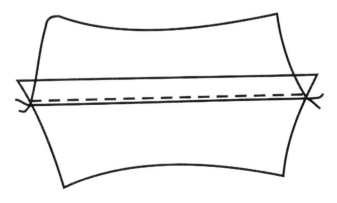

FIGURE 10.6
Understitch the seam.

seam is graded and notching the rounded collar is done wherever necessary (Nayak and Padhye 2015).

10.1.2 Types of Collars

Numerous kinds of collars are commercially available in the garment industry. Three elementary kinds of collars are flat, standing and rolled collars. The elementary collar types and other kinds of collars are given in Tables 10.1 and 10.2, respectively.

10.1.3 Selection of Interfacing for Collars

The type of collar and its applications determine the interfacing material to be used. The following criteria should be met while selecting an interfacing fabric for the collar, which provides stability and shape to the collar (Abernathy and Dunlop 1999; Carr and Latham 2006; Solinger 1988):

- The interfacing materials should be of the same or lower weight than the fashion fabric. For knitted fabrics or stretch fabrics, the interfacing fabric should be chosen to provide stability when the fabric is flexible (Bheda 2002; Kunz 2004; Clayton 2008). Figure 10.7 shows the influence of the interfacing material on the stability/stiffness nature of the collars.

TABLE 10.1

Basic Types of Collars

Type of Collar	Description	Collar Diagram
Flat	Lies flat and close to the garment along the neckline. When the corners are rounded, they are called Peter Pan.	Flat
Full roll	The fall and stand of this collar are the identical height at the center back.	Full roll
Partial roll	These collars have less stand and more fall.	Partial roll

TABLE 10.2

Other Type Collars

Type of Collar	Description	Collar Diagram
Convertible	This is analogous to a full roll collar, but cuddles the neckline closer at the sides of the neck.	Convertible
Shawl	This is identified by its center back seam. The under collar is cut as part of the bodice.	Shawl
Mandarin	A stand-up collar (complete stand with no fall).	Mandarin

(*Continued*)

TABLE 10.2 (*Continued*)

Other Type Collars

Type of Collar	Description	Collar Diagram
Shirt	A distinct neckband serves as the stand.	Shirt
Styles	There are numerous varieties of collar styles like Chelsea, sailor, bertha, Puritan, stovepipe, tie and Peter Pan.	Puritan Sailor Peter pan Tie end
Determination of collar stand	The form of the neckline edge decides the collar stand. If the neck edges are straighter, the collar stand will be more. If the curve of the neck edge is more, then the collar stand will be less.	Less stand More stand More curve Less curve

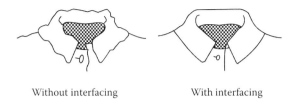

Without interfacing   With interfacing

FIGURE 10.7
Collar without and with interfacing.

- The design of the base fabric determines the selection of the interfacing fabric, that is, whether the interfacing is attached to the under or upper collar. Usually, the interfacing is attached to the under collar. However the fusible interlinings used in the case of see-through fabrics and heavy fabrics are attached to the upper collar (Mathews 1986; Mehta 1992).

10.1.4 Basic Standards for Collars

The basic standard requirements for collars are shown in Figure 10.8.
  A well-applied collar should have the following characteristics.

- Should have a flat and even surface without any wrinkles. The outer edge seam should not be noticeable from the right side.
- The curves should be smooth or have sharp points depending on the type and style of the collar.
- Proper fit should be ensured in the neckline area without unattractive gaps or wrinkles.
- It should be interfaced properly to retain shape.
- Under stitching along the outer seam edge should be done to facilitate the seam to roll to the underside.
- The collar should be pressed well.
- The seam should be enclosed and graded to reduce thickness.

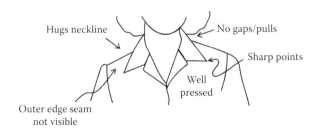

Hugs neckline   No gaps/pulls

Sharp points

Well pressed

Outer edge seam
not visible

FIGURE 10.8
Basic standards for collars.

References

Abernathy, F.H. and J.T. Dunlop. 1999. *A Stitch in Time – Apparel Industry*. Blackwell Scientific Publications, Oxford, UK.

Bheda, R. 2002. *Managing Productivity of Apparel Industry*. CBI Publishers and Distributors, New Delhi.

Carr, H. and B. Latham. 2006. *The Technology of Clothing Manufacture*. Blackwell Science, Oxford.

Chuter, A.J. 1995. *Introduction to Clothing Production Management*. Blackwell Scientific Publications, Oxford, UK.

Clayton, M. 2008. *Ultimate Sewing Bible – A Complete Reference with Step-by-Step Techniques*. Collins & Brown, London.

Fairhurst, C. 2008. *Advances in Apparel Production*. The Textile Institute, Woodhead Publication, Cambridge.

Kunz, G. and R. Glock. 2004 *Apparel Manufacturing: Sewn Products Analysis*. Prentice Hall, Englewood Cliffs, NJ.

Mathews, M. 1986. *Practical Clothing Construction – Part 1 and 2*. Cosmic Press, Chennai.

Mehta, P.V. 1992. *An Introduction to Quality Control for Apparel Industry*. CRC Press, Boca Raton, FL.

Nayak, R. and R. Padhye. 2015. *Garment Manufacturing Technology*. Woodhead Publication, Cambridge.

Shaeffer, C. 2000. *Sewing for the Apparel Industry*. Woodhead Publication, Cambridge.

Solinger, J. 1998. *Apparel Manufacturing Handbook – Analysis Principles and Practice*. Columbia Boblin Media Corp, New York, USA.

11

Plackets and Pockets

11.1 Plackets

A placket is an opening in the waist portion of trousers or in the neckline of a skirt. They can also be found at the cuff of a sleeve in a garment. Plackets facilitate easy usage of clothing and are sometimes used as a component for enhancing the design. Facings or bands are attached in a modern placket to incorporate buttons, snaps or zippers. In designer garments, a placket is a double layer of fabric that contains buttons and buttonholes in a skirt (Chuter 1995; Shaeffer 2000).

Plackets are made by interfacing more layers (generally more than one) of a fabric to impart strength and support to the garment as it is subjected to stress when worn (Chuter 1995). To protect the wearer from fasteners coming in contact with their skin and to hide underlying clothing, the two sides of the plackets are overlapped. Figure 11.1 shows a diagram of a shirt placket.

11.1.1 Continuous Lap Sleeve Placket

A continuous lap sleeve or bound placket is a common finish that facilitates dressing ease in blouses and shirts. The opening is similar to the legs of a dart, which looks like a long thin triangle and the dimensions of the placket are normally mentioned on the sleeve pattern (Shaeffer 2000). Figure 11.2 shows the view of continuous lap sleeve plackets.

11.1.1.1 Construction of Continuous Bound Plackets

11.1.1.1.1 Band Preparation

During cutting operation, a piece of placket binding had to be cut from the same base fabric used for the main garment panel. Based on the customer requirement, the placket can be on the straight grain or on the bias. In case of a decorative effect, the plaids and stripes are cut on the bias. The placket band should be twice as long as the placket opening marking, with an addition of 1″ and 1¼″ wide. Under one long edge of the band of ¼″ the pressing

FIGURE 11.1
Shirt placket.

operation is carried out. Figure 11.3 shows a sample view of a continuous bound placket.

11.1.1.1.2 Placket Opening

In the garment, the placket marking is made by using a tracing paper or soluble marker that is easily removable. During the binding operation, the upper opening of the placket and the pivot point are indicated by dots. In order to make an opening, stitching has to be done along the triangle legs

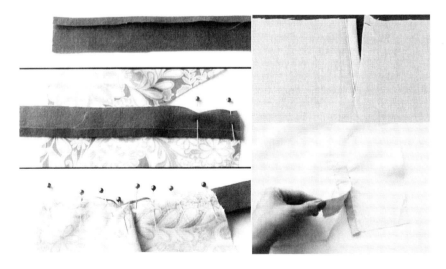

FIGURE 11.2
Continuous lap sleeve plackets.

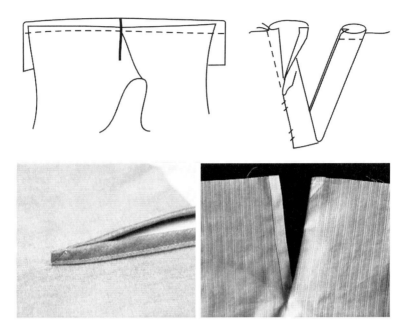

FIGURE 11.3
Continuous bound placket.

and while approaching the point the stitches per inch could be reduced to 15. After reaching the point, the presser foot should be raised for turning the stitch while the needle remains in the fabric. Then, with the 1″ stitch length, stitching has to be carried out from the point to the other side.

Slashing operation should be carried out without locking of threads up to the point and between the stitching lines. The stitching should coincide with the band ¼″ seam allowance, but tapering to near nothing in the centre. Without creating a tuck at the placket point, the band is assembled to the sleeve opening using an even ¼″ seam allowance. The seam is to be pressed flat during stitching. To get a proper finishing to the placket, the seam should be flat pressed during stitching and the placket band should be folded beside the sleeve to cover the stitching line (Joseph-Armstrong 2004; Carr and Latham 2006).

11.1.2 Two-Piece Placket

Figure 11.4 shows the reference diagram of a two-piece placket. This kind of placket is mostly used as an opening in the left side of skirts or in petticoats and sometimes in the back side of dresses. A binding and the overlap with a facing is given as a finishing in the under lap side of this placket. For the overlap a fabric of width 1½″ is used. Another fabric having a wider width of 2½″ is used for the under lap. Both of these separate fabrics should be 1″

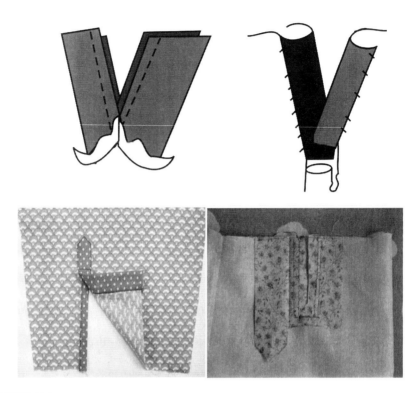

FIGURE 11.4
Two-piece placket.

longer than the placket opening. In the front part of a fabric (overlap side) narrow strip has to be stitched and the wider strip has to be stitched on the back side of the fabric (under lap side) (Hulme 1944).

In order to get the stitching line of the placket and the stitching line of the seam of the garment in line with each other, the seam for the placket to be secured should be equal to the seam allowance used for fixing. An additional length of the fabric strip should be spread out underneath the placket opening. For under lap finishing, a crease mark has to be created with the strip over them and then the free edges which are present at the back side of the panel should be turned under ¼" and hemming of fold to the stitching line should be done to form the bound side of the placket, which is around ½" to ¾" wide (Knez 1994).

Similarly for finishing an overlap, a ¼" or wider fold should be prepared to another side of its free edges so that the width of the binding in the under lap is equal to the distance from the stitching line to the fold line. The strip should be turned over to the back side of the garment and hemming of fold to be done. Then, a line of stitches is worked at the base of the placket holding the under lap and overlap together. This must be done by hand from the wrong side of the fabric so that no stitches are visible on the right side (Kunz and Glock 2004).

11.1.3 Miter Placket

A miter placket or tailored placket provides an attractive look to the garment and also to enhance the strength (Figure 11.5). A miter placket is very often used in the sleeve opening area of men's shirts, children's garments, and in jibbas where a neck opening is needed. It may be utilised for decorative purposes in children's and ladies dresses, where the colour of the placket may be contrasting in order to enhance the aesthetic value of the garment (Kunz and Glock 2004; Le Pechoux and Ghosh 2004).

11.1.3.1 Construction of Miter Placket

Prepare a strip whose width is 1½" to match the exact length of the slit (5") for making the under lap. Having a right side facing, one end of the fabric strip is kept exactly in line with the end of the slit. Now seam joining for about ¼" is stitched and ¼" of the fabric strip is folded to the underside of the free edge and is hemmed along the stitching line. Therefore, the under lap stitch exists on the back side of the garment.

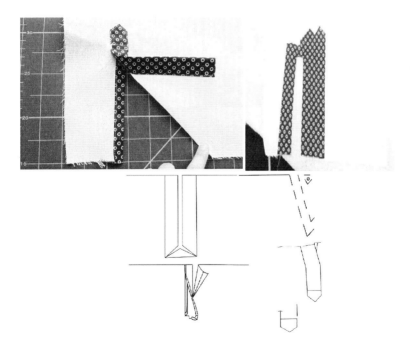

FIGURE 11.5
Miter plackets.

In case of overlap, the 1″ wide plackets are usually shaped at the tip during finishing to give an attractive look. Similar to the under lap, for overlap a strip of 2¼″ wide, which is 1¼″ longer than the slit, is cut and then the seam lines are marked using a dotted line ¼″ inside the outer edge. The right side of the fabric strip is placed facing the back side of the garment. The small side of the fabric strip should be tacked at the free side of the placket opening on the seam line. After machine stitching the strip, the overlap fabric strip should be brought over the right side of the garment and beneath the seam allowance. In this position, top stitching of the fabric strip to the garment should be done and stitching downward until the sleeve opening also has to be carried out. Stitch twice across the overlap to hold the under lap firmly in position (Le Pechoux and Ghosh 2004).

11.1.4 Zipper Placket

Zippers are available in different sizes and are usually assembled to the garment panel using tape. Selection of the zipper depends on the size and the colour of the placket required. Zipper plackets used in skirts, frocks, shirts, handbags, decorative purses and other garments are shown in Figure 11.6 (Beazley and Bond 2006).

11.1.4.1 Construction of Zipper Placket

An opening should be made in the garment and the zipper is chosen according to the opening size. A short slit of width ¼″ should be cut at the end of the opening, diagonally on both sides and then the formed edges are flipped to the back side and a tack should be done. A stitching as well as hemming of a square piece of tape is done at the end of the placket opening. The zipper is placed over this and attached to the fabric edge. Another square piece of tape with neatly finished edges is positioned such that it is covering the zipper edge and finished with hemming (Clayton 2008).

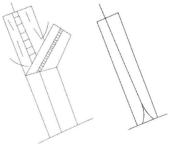

FIGURE 11.6
Zipper plackets.

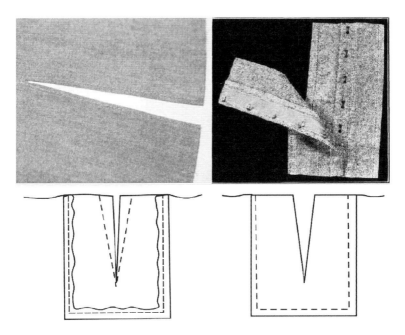

FIGURE 11.7
Faced placket open.

11.1.5 Faced Placket Open

A simple faced placket open is illustrated in Figure 11.7. It is typically a simple neck finish used on infants' and children's clothes and night dresses. To make this, slashing of opening down from the neck at the centre front or centre back should be done and fitted facing is applied to the opening. Place facing piece right side facing the garment, do a row of stitch catching the garment more. Turn the facing to the wrong side and top stitch. Finish the facing by turning the edge and hemming (Clayton 2008).

11.2 Pockets

A pocket is an opening or slot with a closed end that is usually sewn in or over the garment. Pockets may have a decorative or functional purpose (sometimes pockets serve both purposes). Basically, the pocket is utilised as a depository or as a holding provision for items or hands. A pocket opening should be sufficiently wide and deep to accommodate the hands and to prevent objects from falling out. All types of garments can be designed with pockets. Pockets primarily serve a utility purpose rather than a decoration in men's garments. In women's clothing, pockets enhance the aesthetic value drawing attention to the design of the dress. Pockets of varied shapes, sizes, and locations with decorative details such as bias binding, lace, ruffles, tucks, pleats, applique, embroider, etc. can be attached to children's garments to make them attractive (Mathews 1986; Mehta 1992). Pockets can be classified into three types:

- Applied pockets – outside pocket
- In-seam pockets
- Set-in pockets – welt, flap and bound or corded pocket
    - *Applied pockets/outside pockets* – The applied pockets are the pockets that are sewn onto the garment with top stitching. This mainly involves attachment of patch pockets to the exterior/on the surface of the garment.
    - *In-seam pocket/structural pockets* – In-seam pockets are characterised by their unobtrusive appearance as they remain concealed within the seams of the garment. Lining fabric or lightweight self-fabric is used to make these pockets. For the support of the pocket opening and to seal the lining area, generally the facing is extended. These pockets are well suited for trousers, half pants and skirts.
    - *Set-in pockets/inserted pockets* – In this type, the pocket is set into the garment through a narrow opening and made to hang inside. It is difficult to correct stitching errors in these pockets. Accurate marking, stitching, cutting and pressing are mandatory to achieve quality construction. These are used mainly on tailored garments wherever neat appearance is required. Set-in pockets are further subdivided into
        - Welt pocket
        - Flap pocket
        - Bound/corded pocket

11.2.1 Selection of Pocket Design

The most important aspect in selection of pocket design is to ensure that the pocket design blends well with the fabric design, garment design and its components like collar, sleeve, cuff, etc. For example, rectangular shaped pockets cut on crosswise grain and finished with bias edging will go in harmony with striped dresses. The scalloped pocket will be well suited for garments with a scalloped collar (Bray 2004).

The designs of pockets are also influenced by factors such as age, sex, shape and personality of the wearer. For girl's dresses, scalloped and rounded pockets are most preferred. Straight line pockets are more suitable for men's and boy's garments. Inconspicuous designs like set-in pocket are used in garments for older women and short women. The pocket designs should be selected such that it is appropriate for the particular style and end-use of the garment (Shoben and Taylor 1990). For example, simple straight line pockets are suitable for school uniforms and casual wear garments, whereas concealed pocket styles with decoration trimmings are used in party wear dresses.

11.2.2 Patch Pocket

The patch pockets can be cut in the desired shape and are fastened to the outside of the garment. Figure 11.8 illustrates various patch pockets on a garment. The patch pocket may be furnished with a flap that holds it shut. Alternatively, the top of the pocket can be trimmed with a shaped band that looks like a working flap. Flaps can be used purely for aesthetic purposes when they are attached without any pocket (Shoben and Taylor 1990).

Patch pockets are constructed using three layers: the first layer is the pocket itself; a middle layer is an interfacing; and the third layer is a lining matching with the garment lining. Pockets are usually provided with an interfacing and lining when transparent or open weave fabrics are used for the pockets. The lining and interfacing hides the construction details and also matches the pocket with other sections of the garment. When the pocket shape and size are known prior to fitting, the pockets are generally finished and basted in the appropriate place for the fitting. In case the pocket dimensions and pocket location are not confirmed or in case of pockets that require matching with the garment design, a temporary pocket shape is cut out from some fabric scrap and basted onto the garment for the fitting (Nayak and Padhye 2015).

FIGURE 11.8
Patch pockets.

11.2.2.1 Construction of Patch Pocket

11.2.2.1.1 Pocket Cutting and Marking

The steps involved in the attachment of a patch pocket with a separate lining and interfacing is given in detail below:

- A muslin fabric is patterned according to the size and shape of the finished pocket without including seam or hem allowances.
- All the seams in the garment under the patch pocket should be completed and the location of the pocket on the garment panel should be thread-traced.
- With the front side of the garment panel facing up, the pocket pattern is positioned over the thread tracing that marks the pocket's location. The continuation of the grain or the colour bars of the fabric pattern are drawn onto the muslin pocket. The pattern at the edge of the pocket patch, towards centre front is matched with the garment. This process can be made simple by pinning a fabric scrap on it to use as a guide when cutting.
- Right side facing up, the muslin pocket pattern is laced on a large scrap of the garment fabric. The design and grain on the pattern should be matched with those on the fabric scrap. After proper matching, a chalk-mark is made around the pattern, followed by thread tracing.
- Finally, the pocket is cut out along with the seam and hem allowances.

11.2.2.1.2 Interfacing the Pocket

The interfacing assists in retaining the shape of the pocket and the interfacing material for a pocket should be crisper. The weight and drape of the base fabric should be considered while selecting the interfacing material. Commonly used interfacing materials are muslin, linen, hair canvas and crisp lining fabrics. The interfacing can be cut on the length grain, cross grain or on the bias depending on the requirement (Kunz and Glock 2004). Cutting the interfacing along the cross grain will have minimum stretch in the opening of the pocket and eliminates the need for stabilising the pocket opening. An interfacing cut along the bias will have more flexibility and shapes better to the body. Generally, the size of the pocket decides the size of an interfacing. Fusible interfacings can be cut to the size of the finished pocket or can be extended up to the seam allowance. When the interfacing extends into the seam or hem allowances, the pocket edges will be slightly rounded instead of sharply creased.

- The pocket should be kept with the bottom side up so that the interfacing can be placed over the pocket such that the top of the interfacing material matches with the bottom of the pocket seam line. The centres are basted together using large diagonal basting stitches.

- The pocket hem and seam allowances are folded and pinned to the edges of the interfacing. During this, it must be ensured that thread-tracings fall along the extreme edges of the pocket and are invisible from the right side of the pocket. If not, the pins are removed and the interfacing edges are trimmed very slightly. It is repined and checked again.
- The pins are released and the pocket is made flat. Now the pocket is topstitched. Prior to topstitching, a guide is thread traced at the desired distance from the finished pocket edges. Usually, a soft basting thread that breaks easily is used for this purpose. The basting thread can be removed and stitching can be carried out.
- Catch stitches should be used to sew the edges of the interfacing to the pocket in the wrong side. This step can be skipped if the pocket is topstitched.
- The opening of the pocket could be stabilised with a silk organza fabric with seam binding. Then the pocket is steam-pressed.

11.2.2.1.3 Finishing of Pocket Edges

For pockets with curved edges, the seam allowance is ease-basted twice at the curves, generally about ⅛" from the interfacing and again ¼" away. The ease basting is pulled up to make the seam allowance fit elegantly against the pocket, and the excess fullness is shrunk out. A piece of brown paper can be inserted between the seam allowance and the interfacing as a precautionary step to avoid shrinking of the pocket (Sumathi 2002).

- For square cornered pockets, the seam allowance at the bottom is folded underneath. Basting is done at ¼" from the seam line and the pocket is pressed. Likewise, the side seam allowances are folded and basted.
- After folding the bottom and side seam allowances, the excess bulk at the corners are trimmed and the edges are pressed again. The corners of bulky fabrics must be spanked with a clapper to flatten them.
- The top hem should be folded to the back side and all the edges of the pocket are basted at about ¼" from the edges.
- With the back side of the pocket up, the seam allowances are trimmed close to the basting to reduce the bulk. Again, the edges are pressed to shrink out any excess fullness.
- The right side is flipped up, the pocket is positioned on a tailor's ham or pressing pad that simulates the body's curve in the pocket area. The pocket is covered with a press cloth, and shaped to the natural curve of the body.
- When making a pair of pockets, both pockets must be identical. While making asymmetrical garments, where one half of the hip

is larger than the other, the pocket for the larger side can be made slightly larger (up to ¼"), but when worn, the changes in the two pockets should be imperceptible.

11.2.2.1.4 Lining of Pocket

This lining technique with the wrong sides together is applicable for flaps, facings and waistbands.

- A lining fabric is cut in the form of a rectangle on the same grain as the pocket and at least ¼" larger than the pocket on all sides.
- A fold of 1" is made at the top of the lining and it is pressed.
- With the back sides facing each other, the lining material is kept over the pocket with the folded edge lying ¾" below the pocket top. The pocket and the lining are basted together with diagonal basting stitches after matching the centres.
- The remaining raw edges of the lining are folded underneath at about ⅛" to ¼" from the edges. The lining is trimmed to remove excess bulk. The lining is pinned and then basted to the pocket. The edges are pressed slightly.
- The lining is fell stitched in the appropriate place. All bastings are removed and the completed pocket which is ready to be attached to the garment is pressed thoroughly using a damp press cloth.
- This is the last instance where the pocket can be topstitched.

11.2.2.1.5 Set the Pocket

The garment may be stabilised with interfacing under the pocket or just under the opening prior to the setting of the pocket. If the pocket serves a decorative purpose or if the entire front is backed with interfacing, an interfacing is not required. But if the pocket is designed for occasional use and the front is not entirely interfaced, staying the opening becomes essential (Sumathi 2002).

- An interfacing stay of 2" width is cut on the lengthwise grain such that it is long enough to be sewn to the interfacing at the front opening and to a dart or seam at the side.
- Wrong side facing up, the stay is basted over the thread-traced pocket opening. While sewing the pocket, the stay should also be sewed to secure it.
- The face side of the pocket should be turned up and the edges of the pocket are aligned on the garment. A large 'X' is basted at the centre of the pocket following which basting is done at about ¼" from the edges using uneven basting stitches.

- The fit of the pocket is examined. The pocket should fit smoothly or stand away from the garment slightly but it should not be tight. Rebasting is done if the pocket is too tight.
- The garment is turned over to the wrong side and the pocket is permanently secured using short running stitches or diagonal stitches. The basting acts as a guide to sew ⅛″ away from the pocket edges. The stitching should be done in such a way that the stitches do not show up on the pocket face. For regular use pockets, two rows of stitching are done around the pocket.
- At the upper portion of the pocket, with the back side up, numerous cross stitches should be sewn at each side of the pocket for better reinforcement.
- The last step includes removal of the bastings and pressing of the attached pocket with a press cloth.

11.2.3 In-Seam Pocket

A pocket in which the opening falls along a seam line of the garment is known as an 'in-seam pocket'. This type of pocket can be found in pants, skirts, trousers, shorts, kids' wear, kurtas and pyjamas. Figure 11.9 shows the in-seam pocket design.

11.2.4 Slash Pocket

Slash pockets lie inside the garment and the pocket opening is a slash of some type. The slash pocket is subdivided into three types, namely, bound, welt and flap. When each edge of the slash is finished with binding of even

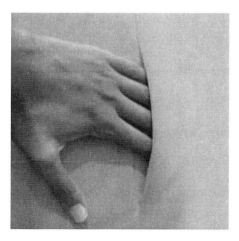

FIGURE 11.9
In-seam pocket.

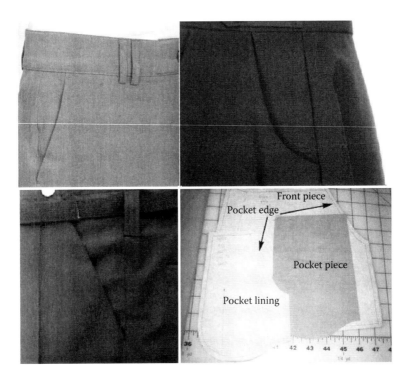

FIGURE 11.10
Slash pockets.

width, it is termed a bound pocket. If one end of the pocket is wider, called the welt, and extends over the pocket opening, it becomes a slash pocket. The flap pocket is provided with a flap of extension turned down over the opening. Figure 11.10 illustrates the slash pocket on men's formal pant.

11.2.5 Flapped Pockets

The side pockets utilise flap pockets, which consist of an extra lined flap of matching fabric to cover the top of the pocket. This flap present over the pocket prevents the contents inside the pocket from getting wet during rain. At other times, the flap can be tucked into the pocket. However, this fact is now often ignored. Nowadays, the flap is left out as it is considered to make a style statement, even during formal events (Aldrich 1999). Figure 11.11 shows the design of a flapped pocket for an outer coat jacket. In general, any type of pocket in any garment that has an overhanging part is called a 'flapped pocket'.

11.2.6 Besom Pockets

Besom pockets are nothing but hidden or secretive pockets. These pockets are not easily visible and have only one slash evident on the front of the

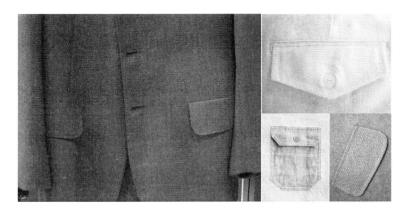

FIGURE 11.11
Flapped pocket.

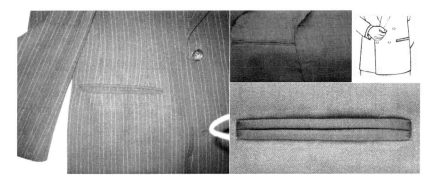

FIGURE 11.12
Besom pocket.

jacket. Moreover, the edges of the slash have narrow stitched folds or 'welts' along the seams, which makes it difficult to distinguish from the garment. Figure 11.12 shows the design of a besom pocket in an outer coat jacket.

11.2.7 Bellows Pockets

Bellows pockets are sporty pockets. They have folds along the three sewn sides of the pocket, which makes them expandable. These pockets can accommodate bigger objects and were typically designed for hunting jackets. The design of a bellow pocket for an outer coat jacket is illustrated in Figure 11.13.

11.2.8 Ticket Pockets

Ticket pockets are basically very small pockets, with or without a flap. These pockets are located on the top of the regular right-hand pocket of a jacket. These

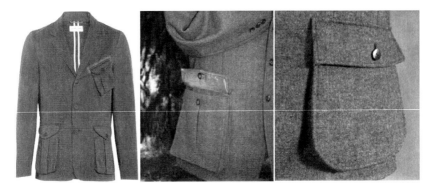

FIGURE 11.13
Bellow pocket.

Ticket pocket is narrow and above the regular pocket

FIGURE 11.14
Ticket pockets.

pockets add to the style and help in convenient usage of the jacket. Ticket pockets are also referred to as 'change pockets'. Figure 11.14 illustrates a ticket pocket on an outer coat jacket.

11.2.9 Variety in Shape, Size, Location and Number

Pockets can be designed in a lot of variants by altering the shape, style, location and number. Different shapes such as rectangular, triangular, heart shaped, oval shaped, scalloped and round can be used in the design of pockets. They can be placed in distinct positions in the garment. For example, a pocket design can be designed with two breast pockets and two hip pockets. In general, the size of the hip pocket will be about 1½ times more than the size of the breast pockets.

FIGURE 11.15
Pocket design variety in material and grain.

11.2.10 Variety in Material and Grain

When fabrics that are different in colour, design, texture or grain from the garment material are used for the pockets, a huge number of fascinating designs can be produced. A simple design of this kind would be a pocket flap (also the collar and buttons) that is made of contrasting coloured material. Figure 11.15 highlights a design with printed pockets on a plain garment and vice versa. A striped checked dress design with pockets cut on a crosswise grain and finished with bias edging can also be seen in Figure 11.15 (Yarwood 1978).

11.2.11 Variety in Decorative Details and Trimmings Used on the Pocket

Decorative detailing and trimming also add some style to the pocket. The outer edges of the pocket can be finished with ruffles, bias binding lace decorative stitches, appliqué, tucks, pleats, embroidery, patch work, etc.

References

Aldrich, W. 1999. *Metric Pattern Cutting*. Blackwell Science Ltd, UK.
Beazley, L. and T. Bond. 2006. *Computer Aided Pattern Design and Product Development*. Blackwell Publishing, UK.
Bray, N. 2004. *Dress Fitting*. Blackwell Publishing Company, UK.
Carr, H. and B. Latham. 2006. *The Technology of Clothing Manufacture*. Blackwell Science, Oxford.

Chuter, A.J. 1995. *Introduction to Clothing Production Management.* Blackwell Scientific Publications, Oxford, UK.

Clayton, M. 2008. *Ultimate Sewing Bible – A Complete Reference with Step-by-Step Techniques.* Collins & Brown, London.

Hulme, W.H. 1944. The *Theory of Garment-Pattern Making: Textbook for Clothing Designers, Teachers of Clothing Technology, and Senior Students.* The National Trade Press Ltd, London.

Joseph-Armstrong, H. 2004. *Pattern Making for Fashion Designing.* Prentice Hall, Englewood Cliffs, NJ.

Knez, B. 1994. *Construction Preparation in Garment Industry* (in Croatian). Zagreb Faculty of Textile Technology, University of Zagreb, Croatia.

Kunz, G. and R. Glock. 2004. *Apparel Manufacturing: Sewn Products Analysis.* Prentice Hall, Englewood Cliffs, NJ.

Le Pechoux, B. and T.K. Ghosh. 2004. *Apparel Sizing and Fit.* Textile Progress 32. Textile Institute, Manchester.

Mathews, M. 1986. *Practical Clothing Construction – Parts 1 and 2.* Cosmic Press, Chennai.

Mehta, P.V. 1992. *An Introduction to Quality Control for Apparel Industry.* CRC Press, Boca Raton, FL.

Nayak, R. and R. Padhye 2015. *Garment Manufacturing Technology.* Woodhead Publication, Cambridge.

Shaeffer, C. 2000. *Sewing for the Apparel Industry.* Woodhead Publication, Cambridge.

Shoben, M.M. and P.T. Taylor. 1990. *Grading for the Fashion Industry.* Stanley Thomas Publishers Ltd, New Jersey, USA.

Sumathi, G.J. 2002. *Elements of Fashion and Apparel Designing.* New Age International Publication, New Delhi, India.

Yarwood, D. 1978. *History of Brassieres. The Encyclopedia of World Costumes.* The Anchor Press Ltd, UK.

12

Sleeves and Cuffs

12.1 Sleeves

The perfection of the sleeves in any garment is important for both functional and aesthetic performance. In a garment, the sleeve is the portion around the arm area of the wearer where it should not arrest the mobility of the wearer. The arm primarily functions in a forward motion, but in reality it can move in every direction. To impart this kind of flexibility, the fit of the sleeve is important. So the sleeve must be designed with a perfect fit, with proper amount of ease for movement of the arm. Before a sleeve is stitched to the bodice, it is necessary to know whether the cap ease is sufficient and is equally distributed between the front and back armhole. A standard sleeve with a perfect fit aligns with or is slightly forward of the side seam of the form (Tyler 1992; Chuter 1995).

One of the characteristics of fashion in dresses could be the use of different patterns of the sleeve which varies in every country and period. By having a single basic sleeve silhouette, any number of sleeve designs can be developed. Once a basic sleeve has been produced, it may be faced, piped and trimmed in countless ways in order to enhance the fabric aesthetically (Fairhurst 2008).

12.1.1 Classification of Sleeves

The different styles of sleeve have definite names like bishop sleeve, bell-shaped sleeve, etc., but most of the sleeves are defined based on their fit levels (tight, loose, shaped), their length (short, three-quarter, elbow-length, etc.), their outline or shape (full at the shoulder, widening around elbow, narrowing toward the wrist etc.) or to style details such as cuffs, openings and trimmings (Jacob 1988; Fairhurst 2008; Ruth and Kunz 2002). There are two basic types of sleeve:

1. Straight
2. Shaped

12.1.1.1 Straight Sleeve

The straight sleeve does not drape along the natural curve of the arm and hence this kind of construction does not allow the bending of the elbow and does not fit neatly. This kind of sleeve hangs loosely; however, it provides an easy and comfortable fit. Its comfort depends on its loose fit so that the whole arm moves freely inside it. The straight sleeve is used for a variety of styles – short, mid-length and long and in a variety of garments, such as blouses, dresses, nightwear, overalls and even coats (Gilmore and Gomory 1961). Figure 12.1 shows the view of a straight sleeve attached to a ladies top.

12.1.1.1.1 Basic Sleeve or Set-in Sleeve

The basic sleeve is a mounted sleeve, and it is drafted to fit a basic armhole. The basic sleeve is used as a foundation upon which an understanding of all other sleeves can be made. It can be made in different lengths.

- *Short sleeve:* This is a sleeve covering the arm up to the middle of the biceps and triceps area.

FIGURE 12.1
Straight sleeve.

- *Mid length:* A sleeve from the shoulder to a length mid-way between the elbow and the wrist.
- *Long sleeve:* A sleeve extending up to the wrist level, that is, it covers the full arm.

12.1.1.2 Shaped Sleeves

Shaped sleeves are constructed such that they follow the natural shape of the arm and help in the bending of the arm at the elbow. This kind of sleeve has some 'dart fullness,' which controls its shape, to fit the sleeve to a curve. These sleeves should have equivalent fullness or a dart, and this 'dart fullness' can be moved from one position to another. The straight sleeve is the primary sleeve pattern from which all types of shaped sleeves are designed. Shaped sleeves are further classified as sleeves with armscye and sleeves without armscye (Gilmore and Gomory 1961).

12.1.1.2.1 Sleeves with Armscye

- *Puff sleeves:* Sleeves that have extra fullness in certain parts (hem, cap or both) of the sleeve. These sleeves can be of any length and fullness. They are designed in the form of gathers by taking in more fabric. The base used for developing this sleeve is the dartless half pattern. The puffed sleeve is most popular among children and young students as it gives a youthful look.
- *Bell sleeve:* The name bell sleeve is derived from the basic pattern 'silhouette.' A bell sleeve has a smooth cap and a hemline flaring out in the shape of a bell. These sleeves can be designed to any length and flare as required.
- *Petal sleeve:* Petal sleeves are similar to the shaped sleeves. The only variation is that this sleeve resembles a petal as the sleeve sections cross over each other at the cap. The sleeves are developed at varying lengths in a number of ways by using a full dartless sleeve block.
- *Lantern sleeve:* Lantern sleeve has two sections where the sleeve widens itself from the cap and hemline to a style line within the sleeve of varying length.
- *Bishop sleeve:* These are gathered into a cuff of a long sleeve, fuller at the bottom than the top.
- *Cap sleeve:* These type of sleeves are often referred as 'sleeveless'. These sleeves are designed to cover the shoulders as they are very short. They do not go below the armpit level.
- *Leg of mutton sleeve:* The sleeve that extremely flares out at the upper arm and narrows down from the elbow to the wrist.

12.1.1.2.2 Sleeves without Armscye

The upper parts of any garment like jackets, blouse, coats, and shirts can be attached to the sleeve in a variety of ways. The basic sleeve pattern can be used as a base to develop numerous designs by making small modifications in their special characteristics or by changing their style (Carr and Latham 2006). These sleeve-bodice combinations are categorised as follows:

- *Kimono designs:* The sleeve merges well with the top of the garment and is developed by combining the sleeve length with the bodice or top.
- *Raglan designs:* When compared to the above design in raglan design the sleeve combines with the armhole and the shoulder area of the garment. This design can be imparted in any garments especially bodice, dress, blouse, jacket, or coat. In order to improve comfort to the wearer, the armhole is lowered at varying depths.
- *Drop shoulder designs:* Only a part of the sleeve cap combines with the garment. It doesn't if the garment is stitched without the lower sleeve.
- *Deep-cut armhole:* The entire portion of the armhole combines with the sleeve.
- *Dolman sleeve:* This sleeve is most preferred when a high arm lift is required. Here the sleeve is designed with a deep armhole. It is lowered under the seams with exaggerated folds under the arms.

12.1.2 Procedure for Construction of Sleeves

- *Grain line:* The straight grain of the fabric acts as the centre of the sleeve along its entire length.
- *Biceps level:* It is the widest portion of the sleeve, which distinguishes the cap from the lower sleeve.
- *Sleeve cap:* It is the portion above the biceps line and is curved.
- *Cap height:* This is the distance between the biceps to the top along the grain line.
- *Cap ease:* It ranges from 1¼" to 1½" depending on size.
- *Elbow level:* It is exactly located at the joining point of the arm and the elbow dart.
- *Wrist level:* It is the place where the hand enters the sleeve.
- *Notches:* The presence of notches eases the attachment of the sleeve to the garment and also helps in the identification of the front and back portion of the sleeve. The front sleeve is identified by the presence of a single notch, and two notches mark the back sleeve. Gathers originate from the notches and also end at the notches.

12.2 Cuffs

A cuff is an added piece of fabric present at the lower edge of the sleeve to cover the arms. The cuff precludes the fraying of the garment and in case of fraying, the garment can be rectified just by replacing the cuff. A simple technique of making cuffs is to fold back the sleeve material at the lower edge. Alternatively, a separate band of material can be sewn onto the sleeve or can be worn separately by means of buttons or studs. A cuff can be made decorative by attaching an ornamental border, lace or other trimming (Joseph-Armstrong 2004).

12.2.1 Shirt Cuffs

In most cases, the cuffs in shirts are divided down along their length and can be fastened together when the shirt is worn. This facilitates convenient wearing of the shirt and allows the cuff to fit perfectly around the wrist when it is fastened. In garments like sweaters and athletic garments, the cuffs are designed with elastic to enable them to stretch around a hand or foot (Hulme 1944). Depending on the method of fastening, divided shirt cuffs can be classified into

- Button cuffs, also known as barrel cuffs, are provided with buttonholes in one side on the cuff and buttons are sewn correspondingly on the other side. (The fit can be made adjustable by providing more than one button.)
- Link cuffs have buttonholes on both sides. They are fastened by means of cufflinks or silk knots. The 'kissing' style, where the inner surfaces of both the sides are pressed together, is the widely used fastening method. In rare cases, the outer face touches the inner face, during fastening. Link cuffs come in two kinds:
  - Single cuffs, which are the typical linked cuffs, are usually worn along with a white tie or a black tie. Some traditionalists wear the single cuff in combination with lounge suits also.
  - French (double cuffs) cuffs can be identified easily by their length. These cuffs are twice as long and need to be folded back while wearing. In earlier days, double cuffs were used to give a more formal look than button cuffs. Previously, French cuffs were required to be worn with a lounge suit or more formal clothing. Today, these cuffs are widely used with many dresses, sometimes even without a tie or jacket. They are highly preferred for semi-formal, black tie events.
- Convertible cuffs can be closed with either buttons or with cufflinks.

12.2.2 Barrel Shirt Cuffs

A two-button barrel cuff is the most widely recognised style among all off-the-rack shirts. Single-button barrel shirt cuffs are recommended for shorter people to create the illusion of longer arms. Likewise, triple-button barrel cuffs are suitable for very tall individuals (Laing and Webster 1998; Beazley and Bond 2006).

Further, barrel cuffs can come in straight cuts or angled cuts or as rounded barrel cuffs making the shirts more fashionable. Barrel shirt cuffs are worn in many instances like regular work, informal occasions. Sometimes they are even worn for formal occasions though they do not provide a very formal look (Le Pechoux and Ghosh 2004; Naya and Padhye 2015).

12.2.2.1 Single Button Barrel Cuff

It provides the perception of longer arms and is well suited for shorter persons (Beazley and Bond 2006). Figure 12.2 shows a shirt with a single button barrel cuff. In actual practices, the single button barrel cuff is considered to be less formal. The major drawback of this cuff style is that the single button often behaves like a pivot and causes the cuff gap to open. Hence, the firmness of the cuff is important in controlling the cuff opening.

12.2.2.2 Double Button Barrel Cuff

In the double button barrel cuff, the two buttons are aligned vertically along the cuff. Horizontal alignment of the buttons is preferred when the cuff is to be tightened around the wrist. In case of a perfect fit, the horizontal button

FIGURE 12.2
Single button barrel cuff.

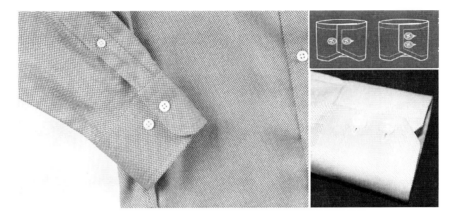

FIGURE 12.3
Double button barrel cuff.

alignment is not required in the cuffs (Mathews 1986). Figure 12.3 illustrates the double button cuff which is a more popular off-the-rack dress shirt style in European nations. These cuffs possess a formal look and are worn with suits. The striking advantage of having two buttons is that the cuff will be retained in place by a single button even if one of the two buttons falls off (Mathews 1986; Mehta 1992).

12.2.3 French Shirt Cuffs

French shirt cuffs provide the best approach to make a shirt more formal. Traditionally, these cuffs were worn with jackets. But now, they are worn with other dresses also. French cuffs can be rendered in numerous styles by varying the type of cufflinks and the cutting styles like straight cut, rounded cut or angled cut (Bray 2004).

12.2.3.1 Kissing French Cuffs

The kissing French cuff style is clearly shown in Figure 12.4. French cuffs, being a part of formal clothing, must be worn in this style. The cuffs are aligned such that their ends kiss each other. The holes should also be in line so that the cufflink can be passed through conveniently (Taylor and Shoben 1990; Sumathi 2002).

12.2.3.2 Undone French Cuffs

Figure 12.5 illustrates undone French cuffs. In this particular style, it is not always mandatory to wear cufflinks. Some people fold back the cuff up to the middle and do not make use of a cufflink. It is obvious that this

FIGURE 12.4
Kissing French cuffs.

FIGURE 12.5
Undone French cuffs.

style will not give a pleasant appearance compared to other cuff styles. However, undone French cuffs are preferred while wearing some clothing like a sweater over a shirt. The additional garment over the shirt ensures that the cuff is retained in a particular position (Pheasant 1986; Taylor and Shoben 1990).

12.2.3.3 Barrel French Cuffs

Barrel French cuffs are an effective alternative to kissing cuffs. In this style, the cuff ends are placed one end below the other such that the holes are in-line and the cufflink is passed through them. Figure 12.6 shows a variant where one end of the cuff end is slid over the other and then passed through a cufflink.

FIGURE 12.6
Barrel French cuffs.

References

Carr, H. and B. Latham. 2006. *The Technology of Clothing Manufacture.* Blackwell Science, Oxford.

Chuter, A.J. 1995. *Introduction to Clothing Production Management.* Blackwell Scientific Publications, Oxford, UK.

Bray, N. 2004. *Dress Fitting.* Blackwell Publishing Company, UK.

Beazley, L. and T. Bond. 2006. *Computer Aided Pattern Design and Product Development.* Blackwell Publishing, UK.

Fairhurst, C. 2008. *Advances in Apparel Production.* The Textile Institute, Woodhead Publication, Cambridge.

Gilmore, P.C. and R.E. Gomory. 1961. A linear programming approach to the cutting stock problem. *Operations Research* 9:349–359.

Glock, R.E. and G.I. Kunz. 2002. *Apparel Manufacturing – Sewn Product Analysis.* Prentice Hall, Englewood Cliffs, NJ.

Hulme, W.H. 1944. *The Theory of Garment-Pattern Making: A Textbook for Clothing Designers, Teachers of Clothing Technology, and Senior Students.* The National Trade Press Ltd, London.

Joseph-Armstrong, H. 2004. *Pattern Making for Fashion Designing.* Prentice-Hall, New York.

Laing, R.M. and J. Webster. 1998. *Stitches and Seams.* Textile Institute Publications, UK.

Le Pechoux, B. and T.K. Ghosh. 2004. *Apparel Sizing and Fit.* Textile Progress 32. Textile Institute, Manchester.

Mathews, M. 1986. *Practical Clothing Construction – Part 1 & 2.* Cosmic Press, Chennai.

Mehta, P.V. 1992. *An Introduction to Quality Control for Apparel Industry.* CRC Press, Boca Raton, FL.

Nayak, R. and R. Padhye. 2015. *Garment Manufacturing Technology.* Woodhead Publication, Cambridge.

Pheasant, S. 1986. *Body Space: Anthropometry Ergonomics and Design.* Taylor & Francis, London, Philadelphia.

Readers Digest. *Sewing Guide.* The Readers Digest Association.

Solinger, J. 1988. *Apparel Manufacturing Handbook – Analysis, Principles and Practice.* Columbia Boblin Media Corp., New York, USA.

Sumathi, G.J. 2002. *Elements of Fashion and Apparel Designing.* New Age International Publication, New Delhi, India.

Taylor, P.J. and M.M. Shoben. 1990. *Grading for the Fashion Industry.* Stanley Thomas Publishers Ltd, New Jersey, USA.

Tyler, D.J. 1992. *Materials Management in Clothing Production.* Blackwell Scientific Publications, Oxford, UK.

13

Apparel Accessories and Supporting Materials

13.1 Closures

Closures or fasteners are normally used to make permanent and semiper-manent connections between the garment panels and joints that could be unfastened and fastened. They are purely functional as well as decorative sometimes (Abernathy and Dunlop 1999).

13.1.1 Zippers

The 'zipper' was introduced in 1893 at the Chicago World's Fair then with the name of 'Clasp Locker'. Initially, during the 1930s zippers were elements in children's clothing for aiding them to dress themselves more easily. Zippers came into public interest in 1937 through the fashion designers in France who used them on men's trousers.

13.1.1.1 Objectives of a Zipper

The main objectives of zippers are

- To increase or decrease the extent of an opening to restrict or permit the passage of items.
- To join or isolate two panels of a garment, as in front of a dress or skirt.
- To detach or attach a detachable panel of the garment from another, as in the adaptation between trousers and shorts.
- To decorate an item.

13.1.1.2 Components of a Zipper

The construction and components of a zipper are shown in Figure 13.1.

The bulk of a zipper includes tens or hundreds of precisely shaped plastic or metal teeth which are attached to two pieces of fabric tape. These teeth could be either individual or fashioned from an endless coil, known as *elements*. The slider moves alongside the rows of teeth which have to be operated manually. The Y-shaped network inside the slider interlocks or separates the opposing rows of zipper teeth, based on the direction of the slider's movement (Kunz and Glock 2004). In zipper construction, either chain zippers where two sets of interlocking teeth are used or coil zippers where coils are used which are attached to a band of fabric tape.

13.1.1.3 Types of Zippers Based on Construction

The different kinds of zippers based on the construction are shown in Figure 13.2.

- *Coil zippers:* This is the most frequently used type of zipper. It runs on two zipper coils on two sides of the fabric tape. The two kinds of coils zippers are spiral coil zipper, with a cable reinforced inside the coil, and a ladder zipper known as 'Ruhrmann type'.
- *Invisible zippers:* It has teeth at the rear side of the tape and the colour of the teeth normally matches with the garment colour. The slider gives an invisible appearance. These zippers are normally coil zippers and are mainly utilised in skirts and dresses.

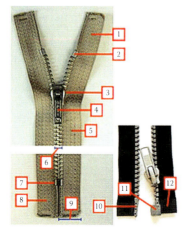

1 - Top tape extension	7 - Bottom stop
2 - Top stop	8 - Bottom tape extension
3 - Slider	9 - Single tape width
4 - Pull tab	10 - Insertion pin
5 - Tape	11 - Retainer box
6 - Chain width	12 - Reinforcement film

FIGURE 13.1
Components of zipper.

FIGURE 13.2
Types of zipper.

- *Metallic zippers:* These zippers are generally made of stainless steel, zinc, brass, nickel or aluminium alloy and are mostly found in jeans. The metal pieces are shaped into the form of teeth and are positioned in a zipper tape at uniform intervals. The metal zippers could be coloured in a variety of colours to match the colour of the garment.

- *Plastic moulded zippers:* It is like metal zippers excluding the fact that it is made of plastic material. These types of zippers can be manufactured in a range of colours to match the garment colour. Polyethylene resins and particularly polyacetal resins are commonly used to produce plastic zippers.

- *Open-ended zippers:* These zippers are generally found in jackets and have a 'box and pin' type mechanism to interconnect the two sides of the zipper. It could be coil, metallic, invisible or plastic zippers. In open-end zippers, both ends are separated from each other as shown in Figure 13.3.

- *Closed-ended zippers:* This kind of zipper is used regularly in baggage and is closed at both ends. The close-ended zippers are nonseparating and are usually opened and closed by means of a slider as shown in Figure 13.4. These zippers are used on trousers, jeans, jacket or shirt pockets, etc.

13.1.1.4 Types of Zippers Based on Material

The kinds of zippers based on material are given below.

1. Polyester Zippers

    Polyester zippers are classified as

    a. CFC – coil filler cord type (Figure 13.5)

    b. CH – coil without cord type

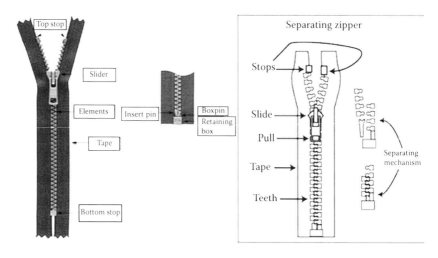

FIGURE 13.3
Open-end zipper.

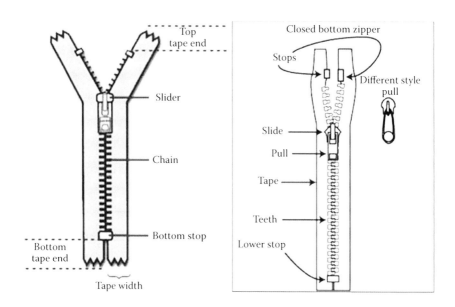

FIGURE 13.4
Close-end zipper.

    c. LFC – ladder type coil, used on trousers

    d. Invisible – concealed coil, used for ladies garments

2. Moulded Zippers

    The teeth of these zippers are produced by injection moulding using high-grade engineering plastic materials.

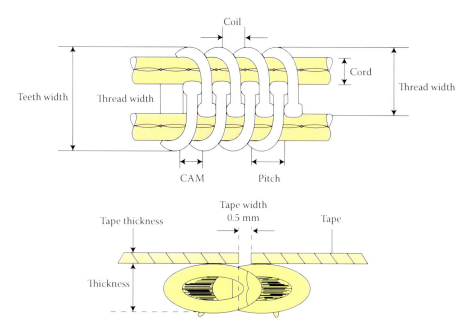

FIGURE 13.5
Construction of a typical CFC coil.

3. Metal Zippers

The teeth of the zippers are stamped out of metal strips such as brass, bronze, aluminium or nickel alloys. The constructions of metal zippers are shown in Figure 13.6.

4. Innovative Zippers

Due to the specific performances of textile and clothing, requirement of specialty zippers are also necessary to meet the functional performance of the garments. As a result, innovations were made to produce novel types of zippers such as flame-resistant zippers, airtight and watertight zippers, chemical-resistant zippers, zippers having electrical conducive yarn, zipper tape printed with ink-jet, environmentally conscious zippers, etc.

13.1.1.5 Manufacturing Process of Zippers

The process sequence of zipper manufacturing is shown in Figure 13.7.

- *Weaving of textile fabric:* The cotton or some other blended yarn in both warp and weft are normally used for weaving a narrow fabric in a needle loom to produce a woven edge braiding. The zipper must be processed through weaving, cutting and winding processes before the complete edge braiding of the zipper.

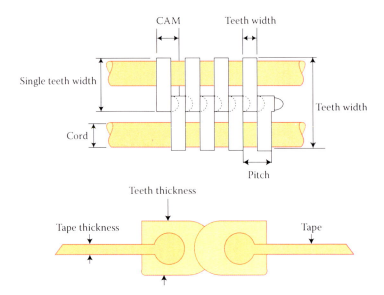

FIGURE 13.6
Construction of a metal tooth.

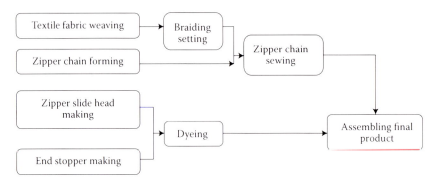

FIGURE 13.7
Manufacturing process sequence of zipper.

- *Zipper chain forming:* The resin (if it is polyester) is fed into the injection moulding machine to form a zigzag line.
- *Zipper slide head and end stopper:* Metal or aluminium alloy is fed into the die casting machine to produce a slider and end stopper with required size and shapes.
- *Sawing and fitting:* The fabric tape and zipper chain are attached together using a sewing machine. After that, the top and bottom stopper of the zipper and the slider are fitted and adjusted.

13.1.1.6 Zipper Size

Zippers are available in a range of sizes such as 3, 4, 5, 6, 7, 8, 9, 10, 12, etc., which are industry standards and does not imitate any dimensions of zipper, however, the larger the zipper size, the stronger the zipper (Park and Stoel 2002). Zipper length denotes the length of the zipper among the topmost point of the top stop and the bottommost point of the bottom stop.

13.1.1.7 Applications of Zippers

The common applications of zippers in different areas are given below.

1. *Ladies' and children's garments:* Coil filler cord (CFC), coil without cord type (CH) or invisible zippers; mostly closed-end zippers are used.
2. *Jackets and overcoats:* Metal or moulded open end or two-way separating zippers.
3. *Trousers:* Three ladder type coil (LFC), CFC with auto lock, closed end zippers.
4. *Denim and casual wear:* Robust closed end metal zippers.
5. *Luggage items:* CFC is most commonly used. Zippers in long chain rolls and sliders are sold separately.

13.1.2 Buttons

Buttons are the most commonly used type of fastener, comprising a disk, ball, or dome-shaped fastener secured to one panel of fabric and joined to another panel of fabric by means of drawing it through a buttonhole. It could be manufactured from an extensive variety of materials, such as natural materials like antler, bone, horn, vegetable ivory, ivory, shell and wood; or synthetics like celluloid, Bakelite, glass, metal and plastic. Hard plastic is the most used raw material for manufacturing of buttons (Carr and Latham 2006).

13.1.2.1 Types of Buttons

- *Shank buttons:* These kinds of buttons have a small bar or ring construction with a hole known as a shank jutting from the rear side of the button, through which sewing thread is sewn to join the button as shown in Figure 13.8.
- *Covered buttons:* In this type, the buttons are generally covered with fabric and have a distinct back panel that protects the fabric over the knob as shown in Figure 13.9.
- *Flat or sew-through buttons:* These buttons have two or four holes in the button through which the sewing thread is sewn to secure the

FIGURE 13.8
Shank button.

FIGURE 13.9
Covered button.

button as shown in Figure 13.10. Flat buttons could be secured by means of a button.

- *Worked or cloth buttons:* These are produced by embroidering or crocheting tight stitches over a knob called a form as shown in Figure 13.11.

FIGURE 13.10
Flat button.

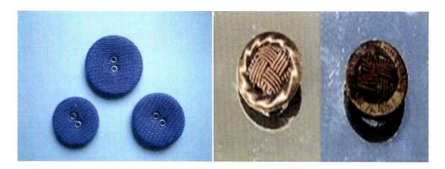

FIGURE 13.11
Worked button.

FIGURE 13.12
Mandarin button.

- *Mandarin buttons:* These kinds of buttons resemble knobs and are made of complicated knotted strings as shown in Figure 13.12. These buttons are a main component in a Mandarin garment, where they are closed with loops.

13.1.2.2 Button Sizes

The button size differs depending on its usage. Shirt buttons are normally smaller in size and the distance between the buttons are less, but coat buttons are comparatively bigger in size and spaced apart. Button sizes are normally expressed in 'lignes' (40 L = 1″). For instance, a formal men's shirt has a button size of 16 L (10.16 mm) and a men's suit jacket has a button size of 32 L (20.32 mm).

13.1.2.3 Buttonholes

Both button loops and buttonholes may be found individually and in sets. Button loops normally extend beyond the edge of the fabric, while buttonholes are cut in the fabric panel itself. There are three standard buttonhole shapes such as rectangular, oval and keyhole. Normally, the number of buttonholes is equal to the number of buttons, but shirt cuffs often have several buttonholes so the wearer could choose the button hole that provides the better fit. Most of the buttonholes have a perpendicular bar type of stitching at both ends which is referred to as a 'bartack', which strengthens the ends of a buttonhole (Clayton 2008).

13.1.3 Hook and Loop Fasteners

Pressure-sensitive tapes, for example, Velcro®, consist of two nylon woven tapes, one of them is covered with fine hooks and the other one with fine loops. When the two pile surfaces are pressed together, the hooks interlock with the loops producing a closed surface that is equal to the size of the tape as shown in Figure 13.13.

Since any one part (hook or loop) of one side will fasten to any part of the other, it is utilised for adjustable fastenings like closing cuffs and ankles on waterproof garments. It can be opened and closed easily, even while wearing gloves.

13.1.4 Eyelets and Laces

Garments often need small holes in the form of eyelets (Figure 13.14) for numerous purposes like for the prongs of buckles on belts, for the emergence of drawstrings at the waist or around hoods, for ventilation, and for use with lacing as a fastener. Diverse sizes of eyelet exist as suitable to the end use and garment type and it is crucial that the material to which they are attached is sufficiently substantial (Mehta 1992).

FIGURE 13.13
Hook and loop fasteners.

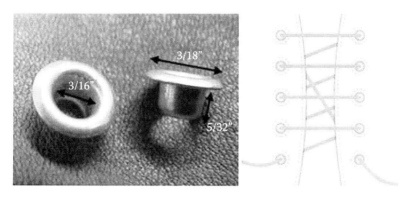

FIGURE 13.14
Eyelets and laces.

13.2 Supporting Materials

13.2.1 Linings

Linings are usually a functional part of a garment rather than a decorative one, being utilised in various shapes. The main objectives of lining materials in garment construction are

- To maintain the shape of garments.
- To improve the comfort as well as drape (hang) of the garment by letting it slip over other garments.
- To add insulation.
- To conceal the inner side of a garment panel of intricate construction to make it neat.
- They are selected to match the garments to be inconspicuous.
- To add to the design of garments.

Lining material could be utilised for small garment sections like in pockets and for complete garments, that is sewn down all the way round. In case of small sections like patch pockets or pocket flaps, it is essential for the lining material to remain concealed. For coats, jackets, and raincoats, the outer garment should not be inhibited in any way by tightness in the lining and in these garments there is normally extra lining fabric in the body and the sleeve (Mehta and Bhardwaj 1998). In skirt and trouser linings, the stability of the outer garment in wear may be aided by the lining being slightly smaller than the garment panel.

13.2.1.1 Fibre Types and Properties

Natural fibres are seldom used to make lining materials owing to their high cost and complicatedness while applying a specific finish to fabrics. Synthetic fibres are presently commonly used material for garment linings.

Viscose: Linings developed from viscose fibres have sufficient strength, softness, luster, and affinity for dyes.

Rayon: Rayon linings have properties like those of viscose but are weaker.

Polyamide: Polyamide linings give exceptional tensile strength and a comparatively high degree of elongation, and good affinity for dyes. Some solvents used for dry cleaning have harsh effects on the fabric which is the main limitation of polyamide fibres.

Polyester: Polyester fibres are like polyamide linings with respect to their properties.

Apart from a few lining materials made of polyamide, lining material from other synthetic fibres are not affected by dry cleaning and can be pressed up to a temperature of 170°C.

13.2.1.2 Function and Consumer Appeal

Function:
The vital objective of lining materials used in garments is to cover all or a portion of the interior surface of a garment. Other functions of lining materials are

1. It aids to protect the shape of garments especially in skirts and trousers, which are made from loosely woven or stretchy materials.
2. Garments that are made from very transparent material in skirts and trousers need 'cover up' areas where linings are utilised.
3. Several kinds of outerwear garments have an inclination to adhere to the body, which could spoil the outline of the garment. A layer of lining fabric kept between the body of the wearer and the main garment cloth could resolve this problem.
4. Linings are frequently used to aid in the development of design features on garments.

Consumer appeal:
The lustre and surface characteristics of the lining materials have a significant influence on consumer appeal. The main characteristics for linings most commonly used are

1. *Taffeta:* It is a crisp woven fabric with a faded warp pattern that yields a shiny surface.

2. *Crepe:* It is generally produced from specially processed yarns, usually from viscose acetate; the finished surface of this lining material has a uniformly crinkled form.

3. *Satin (sateen):* This lining is distinguished by the highly lustrous and smooth face surface and dull back surface.

13.2.1.3 Selection of Linings

There are diverse mixtures of fabrics appropriate for use as a lining. The deciding factors for the selection of lining include

- Type of fashion fabric
- Style of garment
- Type of lining; whether it is partial or complete

Lining fabrics could be woven or knitted but should be capable to provide and recover as essential to accommodate the body movement. It must be durable, colourfast to perspiration and the same care method as the main fabric. It is vital that the lining material should be the same weight or lighter weight and softer than the main fabric so that it does not dominate the garment. Lining fabric should be preshrunk prior to use (Mupfumira and Jinga 2014).

13.2.1.4 Making Up and Testing of Linings

Lining fabrics normally unravel easily and should be secured or stitched with a four-thread safety stitch machine though the thread consumption is higher. Regardless of whether lining materials are pressed or not before setting of seams, all vertical seam pieces should be pressed to one side. The shrinkage of lining fabric and base fabric with respect to length as well as width of fabric should be matched with each other. The wash-n-wear garment should be washed in a washing machine with correct programme settings and the results should be confirmed after the garment has been dried and pressed.

13.2.1.5 Lining Component Patterns

Follow-up of grain line markings for lining materials is equally important as like base fabric since they are for top cloth and fusible. The grain line of lining materials should match with the grain lines of the base fabric panels, even though this could be disregarded for components such as linings for skirts, trousers and sleeve linings. The number of nips on the lining patterns should be kept to a minimum since every nip is a probable weak point on the seams (Brand et al. 2007). Further, if the nips fray out prior to sewing, the stitching operator will have to skirt around the frayed areas by sewing wider seams than those called for.

13.2.2 Interlinings

To retain the shapes of various garment panels, a type of fabric secured between the two ply of fabric in a garment by means of fusing or sewing is known as interlining (Diamond and Diamond 1993). In general, interlinings are soft, flexible and thick. They are made of cotton, polyester, nylon, wool and viscose.

13.2.2.1 Functions of Interlining

- To support the garment.
- To retain the contour of the garment.
- To strengthen the garment components.
- To make the garment stronger and more attractive.
- To enhance the overall performance of the garment during wear.

13.2.2.2 Uses of Interlinings

Interlining materials are generally used in collar, waist band, cuffs, jackets, outerwear plackets, blazers, etc. The intricacy of interlinings used differs largely between tailored wear and other kinds of garments. In the nontailored garments like blouses, dresses, skirts and lightweight jackets and coats, interlinings are occasionally laid into a garment panel such as a collar or cuff and sewn around the edges when the part is constructed. In tailored garments, interlinings play a very crucial role in creating the profile of the garment and smoothing out the contours of the body.

Interlinings are available in a wide range of weights and constructions to match the characteristics of the base fabrics they will support. It can be woven or nonwoven construction, both of which can be constructed to give a different softness or resilience in different directions (Nayak and Padhye 2015; Sayed 2014). Woven interlinings are generally of plain weave construction. In the lighter weight interlinings, they may have a cotton warp and weft yarns which will give a soft handle in both directions, or a cotton warp and a viscose weft to give better crease recovery and retention of shape.

Nonwoven interlinings are made from fibres and are bonded using mechanical, thermal or chemical means or by a combination of these methods. The fibres utilised for nonwoven interlinings could be polyester to provide supple handle, viscose rayon to give a harder handle, nylon to give resilience and bulk, or some other combination of these fibres to give specific physical and mechanical properties. Nonwoven interlinings have diverse characteristics based on the direction in which the fibres are laid in making the material (Diamond and Diamond 2008). The fibres may be laid at random fashion for all-round stability of a product, parallel laid to give stability in one particular direction with extensibility in the direction at right angles, cross laid to give extensibility in both directions, and composite direction to give combinations of properties for general purpose interlinings.

13.2.2.3 Types of Interlinings

The main classification of interlining is shown in Figure 13.15.

13.2.2.3.1 Sew-In Interlining

The interlining that is placed between two layers of fabrics and secured by means of stitching them along with the fabric layers is known as sew-in or nonfusible interlining. Before attaching the interlining material with the layers of fabric, the interlining material should be treated with starch and dried (Tyler 2008). The applications of sew-in interlining are given below:

1. Used as interlining material in flame retardant garments especially for fire service people.
2. Protective garments for people working in rerolling mills.
3. Specially used in embroidery machines.

Merits of sew-in interlining:

- To make flame retardant garments.
- Simple and easy technique.
- No specialised machine is required.

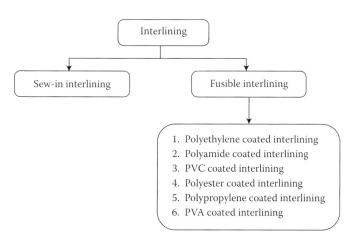

FIGURE 13.15
Types of interlining.

Demerits of sew-in interlining:

- Quality is not good as compared to fusible lining.
- Not appropriate for bulk production.
- No availability of interlining in the market.
- Time-consuming process.
- It involves higher work load and labour cost.

13.2.2.3.2 Fusible Interlining

In case of fusible interlining, the interlining material is kept between fabric layers and are attached to them through fusing by applying pressure and heat for a particular temperature and time period. In this system, the base component fabric is coated on one side with a thermoplastic resin, which is then bonded to another fabric by means of application of heat and pressure (Fairhurst 2008). These kind of interlinings improve the look of finished garments through the following:

- Control and stabilisation of critical areas.
- Reinforcement of specific design features.
- Least change in handle of the base fabric.
- Conservation of a crisp and fresh look of fabric.

Advantages of fusible interlining

- To get resemblances amongst the apparel.
- Application process is very easy.
- It has high productivity.
- Fusing time is less.
- It is cheap.
- Superior performance compared to sew-in interlining.

Disadvantages of fusible interlining:

- High temperature is required.
- Special attention is required during attachment of interlining material.

13.2.3 Difference between Lining and Interlining

The comparison of lining and interlining materials is given in Table 13.1.

TABLE 13.1

Comparison of Lining and Interlining Materials

Sl. No	Lining	Interlining
1	It is used inside of garments or garment components.	It is used between two layers of fabric.
2	It is attached by sewing.	It is attached by sewing or application of heat and pressure.
3	Finishing is not necessary.	Sometimes finishing is necessary to improve its properties. For example, shrink resist finish. Crease resists finish.
4	No coating is used.	Coating is used.
5	It is used in coat, rain coat, over coat, pocket flap, kids garments, jacket, etc.	It is used mainly in collar, cuff and front of jacket, waistband and front part of coat.
6	No classification.	It is of two types: a. Sewn interlining. b. Fusible interlining.
7	It is used to increase hang and comfort of garments.	To support, reinforce and control areas of garments and to retain actual shape.

13.2.4 Interfacing

It is an interior construction fabric that is positioned between plies of fashion fabric. It gives strength, shape and body to the garment. Interfacing is an extra layer secured to the inside of garments, to add shape, firmness, structure, and support to areas such as collars, cuffs, waistbands and pockets; and to stabilise areas like shoulder seams or necklines. A suitable interfacing should

- Be suitable to the fashion fabric with respect to fabric construction, fibre content, care and method of application (sew-in versus fusible).
- Have the same grain line as the fashion fabric.
- Harmonise in colour as that of the fashion fabric.
- Give necessary reinforcement required to enhance the contour of the garment.
- Not alter the drape characteristic of the fashion fabric.

13.2.4.1 Purposes of Interfacing

The objectives of interfacing are to

- Stabilise the fabric by avoiding sagging and stretching
- Strengthen the specific garment region
- Support facings

- Stabilise waistbands and neckline areas
- Soften the edges and provide smooth as well as stability of base fabric
- Maintain shape to garment areas like shoulders, hems, collars and cuffs

13.2.4.2 Types of Interfacing

Interfacings are available in two main types (fusible or sew-in), in three weave construction (nonwoven, woven and knit) and in diverse weights (light, medium and heavy weight). It is vital to select the appropriate type of interfacing for the particular garment. Woven interfacings have warp grain and weft grain similar to the base fabric to retain the drape of the garment (Baker 2006). Interfacings are primarily utilised on knitted fabrics to stabilise and to prevent excessive stretching. There are several kinds of nonwoven interfacings.

- *Stable:* It has little grain in any direction and is excellent for shoulder pads.
- *Stretch:* It is a crosswise stretch but is stable lengthwise.
- *All-bias:* It has stretch in all directions.

Knit interfacings are generally softer and more flexible due to their excellent stretch properties. Weft-insertions and warp-insertions are created on a knitting machine, and then either a warp or weft yarn is inserted. The inclusion of the additional yarns provides a more stable and secure knitted interfacing. The weft insertion has the higher stretch on the bias direction whereas the warp insertions have the higher stretch in the crosswise direction and they can be fused at a lower temperature compared to other fusibles.

13.2.5 Shoulder Pads

Occasionally garments, particularly shirts, sweaters, suits and dresses, can benefit from extra shaping. Shoulder pads (Figure 13.16) are generally used

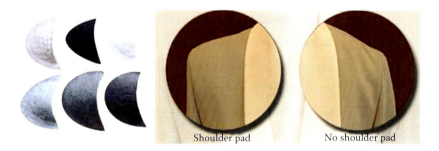

Shoulder pad   No shoulder pad

FIGURE 13.16
Shoulder pads.

for this function and are secured to the inside portion of the garment at the shoulder region. These shoulder pads are generally made of foam and covered with fabric. For unlined garments, a shoulder pad covered with nylon is commonly used, and for lined garments there is no necessity to cover the pad. The pads for unlined garments can be secured by means of a series of tacks made by a blind stitch machine. In case of tailored garments, the shoulder pads could be sewn in or fused to the shoulder area with thermoplastic resin positioned on the top layer of the pad (Soto 2001).

13.2.6 Waddings

Wadding or batting is an insulation layer normally used in quilted fabrics between a patchwork top layer and a backing material of bottom layer. The fibers such as cotton, wool, polyester and its blends are commonly used for batting (Stillwell 2014). It is frequently made up of a variety of fibres held together using several methods. The common techniques for holding the fibres together are

1. *Bonding:* In this method, the fibres are bonded using thermal or resin bonding. Thermal bonding has a mix of low melt polyester fibre and normal polyester to hold it together. Fibres such as polyester, cotton and wool are used in resin-bonded batting.

2. *Needle-punching:* In this method, the fibres are interlocked mechanically by means of punching them with needles, which make the batting denser as well as stronger while being lower loft.

3. *Scrim:* It is a low-weight fabric, which is needle-punched into the batting with the objectives of stabilising the batting, enhancing loft and strength of batting and to avoid stretching and distorting.

References

Abernathy, F.H. and J.T. Dunlop. 1999. *A Stitch in Time – Apparel Industry.* Blackwell Scientific Publications, Oxford, UK.

Brand, J., J. Teunissen and De Muijnck, C. 2007. *Fashion & Accessories.* TERRA, ArtEZ Press, Lannoo International, CA, USA.

Carr, H. and B. Latham. 2006. *The Technology of Clothing Manufacture.* Blackwell Science, Oxford.

Clayton, M. 2008. *Ultimate Sewing Bible – A Complete Reference with Step-by-Step Techniques.* Collins & Brown, London.

Diamond, J. and E. Diamond. 1993. *Fashion Apparel and Accessories.* Delmar Publishers Inc, New York.

Diamond, J. and E. Diamond. 2008. *Fashion Apparel and Accessories and Home Furnishing.* Pearson Publications, New Delhi, India.

Fairhurst, C. 2008. *Advances in Apparel Production.* The Textile Institute, Woodhead Publication, Cambridge.

Kunz, G. and R. Glock. 2004. *Apparel Manufacturing: Sewn Products Analysis.* Prentice Hall, Englewood Cliffs, NJ.

Baker, M.M. 2006. *Interfacing. UK Co-Operative Extension Service.* University of Kentucky, UK.

Mehta, P.V. 1992. *An Introduction to Quality Control for Apparel Industry.* CRC Press, Boca Raton, FL.

Mehta, P.V. and S.K. Bhardwaj. 1998. *Managing Quality in the Apparel Industry.* New Age International, New Delhi.

Mupfumira, I.M. and N. Jinga. 2014. *Clothing Care Manual.* Strategic Book Publishing, India.

Nayak, R. and R. Padhye. 2015. *Garment Manufacturing Technology.* Woodhead Publication, Cambridge.

Park, J.H. and L. Stoel. 2002. Apparel shopping on the internet: Information availability on US apparel merchant web sites. *Journal of Fashion Marketing & Management* 6(2):158–176.

Sayed, A. 2014. An overview on garment lining. http://textileapex.blogspot.in/2014/03/garments-lining.html (accessed on November 21, 2015).

Soto, A.M. 2001. *Simplicity: Simply the Best Sewing Book.* Simplicity Pattern Co, UK.

Stillwell, T. 2014. A Guide to Batting. http://www.connectingthreads.com/tutorials/A_Guide_to_Batting__D88.html (accessed on October 23, 2015).

Tyler, J.D. 2008. *Carr and Latham's Technology of Clothing Manufacture.* Blackwell, UK.

14

Production Planning and Control

Production planning entails the organisation of an overall manufacturing process to manufacture the end product. In the garment industry, various activities to be carried out in production planning are designing the end product, determining the machinery required and capacity planning, plant layout and material handling, establishment of sequence of operations for a style and nature of the operations to be carried out and specification of certain production quantity and quality levels.

Production planning and control is a vital part of the garment industry. Accuracy in planning equates to timely shipment of orders, better utilisation of operators and guarantees that proper supplies and machineries are available for each style and order (Ramesh and Bahinipati 2011). It includes every process from scheduling of each and every task in the particular process to dispatch of the garment.

14.1 Production Planning

The main aim of production planning is to provide a system along with a set of procedures for effective conversion of raw materials, labour and other inputs into final product (garment). The three key elements determining production planning in industry are

- Volume of production
- Nature of production process
- Nature of operation

14.1.1 Volume of Production

The intensity and quantity of production planning could be determined by the volume and character of the processes and the nature of the production processes. For example, production planning for manufacture of 10,000 garments would be different from the planning for a 1000 garments.

14.1.2 Nature of Production Process

In a job shop, the production planning would be very casual and informal and creating work methods are up to the skill of the individual person. But in the case of high volume production, many garment designers, process engineers and industrial engineers (IEs) are involved.

14.1.3 Nature of Operations

Comprehensive planning is necessary for cyclic operations, for example, in continuous manufacturing of a single standardised style of garment. The alternatives in manufacturing approach are

- Manufacturing to order, which may or may not be repetitive at regular intervals
- Manufacturing for stock and sell – batch or mass production
- Manufacturing for stock and sell – continuous process manufacturing

14.2 Production Planning System

The two interconnected subsystems in a production planning system are

- Product planning system
- Process planning system

A product planning system includes processes related to the development of product based on market necessities. In the case of the apparel industry, it includes fashion forecasting, customer research, development of protocol, etc. A process planning system includes activities that are required for the production of product based on expected demand (Nahmias 1997; Fairhurst 2008). This involves determination of the amount of material required in various processes, the sequence of processes that include the fabric inspection, spreading, cutting, sewing, finishing, packing, etc. The product planning system pays more attention to market requirements and the creation of product design based on the requirement. But process planning is more focused on the activities that are aimed at processes involved in the development of the product.

14.3 Production Control

Production control involves planning of production of the garments and the resources in terms of equipment and the labour available for translating the requirement of the garment production into reality. Due to the continuous monitoring of production flow and the utilisation of resources by the production control department, any deviation from the predetermined plan can be managed; hence, the productivity may run according to the original schedule (Schertel 1998; Russell and Taylor 1999). It manages all the garment production operations by means of gathering the significant information regarding the various types of inputs and outputs, and by making required changes in them. It guides and inspects the progress of the process and closes the records on the completion of the work or order. The functions of production control are to

- Offer the production of component panels, assemblies and garments of requisite quantity and quality and at the target time.
- Coordinate, monitor, and feedback concerned with the production of a particular style to the management, analysing the results of the production activities, understanding their importance and taking necessary corrective action.
- Offer optimal use of all resources.
- Achieve low production cost and reliable customer service.

14.3.1 Elements of the Production Control

Production control is updating and improving the procedure. Based on the requirements of implementation, the worker and machinery assignments, the job priorities, the production routes, etc. may be modified (Solinger 1988; Chuter 1995; Lim et al. 2009). The features of production control include the following:

- *Control of planning:* It guarantees the receipt of up-to-date estimated data from PP (production planning) department, bill of material (BOM) information from product engineering and data regarding routing from process engineering.
- *Control of materials:* It ensures delivery of necessary raw materials to the work floor and movement of materials within the shop.
- *Control of manufacturing capacity:* Establish the availability of machinery and labour skill level and give the practically achievable production schedules.
- *Control of activities:* Release order and information.

- *Control of quantity:* Follow-up of progress of production to ensure that the necessary quantities are processed at each production stage.
- *Control of due dates:* Check on the relation of actual and planned schedules and establish the reasons for delays or stoppages that hinder the weekly schedules of work allocated to each machine or work station.
- *Control of information:* Issue timely information and reports showing deviations from plans; hence corrective action could be carried out.

14.4 Production Planning and Control

It basically comprises planning production in an organisation prior to actual production processes and practicing control activities to ensure that the intended production is achieved with respect to quantity, quality, delivery schedule, and cost of manufacturing (Johnson and Moore 2001; Babu 2006; Mok et al. 2013). The aim of production planning and control (PPC) in the apparel industry consists of the following factors:

- To dispatch the garments at required quality and quantity in time to attain buyer satisfaction.
- To ensure the maximum use of all resources.
- To ensure that quality garments are produced.
- To minimise the product manufacturing time.
- To maintain optimum inventory levels.
- To maintain flexibility in the manufacturing process.
- To coordinate between operator, machines and different departments.
- To eliminate bottlenecks at all phases of production and resolve the issues associated with production.
- To ensure efficient cost reduction and control.
- The vital objective is to increase the profit of the garment industry.

14.4.1 Stages of Production Planning and Control

There are three stages in PPC, which are as follows:

- Planning stage
- Action stage
- Control stage

14.4.1.1 Planning Stage

The manufacturing of a garment begins with the planning for the same. It comprises selection of the best course of action within numerous alternatives. The two stages in the planning stage are preplanning and active planning.

- *Preplanning:* Preplanning process comprises product planning and development, demand forecasting, resource and facilities planning, plant planning and plant layout. Preplanning in the garment industry plays a prominent role. Fashion forecasting is the first and foremost stage in planning for production.
- *Active planning:* It comprises planning for quantity, product mix determination, scheduling, routing, material and process planning.

14.4.1.2 Action Stage

This stage is considered as the execution stage. It involves the dispatching and progressing function. This stage in apparel industry is the stage where the production of the garment is in process according to the requirements of the product. The planning and scheduling for the garment production including the assortment plan, layout plans, cutting and sewing are in progress. It includes all the stages from receiving the fabric to dispatching the garments to the customer.

14.4.1.3 Control Stage

It involves material control, inventory control, quality control, labor control and cost control. This phase is more in terms of controlling the functions in production with an objective of manufacturing the products as planned.

The planning of these three different stages depends on the principles of the production planning, which are as follows:

- Type of production determines the kind of PPC system needed.
- Number of parts involved in the product affects expenses of operating the PPC department.
- Complexity of the PPC function varies with the number of assemblies involved.
- Time is a common denominator for all scheduling activities.
- Size of the plant has relatively little to do with the type of PPC system needed.
- PPC permits 'management by exception'.

- Cost control should be a by-product of the PPC function.
- The highest efficiency in production is obtained by manufacturing the required quantity of a product, of the required quality, at the required time by the best and the cheapest method.

14.4.2 Levels of Production Planning and Control

Production planning takes place at many levels of the industry/organisation and covers different time perspectives. It could be categorised as strategic planning, tactical planning, and operational planning based on the graded altitudes in which it is carried out in the garment industry.

14.4.2.1 Strategic Planning

Strategic planning is an organization's process of defining its strategy, or direction, and making decisions on allocating its resources to pursue this strategy. Generally, it is a long-run plan carried out at the top management level. The long-term plans concentrate on product lines, divisions, markets, and other business units. The factors considered for the long-term planning includes investment capacity of the organisation, life cycle of product, market requirements, etc.

14.4.2.2 Tactical Planning

It is executed for an intermediate term by the middle level management in an organisation. It focuses on comprehensive products instead of individual specific style of products and has a time span of 6–18 months. It indicates the employment plans, utility plans, materials supply plans and expansion plans in the industry.

14.4.2.3 Operational Planning

This is executed for a short range time period by the lower level management in an organisation. It is mainly concerned with the use of existing services or facilities in the industry rather than creation. It comprises adequate utilisation of resources like raw materials, machinery, energy, etc. Short-term planning takes into account existing customer orders, priorities regarding material availability, labour absenteeism rate, cash flows, etc.

14.4.3 Functions of Production Planning and Control

The functions of PPC could be discussed in two individual stages as shown in Figure 14.1.

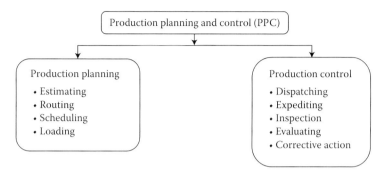

FIGURE 14.1
Functions of production planning and control.

14.4.3.1 Production Planning Functions

The production planning functions include the following:

1. Estimating

    It involves determining the quantity of garments to be produced and associated cost involved for the same based on the sales forecast. Determinations of raw materials and labour required to meet the planned targets and machine capacity are the vital activities prior to budgeting for resources.

2. Routing

    It is the method of determining the chain of operations to be carried out in the production line to complete the assembling of garments. This information is given by a product engineering function and is beneficial to make machine loading charts. A route sheet is a document giving the guidelines and information for conversion of raw materials into finished products. Route sheets contain the following information:

    a. The necessary operations and their sequence.

    b. Machine has to be used for every operation.

    c. Projected set up and operation time per garment piece.

    d. Description of raw materials to be utilised for garment production.

    e. Inspection procedure and tools required for inspection.

    f. Garment packing and handling guidelines during the movement of parts and subassemblies through the operation stages.

3. Scheduling

    It involves standardising the priorities for each work and determining the starting and finishing time for each process or operation.

It gives a time table for production, representing the total time period essential for the production of a specific garment style. The objectives of scheduling are as follows:

a. To avoid unbalanced utilisation of time amid various departments as well as work centres.

b. To utilise labour in an efficient manner such that the target is achieved well within the established lead time to dispatch the order in time and complete production at a minimum total cost.

4. Loading

Loading is the process of transforming the scheduled processes into practical work. Two main concepts of loading are facility loading and machine loading.

a. *Facility loading:* It is the loading of the work centre and deciding which kind of jobs to be allotted to which machine.

b. *Machine loading:* It is the process of allocating specific jobs to machines or workers based on primacies and capacity utilisation. A machine loading chart has to be made to demonstrate the planned utilisation of machines and workers by allocating the jobs to machineries as per priority determined at the time of scheduling.

14.4.3.2 Production Control Functions

Production control functions include the following:

1. Dispatching

It is defined as making the production-related activities in a dynamic manner by issuing the orders and guidelines in agreement with the previously planned time frames. It also gives a means for comparing actual progress of the work with respect to the planned progress. The functions of dispatching are given below:

a. Ensuring the smooth flow of raw material and other accessories from stores to first garment production operation and then from one operation to the next operation until all production processes are carried out. In the garment unit, it comprises the flow of fabric to inspection and then to the spreading room, cutting section, sewing room, finishing, packing and dispatching.

b. Gathering tools like cutting tools, sewing tools, etc., from tool stores and delivering them to the concerned department or operator.

c. Delivering the specification sheets, drawings and route cards to the concerned departments.

d. Giving the job orders and approving the processes in agreement with production schedule and time frame as indicated in schedules or machine loading charts.

e. Getting the schedule of inspection by the buyers or internal inspectors in an organisation and delivering it to the inspection section of the line.

2. Expediting/Follow-up

It confirms that the process is done as per the production plan and the delivery schedules are met. Progressing comprises activities like status reporting, attending to bottleneck processes in the production line and eliminating them, controlling of deviations from the planned performance levels, monitoring and follow-up of progress of work in all stages of production, coordinating with stores, tool room, purchase and maintenance departments and revising the production plans and replanning it if necessary (McBride 2003; Bubonia 2012). The necessity for follow-up could arise owing to the following reasons:

a. Delay in supply of materials.

b. Excessive absenteeism.

c. Changes in design specifications.

d. Changes in delivery schedules initiated by the customers.

e. Breakdown of machines, tools, jigs and fixtures.

f. Errors in design drawings of patterns and process plans.

14.4.4 Requirements of Effective Production Planning and Control

- Better organisational structure with proper guideline delegation of authority and finalisation of responsibility at all levels in an organisation.

- Information feedback system should give reliable and latest information to the concerned persons who are all carrying out PPC functions.

- Standardisation of materials, equipment, labour, workmanship, quality, etc.

- Trained person for handling the special equipment and manufacturing processes.

- Flexibility to accommodate changes and bottleneck circumstances like shortage of raw materials, power failures, machine break-downs and absenteeism of workers.

- Correct management policies concerning production level and inventory cost, product mix and inventory turnover.

- Precise assessment of manufacturing lead time and procurement lead times.
- Plant capacity should be sufficient to achieve the demand as well as flexible enough to respond to the introduction of new product styles, changes in product mix, production rate, etc.

14.4.5 Production Activity Control

The materials requirement planning system states what products are required in how much quantities and when they are needed. The production activity control (PAC) directs when, where and how the products should be made in order to ensure the dispatch of garments as per schedule (Burbidge 1991). Figure 14.2 shows the major concerns of PAC.

14.4.5.1 Objectives of Production Activity Control

- To know the current progress of the job. For example, is the garment production at the stage of sewing, cutting or packing?
- To decide upon what should be the next operation to be processed and in which work centre. For instance, once the fabrics are cut, they need to be sent to the spreading department and hence it would help to organise the same.
- To make sure that the right quantity of materials is in the right place, at the right time and the requisite capacity and tooling are provided.
- To improve the operational efficiency, that is, efficiency of worker and machine utilisation.
- To minimise work-in-progress inventory.
- To minimise set-up costs.
- To maintain control of operations by monitoring job status and lead times, measuring progress and indicate corrective action, when necessary.

14.4.6 Operations Planning and Scheduling

The operations scheduling and control process involves activities like priority as well as detailed scheduling, loading, expediting and follow-up, and input/output control. The several terminologies used in operations planning and scheduling are as follows:

FIGURE 14.2
Production activity control (PAC).

14.4.6.1 Loading

It is the assignment of jobs or processes to various machines or work centres for future processing, giving much attention to the sequence of operations based on the route sheet and the priority sequencing.

14.4.6.2 Sequencing

It is the practice of arriving at the sequence of operation of all jobs at each machine or work centre. It creates the priorities for carrying out the jobs that are waiting in the line at each machine or work centre.

14.4.6.3 Detailed Scheduling

It is the process of defining the starting and finishing time at every work centre, which is possible only after loading and sequencing.

14.4.6.4 Expediting

It is an action necessary to keep the work order to flow through the production line as per the detailed schedule. Delay in production due to equipment breakdown, nonavailability of materials when needed, etc., demands the expediting action for some vital processes.

14.4.6.5 Input–Output Control

The input–output plans and schedules require definite capacity levels at a work centre, but real utilisation could vary from what was planned. Input-output control is a main activity that gives comprehensive information about the real utilisation of a machine's capacity or work centre compared to the planned capacity utilisation.

14.4.7 Scheduling Techniques

The type of scheduling technique utilised in a job shop is based on the quantity of the received orders, the nature of the process and its complexity. The two types of scheduling techniques are

- Forward scheduling
- Backward scheduling

14.4.7.1 Forward Scheduling

In this process, each task or operation is scheduled to happen at the earliest time that the required material will be on hand and capacity will be available. It presumes that procurement of material and operations starts as soon as the buyer/customer requirements are known. Some buffer time could be added to estimate the target date and time for dispatching the order to

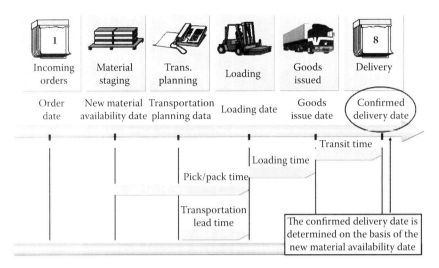

FIGURE 14.3
Forward scheduling.

the buyer. Figure 14.3 illustrates the forward scheduling process (Glock and Kunz 2004). From Figure 14.3 it could be noticed that the total time involved in starting and finishing the specific tasks in each department is calculated on the base of the time scale and then the date for the delivery of the products.

14.4.7.2 Backward Scheduling

This technique is normally utilised in assembly type industries where they commit in advance to specific delivery time. After the determination of the essential schedule dates for key subassemblies, the schedule utilises these dates for each component and works backward to determine the proper dispatch date for each component manufacturing order (Carr and Latham 2006; Ray 2014). The work or jobs start date is calculated by 'setting back' from the finish date the processing time for the job. The backward scheduling is shown in Figure 14.4, where the overall manufacturing lead time is split into the time schedule of different departments on the basis of time scale.

14.4.8 Sequencing

The sequence or order in which the jobs are executed is determined by the sequencing. The sequence of processing each job will be vital when it comes to the cost of idle time at work centres. Hence, establishment of priority for all the jobs waiting in the queue by applying priority sequencing rules has to be done (Collins and Glendinning 2004). The conditions for selection of the right sequencing are

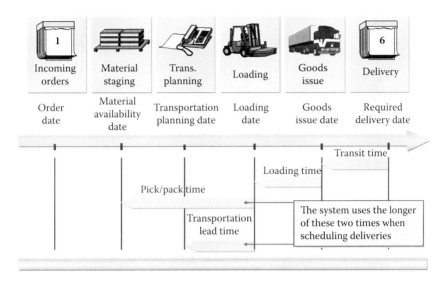

FIGURE 14.4
Backward scheduling.

- Set-up costs or change over costs
- Work-in-progress inventory cost
- Idle time
- Average job lateness
- Average flow time
- Average number of jobs in the system
- Average time to finish a job

There are two types of sequencing that could be selected based on the specific criteria as given below:

- Single criterion priority sequencing rules
- Dynamic sequencing rules

14.4.8.1 Evaluating Sequencing Rules

The most frequently used criteria for evaluation of sequencing rules are as follows:

1. *Average flow time:* It represents the average time period each job occupies in the work centre or shop.

$$\text{Average flow time or} \atop \text{Average Completion Time} = \frac{\text{Total flow time for all jobs}}{\text{Number of jobs in the system}}.$$

2. *Average number of jobs in the system or shop:* It is the average number of tasks or jobs in the shop every day.

$$\text{Average number of jobs in the system} = \frac{\text{Total flow time for all jobs}}{\text{Number of jobs in the system}}.$$

3. *Average job lateness:* It is the average time period that each task or job is delayed compared to its actual due date.

$$\text{Average Job lateness} = \frac{\text{Total Job lateness}}{\text{Number of jobs in the system}}.$$

4. *Change over cost:* It represents the total cost of making the entire machine change over in groups of jobs.

14.5 Production Planning and Control in Garment Industry

14.5.1 Production Strategies in Garment Industry

The commonly used production policies in garment industries are

- Flexible manufacturing strategy
- Value-added manufacturing strategy
- Mass customisation

14.5.1.1 Flexible Manufacturing Strategy

This system ensures quicker and effective manufacturing of a diversity of garment styles in small production runs with the least defects. This is quicker to respond to demands of the consumer especially for small orders and shorter lead limes. The main advantages of implementing this strategy are its flexibility to operate to achieve the consumer demands and the capability to adapt swiftly to changes in the garment market (Jang et al. 2005).

14.5.1.2 Value-Added Manufacturing Strategy

It is a quick response plan that concentrates on the managing of avoidable operations that does not improve the product value but instead leads to delays in production. The underlying principle of this strategy is that every operation or task performed on a garment style should add value. Operations like material handling, sorting, inspection, ware housing, etc., involve additional time, handling, and workers; however, it does not add product value (Nayak and Padhye 2015).

14.5.1.3 Mass Customisation

The strategy here is to produce products that could be made to order rather than made to plan. Product life cycle is short and the strategy requires processing single orders with immediate turnaround. Taking into account the complexity of many garment products and the number of processes that a style could require, the machinery, labour skills, information, and processes must be integrated (Goworek 2010). This may involve single ply culling, single piece continuous floor manufacturing, and integral information technology. Mass customisation reduces the risk associated in trying to anticipate consumer demand months ahead of the point of sale to the ultimate consumer.

14.5.2 Roles of PPC Department in Garment Industry

14.5.2.1 Task Scheduling

It involves planning of Time & Action (T&A) calendar for every order from the receipt of the order to dispatch of the same. The task schedule comprises a list of jobs to be processed for the style of garment. Alongside each task the production planner cites the start time of a task and the scheduled date for completion.

14.5.2.2 Material Resource Planning

It is the planning and creation of material requirement sheet based on sample product and the specification sheet. The consumption of raw material such as fabric, button, sewing thread, and twill tape and their costs are estimated.

14.5.2.3 Loading Production

Production planner delineates which garment style and how the quantity has to be put into the production line.

14.5.2.4 Process Selection and Planning

The operations required to finish an order differs from garment style to style. Based on the buyer requirement, production planning section decides on processes for the orders. Sometimes additional operations or processes are removed to minimise the cost of production.

14.5.2.5 Facility Location

For a garment industry that has multiple factories for production and are set for specific products, the production planner has to identify which facility will be the most appropriate for new orders.

14.5.2.6 Estimation Quantity and Costs of Production

The production planner should estimate the daily production based on garment style work content. Based on the estimated production rates, production runs and operator involvement planner also estimate production cost per pieces.

14.5.2.7 Capacity Planning

The PPC department plays a vital role during booking of orders. They have to provide information regarding quantity of order they could accept based on their estimated production capacity.

14.5.2.8 Line Planning

It is the preparation of comprehensive production line planning along with daily production target for the specific production line. In a majority of cases, line planning is prepared after having a discussion with the production department and the IE in the industry.

14.5.2.9 Follow-Up and Execution

The PPC department ensures that the order is moving in a particular production line as per the production plan.

14.5.3 Standard Allowed Minute

Standard allowed minute (SAM) is the time (including allowances) necessary to produce one finished garment. In the garment industry, particularly in production, SAM is used for assessing the efficiency of work. In the garment industry, the industrial engineering department determines and calculates SAM for assembling processes of garments using a standard calculation method (Vijayalakshmi 2009). The applications of SAM are

- Firm as well as individual operator's performance
- Operator and associated cost ratios
- Operators' payroll and incentive amount
- SAM is one of the key parameters in state-of-the-art production scheduling methods like in line balancing and performance measuring systems

14.5.3.1 Calculation of SAM of a Garment

For the valuation of garment cost, determination of SAM value plays a critical role (Martinich 1997; Sarkar 2011). General sewing data (GSD) has a definite set of codes for motion data for determination of SAM. Other than using GSD and synthetic data, other methods are also available for the calculation of SAM.

14.5.3.1.1 *Calculation of SAM by Synthetic Data*

In this system 'predetermined time standard' codes are utilised to establish 'standard time' of a specific style of garment. The step-by-step procedure for calculation of SAM by this method is given below.

- Selection of any one process or operation for which the SAM has to be determined.
- Study of various motions of the specific process/operation performed by an operator and remarking all movements used by the operator in carrying out one complete cycle of work.
- Enlist various motions performed by an operator sequentially. By referring to GSD and synthetic data for time measurement unit (TMU) values, TMU value (1 TMU = 0.0006 minute) for one operation could be obtained, which is then converted into minutes which is known as basic time.
- SAM = basic minute + bundle allowances (10%) + machine and personal allowances (10%).

14.5.3.1.2 *Calculation of SAM by Time Study*

The step-by-step procedure for calculation of SAM by this method is given below.

- Selection of one process or operation for which the SAM has to be estimated.
- Note down the cycle time (total time necessary to carry out all tasks required to complete one operation) for the specific operation using a stop watch by standing at the side of the operator using the stop watch. It has to be done for five consecutive cycles of that operation and the average has to be determined. Basic time = cycle time × performance rating.
- Establishing the performance rating of an operator after evaluating his or her movement and work speed. Suppose if the performance rating of an operator is 85% and the cycle time is 0.55 minutes, then basic time = $(0.55 \times 85\%) = 0.46$ minutes.
- SAM = basic minute + bundle allowances (10%) + machine and personal allowances (20%). Now, SAM = $(0.46 + 0.046 + 0.092) = 0.598$ minutes.

14.5.3.2 **Functions of SAM Value in Production Planning**

1. *Determination of line capacity* – The systematic method of estimating the production capacity of a line by utilising the SAM of a garment.

2. *Determination of lead time* – Based on the production capacity of a garment unit, order allocation has to be done for different lines.

3. *Order booking* – While booking the orders, available capacity in a particular period of time has to be taken into account. In these circumstances, determination of time required to complete the new order using SAM and comparing the same with production minutes available in the factory for the particular period will be helpful.

4. *Process scheduling* – Time and action calendar of each and every order is carried out by the production planning department based on capacity of each process, which is known by calculating SAM.

5. *Order execution and production monitoring* – SAM facilitates the production planning department to set targets for sewing lines.

6. *Estimation of labour* – For the estimation of labour cost for a particular style, the SAM value will play a vital role.

14.6 Performance Measurement Parameters in Production Planning

14.6.1 Cut to Ship Ratio

This demonstrates the percentage of garments dispatched out of a total number of garments cut for a particular style or order. A ratio around 98% will be considered very good, which means that only 2% of garments are rejected in the style.

$$\text{Cut to Ship ratio} = \frac{\text{Total pieces shipped}}{\text{Total Pieces cut}}.$$

Total pieces shipped – Information collected from final packing list.
Total pieces cut – Information collected from cutting room records.

14.6.2 Labour Cost per Minute

This is a significant parameter while determining the garment cost. It assists in determining the labour cost involved in the production of a garment.

$$\text{Cost per minute} = \frac{\text{Total cost incurred on labour}}{\text{Total available working minutes} \times \text{No. of labour}}.$$

To determine the labour cost of a new style or order, its SAM value is estimated and then multiplied to cost per minute and efficiency to get the actual labour cost.

Example: A garment unit has 500 direct sewing operators and helpers. The cost to the company for the 500 operators is Rs. 3,900,000. The company works for 8 hours per day for 26 days in a month. Therefore,

Total working minutes per operator = 12,480 minutes.

Cost per minute = 3,900,000/(12,480 × 500) = Rs. 0.625 per minute.

14.6.3 Plan Performance Index

This demonstrates the variation between the planned work and the actual work completed.

$$\text{Plan performance index} = \frac{\text{Achieved production}}{\text{Planned produciton}} \times 100.$$

This assists in assessing the effectiveness of the garment factory plan in order to achieve the target dates.

14.6.4 On-Time Delivery

This shows the percentage of deliveries that a garment industry is able to make on-time without any delays. An on-time delivery represents that the garments or products are dispatched on time to the buyer.

$$\text{On time delivery} = \frac{\text{On time delivery}}{\text{Total deliveries}} \times 100.$$

14.6.5 Capacity Utilisation

The degree to which the manufacturing capacity of an industry is utilised in production of garments is known as capacity utilisation. It is the ratio between actual output produced with the existing facilities such as equipment and labour and the potential output.

$$\text{Capacity Utilization} = \frac{\text{Produced garments or Produced minutes}}{\text{Capacity in terms of garments or minutes}} \times 100.$$

Suppose the garment industry produces a standard garment style. Then capacity could be expressed by means of number of garment pieces. But, if there is a vast variation in the products produced, then the unit of measure should be number of minutes produced.

Example: Suppose if the industry has 500 operators and they work at 67% efficiency. The factory runs 24 days in a month, 8 hours each day. The factory made 280,000 pieces in 12 minutes each in the last month.

Produced minutes = 280,000 × 12 = 3,360,000 minutes.

Capacity = 500 × 0.67 × 26 × 480 = 4,180,800 minutes.

Capacity utilisation = 3,360,000/4,180,800 = 80.37%.

14.6.6 Lead Time

It is the time period between the confirmation of an order and the dispatch of the order to the buyer. A shorter lead time is beneficial as the buyer wants dispatch of their products as early as possible. Lead time comprises waiting time before or after actual manufacturing and throughput time.

14.6.7 Overtime %

It is the percentage of overtime utilised apart from total working time.

$$\text{Overtime} = \frac{\text{Total overtime minutes}}{\text{Total working minutes}} \times 100.$$

14.6.8 On Time in Full

OTIF (on time in full) is the primary parameter for logistics performance. It evaluates whether the supply chain was capable to deliver:

1. The expected product
2. In the quantity ordered
3. At the place agreed
4. At the time expected by the customer

$$\text{OTIF} = \frac{\text{Number of deliveries OTIF}}{\text{Total number of deliveries}} \times 100.$$

14.6.9 Absenteeism

It refers to the absence of operators from regular work without permission and is avoidable. Absenteeism in the apparel industry is in the range of 10%–15%.

$$\text{Absenteeism} = \frac{\text{Number of man days lost}}{\text{Number of man days scheduled to work}} \times 100.$$

14.6.10 Attrition Rate

Retention of labourers is a huge issue in an apparel industry. This parameter could tell the manager whether their efforts toward labour retention are harvesting benefits.

$$\text{Attrition rate} = \left(\frac{\text{Number of attritions} \times 100}{\text{Actual employees} + \text{new joined}} \right) \times 100.$$

14.7 Production Planning Software for Apparel Industry

A PPC kit for the industry level management is vital for timely delivery of an order. Some of the software solutions existing for the garment industry are given below.

14.7.1 Evolve by Fast React

Evolve is a vibrant solution that gives highlights of important activities related to production planning, reveals up-to-date performance of an industry and offers a prompt cautionary alert of any critical actions required. The main features of Evolve stated by Fast React comprises

- Multilevel planning in an industry as well as at the machine level.
- During product changeover, it offers efficiency profiles of various operations and start-up allowances.
- It provides management of high-quality machines, WIP and auxiliary operations.
- Due to better production management aspects in the software, it provides better production scheduling and communication with subcontractors.
- The production planning offered by this software is more dynamic, which reflects the up-to-date condition about the progress of the order.
- Materials and critical path priorities are actively 'driven' to support the latest plan.

14.7.2 Plan-IT by Gemserp

Plan-IT is specially made for apparel production planning. It facilitates merchandisers and production departments for decision making. Hence, it reduces last minute urgency in order processing.

14.7.3 PPC Module by APPS

The significant characteristics of APPS PPC module are as follows (Kumar 2008):

- Efficient production line planning
- Production monitoring of vendor
- Provides details about consumption of raw material and critical path monitoring of the same
- Easy analysis of production in an industry by providing quality control reports
- Offers comprehensive summary reports

14.7.4 MAE by Parellax

MAE determines the number of orders that could be taken by the production line at one time. The aim is to load plan; hence, the production line will not be idle at any time. It also has an option for semiautomated line planning capabilities (Keiser and Garner 2012).

14.7.5 STAGE Production Planning Management

The significant characteristics of this software are

- The planning department could plan merchant's orders, cut and paste orders from one production line to another one easily.
- No data entry is necessary on the planning board since all data are extracted from ERP software and outcomes are showed automatically on the planning board.
- Additional features of this software are automatic mailing, integration with T&A and expected completion report.

14.7.6 Pro-Plan by Methods Apparel

Pro-Plan aids to set up capacity for each production department, remove holidays, add overtime, use of current efficiencies and absentee levels and the capacity is instantly available.

References

Babu, V.R. 2006. Garment Production Systems: An Overview. http://www.indiantextilejournal.com/articles/FAdetails.asp (accessed on March 12, 2015).

Bubonia, J.E. 2012. *Apparel Production Terms and Processes.* Fairchild Books, New York.

Burbidge, G.M. 1991. Production flow analysis for planning group technology. *Journal of Operation Management* 10(1):5–27.

Carr, H. and B. Latham. 2006. *The Technology of Clothing Manufacture.* Blackwell Science, Oxford.

Chuter, A.J. 1995. *Introduction to Clothing Production Management.* Blackwell Scientific Publications, Oxford, UK.

Collins, P. and S. Glendinning. 2004. Production planning in the clothing industry: Failing to plan is planning to fail. *Control* 1:16–20.

Fairhurst, C. 2008. *Advances in Apparel Production.* The Textile Institute, Woodhead Publication, Cambridge.

Glock, R.E. and G.I. Kunz. 2004. *Apparel Manufacturing – Sewn Product Analysis.* Prentice Hall, Englewood Cliffs, NJ.

Goworek, H. 2010. An investigation into product development processes for UK fashion retailers: A multiple case study. *Journal of Fashion Market Management* 14(4):648–62.

Jang, N., K.G. Dickerson and J.M. Hawley. 2005. Apparel product development: Measures of apparel product success and failure. *Journal of Fashion Marketing Management* 9(2):195–206.

Johnson, M. and E. Moore. 2001. *Apparel Product Development.* Prentice Hall, Upper Saddle River, NJ.

Keiser, S. and M. Garner. 2012. *Beyond Design: The Synergy of Apparel Product Development,* Fairchild Publications, New York.

Kumar, A. 2008. Production Planning and Control: Lesson 8 Course material. Delhi University. www.du.ac.in/fileadmin/DU/Academics/course_material/EP_08.pdf (accessed on March 14, 2015).

Lim, H., C.L. Istook and N.L. Cassill. 2009. Advanced mass customization in apparel. *Journal of Textile Apparel Technology Management* 6(1):1–16.

Martinich, J.S. 1997. *Production and Operations Management – An Applied Modern Approach.* John Wiley & Sons Inc, New York.

McBride, D. 2003. The 7 Manufacturing Wastes. http://www.emsstrategies.com/dm090203article2.htm (accessed on March 25, 2015).

Mok, P.Y., T.Y. Cheung, W.K. Wong, S.Y.S. Leung and J.T. Fan. 2013. Intelligent production planning for complex garment manufacturing. *Journal of Intelligent Manufacturing* 24(1):133–45.

Nahmias, S. 1997. *Production and Operations Analysis.* Irwin, Chicago, IL.

Nayak, R. and R. Padhye. 2015. *Garment Manufacturing Technology.* Woodhead Publication, Cambridge.

Ramesh, A. and B.K. Bahinipati. 2011. The Indian apparel industry: A critical review of supply chains. In: *International Conference on Operations and Quantitative Management (ICOQM).* Nashik, India.

Ray, L. 2014. Production Planning for Garment Manufacturing. http://smallbusiness.chron.com/production-planning-garment-manufacturing-80975.html (accessed on October 5, 2014).

Russell, R.S. and B.W. Taylor. 1999. *Operations Management.* Prentice Hall, Upper Saddle River, NJ.

Sarkar, P. 2011. Functions of Production Planning and Control (PPC) Department in Apparel Manufacturing. http://www.onlineclothingstudy.com/2011/12/functions-of-productionplanning-andhtml (accessed on October 5, 2014).

Schertel, S. 1998. New Product Development: Planning and Scheduling of the Merchandising Calendar. Master dissertation, North Carolina State University, Raleigh, NC.

Solinger, J.S. 1988. *Apparel Manufacturing Hand Book – Analysis Principles and Practice.* Columbia Boblin Media Corp, New York, USA.

Vijayalakshmi, D. 2009. Production Strategies and Systems for Apparel manufacturing http://www.indiantextilejournal.com/articles/FAdetails.asp (accessed on March 12, 2015).

15

Fabric Utilisation in Cutting Room

With the current competitive and aggressive world market, manufacturers are facing invariable pressure to minimise costs, propose product styles and deliver products quickly. The important aspect of achieving better efficiency in the apparel industry is to minimise the raw material costs, which often go up to 70% of the total manufacturing costs (Glock and Kunz 2004). The method of handling cutting orders and planning economic cutting lays is of extreme significance for better utilisation of materials and for increasing the efficiency of the cutting process (Aldrich 2002; Beazley and Bond 2003). To reinstate a competitive spot in the international marketplace, the garment industry is looking to upgrade its responsiveness to customer needs.

15.1 Cut Order Planning

One of the vital operations that take place in the cutting section is known as cut order planning. This could also be called 'lay plan', 'cut plan', etc. It is optimising principally the cutting operation under certain constraints by following certain parameters. In short, cut order planning is nothing but deciding the arrangement or combination of markers and spread lays for a particular garment style order. This is a benchmark process to be done in every garment industry which has a huge impact on overall savings for the order (Solinger 1988; Tyler 1991).

Cut order planning is the activity of planning the purchase order for the process of cutting, as input into the marker making stage. Therefore, the cutting section receives all the spreading and cutting instructions. This process is a dynamic job that must react to the ever-changing field of many decisive factors such as sales, inventory levels, raw materials, availability of equipment, etc. The various kinds of sizes, styles, fabrics and colours introduce significant complications into the problem (Mathews 1986; Shaeffer 2000).

The cutting section has a greater influence on higher manufacturing costs than any other section concerned with the manufacturing of garments. Several garment manufacturing industries are still using unsophisticated methods or processes, depending on the knowledge of one individual who has the required data and decision making tools only in his or her memory. Commercial software for cut order planning has been developed, but effectual application needs extensive customisation and the necessary hardware for implementation.

The utilisation of fabric in a cutting room is decided by three factors such as pattern engineering, marker making and the selection of markers for the specific production plan. A fundamental prerequisite to attain high fabric utilisation is designing the garment patterns in a logical way that produces a proper garment construction in line with the current fashion trends and comfort requirements and also with fabric utilisation and garment sewing aspects. The second factor that determines utilisation of fabric is the manner in which the garment patterns are arranged on the marker (Carr and Latham 2006). This is a critical stage where fabric wastage could be controlled. The important development in this area was utilisation of computers in terms of CAD for pattern making and marker planning.

In a typical computer-aided marker making system, the information on pattern shapes obtained from a digitiser is stored in the computer memory and is showed on a CRT. A marker planner decides the movement of individual garment patterns by means of a stylus to get the most efficient marker. The software prevents overlapping of any garment patterns, protecting predetermined match of stripes, checks, etc., stores the information about the markers and relays this information to a computer-driven plotter or disk to a digitally controlled cutting device (Chuter 1995).

Fabric utilisation also depends to a huge extent on the manner in which the stock of fabrics is scheduled for the products of different garment styles as well as sizes. Separate markers are produced for a particular garment style and sizes on a material of a given width. The selection of markers is now commonly done by either first match or enumeration method (Fairhurst 2008). In the first match system, the first combination of markers observed that comprise all required style/sizes in the cut order is the one used. This method is perceptibly inefficient. The enumeration method is done by manually listing several marker combinations in order of efficiency. This is time consuming and the solution is most likely not optimal.

Computerised marker planning is the connection between computerised pattern grading and computer controlled cutting systems. It passes the data or information from the pattern grading operation to the cutting machine through a computer controlled plotter. The plotter prints full scale master markers from the arrangement of patterns generated on the CRT and stored in memory (Mehta 1992). Various statistical parameters like the pattern piece area could be determined and printed in every graded pattern.

The integrated marker planning will lead to reduced costs in the pattern making section with reference to labour, supplies and occupancy. Productivity in cutting section increases compared to conventional marker planning system primarily due to the elimination of manually drawing around the patterns, which is replaced by the plotter. The higher productivity provided by the computer-aided planning could be utilised to allow additional effort to improve marker efficiency.

One of the important characteristics of the computer aided marker planning process is the higher material utilisation and minimal fabric wastage.

Fabric utilisation is superior compared to the conventional manual method because less time is necessary to make the original marker and therefore more time can be spent for improving the marker efficiency (Burbidge 1991). Computerised pattern grading is the main source data of manufacturing control. These data are digitised and controlled by the grading rules to be used. This supplementary output could be used for further stages in garment production. Since computerised pattern grading systems carry out the translation from a sample size pattern into a full range of sizes, it mainly influences the pattern making and grading labour costs.

The computerised pattern grading process could improve the capacity to make several garment pattern sets in a given period of time, thereby reducing the cycle time. This is occasionally required due to style changes (Ambastha 2012). The computerised pattern grading is helpful to those industries where frequent changes in garment style and short lead times are requisite. The current advancements in computerised marker planning and grading still need a person at some point to interact with the computer to execute certain functions, like selection and placement of pattern pieces which are needed to build the marker.

15.1.1 Cost Involved in Cut Order Planning

The main costs involved in the cutting section are

- Raw material cost (fabric cost)
- Marker making cost
- Spreading cost
- Cutting cost
- Bundling and ticketing cost
  - *Fabric cost:* The cost of raw material, that is, fabric is around 50%–70% of the total garment cost and total labour cost is only about 10%–15%, out of which cutting room labour cost would be a minimum.
  - *Marker making cost:* With the utilisation of CAD in marker planning, the evaluation of marker making cost has become quicker as well as more efficient.
  - *Spreading cost:* The spreading cost depends on the following parameters:
    - Total fabric laid in a spread
    - Number of plies
    - Number of lays
    - Number of roll changes
  - *Cutting cost:* It is dependent on the number of garment panels to be cut.

- *Bundling cost:* It depends on order quantity, size of bundles, number of parts, etc. Cut order plan does not have any major impact on bundling.

15.1.1.1 Types of Fabric Losses

Analysing the above costs, the fabric cost is the most dominant one and it is the first priority of the cut order plan. In order to reduce the fabric wastage, it is necessary to classify the various kinds of fabric wastes that occur during garment manufacturing and then segregate them into essential and nonessential ones and, finally, develop ways to minimise or eliminate the wastes (Anon 1993; Rogale and Polanovi 1996). The types of fabric wastes are given below.

- *End loss:* It is an extra fabric left at the ends of a ply in a spread to ease cutting. The standard end loss per ply is 24 cm.
- *Fabric joint loss:* Fabric rolls are stitched jointly while going to the manufacturing processes. This results in fabric wastage of the areas having stitch holes or marks. It is known as fabric joint loss.
- *Edge loss:* The width of the marker is always a few centimetres lesser than the width of the fabric roll. This has been considered to accommodate the selvedge of fabric. The loss of fabric on the sides for selvedge accommodation is called edge loss.
- *Splicing loss:* Cutting of fabric crossways along the fabric width and overlapping fabric layers in between the two ends of a lay is known as splicing. The splicing operation could be used for adjustment of fabric fault, which has been observed during the spreading operation. The length of overlapped fabric during the splicing operation is known as splicing loss.
- *Remnant loss:* It is the remnant fabric left after the complete laying of a single fabric roll is thrown aside or used for part change. More precise fabric roll allocation methods are necessary to reduce the remnant loss.
- *Ticket length loss:* Normally, the actual fabric length and the length mentioned in the fabric roll will not be the same. The variation between these two lengths is known as ticket length loss.
- *Stickering loss:* Sometimes the patterns are cut a little extra for pattern marking and stickering and this area may get damaged because of glue or ink. It has to be cut off and is wasted.
- *Cutting edge loss:* It is a minor loss due to uneven and faulty cutting during fabric spreading and cutting. This is caused by faulty cutting methods or faulty cutting machinery.

15.1.1.1.1 Remedial Measures to Reduce Fabric Loss

15.1.1.1.1.1 End Loss

- *Standardisation of end loss in fabric:* The end loss of fabric for one particular spread/lay should be standardised and kept as a minimum value as per the requirement of the spread. The standard end loss set for straight knife cutting of garment panels is around 2 cm.
- *Minimising the number of plies in a lay:* To reduce the overall end loss of fabric, the number of plies in a lay should be reduced. The minimum number of plies required for a particular order could be determined by

$$\text{Minimum number of plies} = \frac{\text{Total Order Quantity}}{\text{Maximum number of pattern Pieces Allowed in one marker}}$$

15.1.1.1.1.2 Edge Loss

- *Markers in cuttable width:* The fabric loss at the width of the fabric could be lessened by preparing the markers whose width is equal to the minimum width (cuttable width) of the fabric.
- *Fabric grouping:* If more variation in width between as well as within the fabric roll is noticed, grouping of fabrics having similar widths could be tried to reduce the edge loss.

15.1.1.1.1.3 End Bits

- *Roll allocation:* Generally, the fabric rolls are picked at random and are spread during the spreading process, which leads to enormous amounts of fabric end bits after the completion of spreading of particular lays. Association of fabric rolls with lays should be done in a manner that minimum end bits are left.
- *Planning of fabric end bit:* Consideration of fabric end bits during the planning of the spread could minimise the end bits.

15.1.1.1.1.4 Ticket Length Loss

- *Complete inspection of fabric rolls:* This ensures no surprises on the cutting floor and effective fabric control.
- *Vendor management:* Fabric received in the factory should be tracked vendor-wise. Vendor-wise tracking enables the management to tab the vendors giving less fabric and make informed decisions for the future.

15.1.2 Fabric Saving Using a Cut Order Plan

On a lay, the fabric used is defined as shown in Figure 15.1.

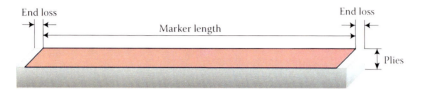

FIGURE 15.1
Fabric utilised in marker.

In order to minimise fabric waste, all three parameters such as marker length, end loss and plies have to be reduced for the overall order.

15.1.2.1 Marker Length

The higher efficiency marker normally gives better savings by reducing the fabric wastage. The vital thing is to concentrate on the overall order instead of individual markers. The reduction of marker length for the entire order could be calculated by the following parameters.

i. *Lay consumption:* It is a realistic method for assessing the overall consumption of fabric for the order as it considers the influence of all the markers over the order in terms of plies.

$$\text{Lay consumption} = \frac{\sum (\text{marker length} \times \text{plies})}{\text{Total Pieces cut}}$$

An accepted measure used in garment industries is marked consumption. However, it only gives information about the quality of markers and ignores their impact on the order.

$$\text{Marked consumption} = \frac{\sum \text{marker length}}{\text{Total Bodies marked}}$$

Example: An order XYZ has the following quantity as shown in Table 15.1

The cut plan for the above order was made as shown in Table 15.2.

TABLE 15.1

Order Quantity

XS	S	M	L	XL	XXL	Total
1	14	23	16	5	1	60

TABLE 15.2

Cut-Plan for the Order Quantity by Lay Consumption

	Plies	XS	S	M	L	XL	XXL	Marker Length	Marker Length × Plies	Marked Efficiency	Marker Efficiency × Pieces
Lay 1	10			1	1			2.5	25	77.32	1933
Lay 2	4		1		1			2.4	9.6	76.91	738.336
Lay 3	2	1		1				1.4	2.8	76.51	214.228
Lay 4	4		1					1.3	5.2	72.16	375.232
Lay 5	2			1	1			1.2	2.4	73.05	175.32
Lay 6	1					1		1.3	1.3	75.94	98.722
Total	23	1	14	23	16	5	1	10.1	46.3		3534.84
						Lay Consumption			1.08	Weighted Efficiency	76.35%

Marked consumption = Total length of marker/bodies marked

$$= 10.1/10 = 1.01 \text{ meters}$$

Lay consumption = (Total length of marker × number of fabric plies)/

Number of pieces cut

$$= 46.3/60 = 0.77 \text{ meters}$$

ii. *Weighted efficiency:* The quality of the marker could be assessed by the determination of the marker efficiency. This gives information about the efficiency of the markers over the whole order weighed according to its number of pieces.

$$\text{Weighted Efficiency} = \frac{\sum(\text{number of pieces in marker} \times \text{Efficiency})}{\text{Total pieces}}$$

An example of calculation of weighted marker efficiency for Table 15.2 is given below:

Weighted Efficiency = (Total marker efficiency of all the lay

× number of pieces)/Total Pieces

$$= 3534.84/60 = 76.35\%$$

Marker length can affect savings in the following ways:

- The marker should have the highest number of pieces allowed in it wherever possible. The more pattern pieces mixed in a marker normally gives higher marker efficiency and fabric savings.
- The marker should have a homogeneous mix of smaller and larger pattern sizes to get higher marker efficiency, which results in fabric savings.
- Use of long length markers provides less number of markers for a particular order leading to reduction of marker making cost.

15.1.2.2 End Loss

It is unavoidable that a little quantity of fabric is wasted at the ends while spreading every ply. This indispensable laying wastage is incorporated in the rating of fabric usage, since it is accepted that spreaders cannot lay-up precisely to the ends of the marker (Ng et al. 1998, 1999). A practical target for ends waste is 2 cm per end or 4 cm per ply. This could be kept within the limits by following the practices below.

- Fabric end loss could be minimised by use of lesser number of fabric plies in a spread.
- Better control on the production line could facilitate to minimise fabric end loss.
- Preparing longer spreads by combining the lays could reduce the fabric end loss.

15.1.2.3 Plies

This is a vital factor that should be considered to obtain the correct mix ratio of the order. For example,

Marker length = 25 m

Cuttable width = 2.5 m

Marker efficiency = 83%

Total ply area = 25 m $\times$ 2.5 m = 62.5 m^2

Fabric area utilised in garments = Marker efficiency $\times$ Total ply area
= 0.83 $\times$ 62.5 = 51.8 m^2

Wastage per ply = 62.5–51.8 = 10.6 m^2

This shows that 51.8 m^2 of the marker will be actually utilised for cutting garments and 10.6 m^2 will be wasted within the marker. It could be observed from the example that the wastage of the fabric increases with an increase in the number of fabric plies. Therefore, to minimise fabric wastage, the number of fabric plies should be as close to the ideal plies, which are nothing but the minimum number of fabric plies required for a complete order under the present constraints.

Example: An order XYZ has the following quantity as shown in Table 15.1. The marker maker can put a maximum of 3 pieces in a marker.

Ideal plies = Order quantity/Max pieces allowed in marker = 60/3 = 20.

Therefore, a cut plan with 20 plies should be the ideal solution.

It is feasible to generate a cut order plan with ideal plies for any order quantity. As the order quantity increases, it becomes tedious to get the ideal number of plies. The number of plies can influence fabric saving in various ways as given below:

- *Wastage in markers:* Fabric wastage in the spread could be reduced by laying the number of plies close to the ideal fabric plies as increase in each ply in the spread could lead to increase in fabric wastage.
- Further, each extra ply leads to increase in cost of spreading.

The characteristics of a good cut order plan are

1. It should utilise an ideal number of plies and only increase a few number of the plies due to remnant markers.
2. It should have the majority of markers as long markers with a maximum number of pieces allowed in a marker. Further, the markers should have a better blend of longer and smaller size patterns.

If the above two objectives are met, the cut plan will give a lower fabric consumption and higher weighted efficiency, minimising the particular order cost. Two cut plans could be compared based on number of fabric plies, consumption of lays and weighted efficiency to decide which one is a better solution.

15.2 Roll Allocation

Fabric roll allocation or roll planning is the critical part because it influences the fabric savings in the apparel industry. It is the process where the order in which rolls are to be utilised is programmed to minimise fabric wastage. During the spreading process, the variation in length of fabric between the fabric rolls could result in fabric wastage. As there are several combinations for the preparation of the fabric roll order for each lay, it is complicated to create a roll plan to reduce the fabric wastage during spreading (Knez 1994).

TABLE 15.3

Lay Information

	Plies	Lay Length (m)	Fabric Required (m)	Pieces in Lay	Pieces Cut in Lay
M1	10	10	100	7	70
M2	10	3	30	2	20

Note: Total fabric required = 130 m.

A better roll allocation shall minimise the fabric remnants, which gives higher fabric utilisation. End bits from the fabric could be produced when shorter lengths of fabric are left over after finishing the laying process. The remnants left over after cutting a spread should be less (Hui et al. 2000). The markers produced for remnant lays generally have less fabric utilisation than the normal production marker.

Example: An order XYZ has the following two lays as shown in Table 15.3. Total fabric required = 130 m.

Two rolls are allocated to the order with a total of 130 m of fabric. Roll 1 has a length of 68 m and Roll 2 has a length of 62 m.

15.2.1 Manual Roll Allocation Method

The normal practice in the garment industry is to randomly select a fabric roll and start spreading until it is exhausted and then select another roll and so on until the completion of the lay. Therefore, if the same process is done the result will be as shown in Table 15.4.

In the above case, we have only been able to lay 9 plies in the second lay, still 1 ply short and 3 m of end bits are left which cannot be used.

TABLE 15.4

Roll Allocation Using Manual Method

Lay 1	Lay Length = 10 (m)			
Roll No.	Roll Length (m)	Plies	Fabric Used (m)	Left (m)
R1	68	6	60	8
R2	62	4	40	22
Total	130	10	100	30

Lay 2	Lay Length = 3 (m)			
Roll No.	Roll Length (m)	Plies	Fabric Used (m)	Left (m)
R1	8	2	6	2
R2	22	7	21	1
	30	5	27	3

15.2.2 Automated Allocation Method

If a roll allocation logic process is utilised, then it could give the result as shown in Table 15.5.

The comparison of two methods are given in Table 15.6.

15.2.3 Important Consideration in Roll Allocation

The main aim of a better roll allocation method is to minimise the raw material (fabric) costs and manage all the practical aspects of an industry. The major concerns regarding the fabric roll allocation are given below:

1. Fabric cost
2. Roll length variation
3. Fabric defects and part change
4. Fabric shade, shrinkage and width variation
5. Spreading costs

TABLE 15.5

Roll Allocation Using Automated Method

Lay 1	Lay Length = 10 (m)			
Roll No.	Roll Length (m)	Plies	Fabric Used (m)	Left (m)
R1	68	5	50	18
R2	62	5	50	12
Total	130	10	100	30
Lay 2	**Lay Length = 3 (m)**			
Roll No.	Roll Length (m)	Plies	Fabric Used (m)	Left (m)
R1	18	6	18	0
R2	12	4	12	0
	30	10	27	0

TABLE 15.6

Comparison of Manual and Automated Method

Metric	Calculation	Case: 1 (Manual)	Case: 2 (Automated)
Fabric used		130	130
Fabric returned		0	0
Pieces cut		88	90
End bits left		3	0
Lay consumption	Total fabric in lay/pieces planned	1.444	1.444
Achieved consumption	Total fabric used/pieces cut	1.477	1.444
Increase in consumption	Achieved – lay consumption	0.033 (2.3%)	0
End bits	End bits left/total fabric	2.3%	0%

15.2.3.1 Fabric Cost

The key area of fabric wastage in roll allocation is the huge quantity of end bits that are left at the end of the order. This could be reduced by proper roll planning and practicing it on the production department.

15.2.3.2 Roll Variation

There could be roll to roll variation with respect to the fabric length. This means that if a roll ticket says it has 100 m of fabric it is possible that the actual length in the roll is a few metres higher or lower than what is stated. This could be due to the following reasons:

- Mill gives extra fabric to compensate for defects in the roll.
- Incorrect determination of roll length.
- Defects and damages being removed from the roll reducing the length.
- Swatches being cut or markings on the roll making that length shorter.
- Joins in the roll making it into two rolls instead of one.

A good roll allocation process should be able to admit all practical changes in the production floor and give a tailored plan according to the present state of fabric.

15.2.3.3 Fabric Defects

A better roll allocation process should manage practical problems related to fabric defects. It will not finish all end bits completely; there will always be a few pieces left over which could be used for part changes. If the fabric has a greater amount of fabric defects, then more end bits could be left for part change based on fabric, process and production requirements.

15.2.3.4 Fabric Shade, Shrinkage and Width Variation

Fabric is grouped in garment units based on their shade, shrinkage level and fabric width to ensure quality and minimise fabric wastage. Hence, a roll allocation system should stand by these practical considerations and also ensure fabric savings.

15.2.3.5 Spreading Costs

A roll allocation system should not raise the spreading cost and it could be reduced by

- Minimal roll changes while spreading
- Providing the spreader a plan for spreading based on roll number and plies saving time for decision making

- Prefilled lay slips provided to the spreader to reduce filling and calculation time

If a defect is identified in the fabric during the spreading process, the fabric is normally cut and removed. The fabric left in the lay is either kept separately or used with splicing.

15.2.4 Characteristics of Roll Allocation

A good roll allocation system should possess the following characteristics:

1. It should internally carry out all the possible options and generate the best option for roll allocation automatically.
2. It should create less end bits in fabric and leave fabric as much as possible in a utilisable roll form.
3. It should permit the user to do practical modifications on the production floor if required.

15.3 Fabric Grouping

Fabric grouping, otherwise called fabric batching, is the main process that will give improvement with respect to material utilisation along with improvement in quality control (Ambastha 2012). Here the fabric rolls are categorised to create fabric groups with identical properties with reference to fabric width, shade and shrinkage, where each of these considerations can have different implications on marker planning, cutting plan and fabric utilisation and have a critical influence on the overall material utilisation.

15.3.1 Fabric Grouping by Shrinkage

Every fabric roll is determined for its shrinkage level and could result in different warp and weft shrinkages. An adequate level of shrinkage is decided based on measurement tolerances and buyer guidelines. This acceptable level is used to generate fabric groups within a certain warp and weft shrinkage acceptance level. The fabric rolls of one group should not be mixed with any other group of rolls. Several garment industries carry out the process of dividing the order quantity in the ratio of shrinkage groups and treat each group as a separate order. This kind of approach is useful only if the quantity in each group is large.

15.3.2 Fabric Grouping by Width

Generally, the variation in width of the fabric will be noticed between the fabric rolls received by the garment industry. The markers should be made with the width equal to the lowest available width of the fabric if the fabric rolls are not grouped according to the fabric width, which leads to substantial fabric wastage. If the marker is created with the lowest width of the fabric in the lot and the patterns remain unaltered, then the fabric rolls having higher width could be mixed with fabrics having lower width. The fabric rolls are separated in different width groups and the major markers are cut with the relevant groups to get the maximum benefit of width. Once the remnant planning is started, these groups can be mixed to minimise the number of spreads and have a better utilisation of labour and time.

Width grouping will happen normally and disregarding this aspect creates possible wastage; for example, if the fabric rolls are received in the garment unit with two different fabric width groups and each has the same quantity of fabric.

Group 1: Fabric width = 1.48 m, total fabric length in a roll = 120 m

Group 2: Fabric width = 1.45 m, total fabric length in a roll = 120 m

- If the fabric rolls are not grouped based on width, then combined cuttable width = least width of the fabric in all fabric roll groups = 1.45 m
- Total cuttable area without fabric grouping = $(1.45 \times 100) + (1.45 \times 100) = 290$ m^2
- Total cuttable area with fabric grouping = $(1.48 \times 100) + (1.45 \times 100) = 293$ m^2
- Hence, fabric wastage because of nongrouping of fabric rolls = 293–290 = 3 m^2
- Fabric wastage (%) = $(3/290) \times 100 = 1.03\%$

15.3.3 Fabric Grouping by Shade

Fabric grouping based on fabric shade is generally the regular method of grouping. Most garment industries have received fabric in different shades. In this situation, there is no change in the pattern or the marker. Hence, these fabric rolls could be spread together in a lay. However, many garment industries prefer to separate them and utilise only one shade in a spread and at this point the method of shade grouping is used (Ambastha 2013b). To obtain the full advantages of shade segregation, the remnants and end bits should be marked with shade to make sure that part change occurs from the required shade only. If this operation is skipped and visual authentication is used for part change, then there is more of a chance of rejections and poor quality product.

15.3.4 Manual Grouping Approach

Normally, fabric grouping is done manually during the production planning operation. One main garment style of fabric grouping splits the total order

TABLE 15.7

Order Quantity as Per the Ratio of Roll Width

	S	M	L
Group 1	68	113	45
Group 2	22	37	15
Total	90	150	60

quantity in the grouped fabric quantity's ratio and then treats them separately. For example, the order quantity in S size is 90, M size is 150 and L size is 60. For instance, if the consumption of fabric is 1 m, then 300 m fabric should be procured. Now, consider that the fabric rolls are received in two width groups.

Group 1: Width of fabric = 1.48 m; fabric roll length = 200 m

Group 2: Width of fabric = 1.45 m; fabric roll length = 100 m

Group 1:Group 2 = 2:1. Therefore, after breaking the order quantity in the same ratio and allocating them to groups, it will be as shown in Table 15.7.

Now, if 3-way markers for the lays are put, then each group will have its own cut plan.

Also, say a lay can be put at a maximum of 80 plies. Therefore, the manual grouped cut plan is shown in Table 15.8.

15.3.4.1 Problems in Manual Grouping Approach

1. *Too many markers*: In the example shown above, both the fabric groups I and II had the same cut plan markers, hence they have to be created twice (total number of markers – 6).

TABLE 15.8

Manual Grouping of Cut Plan

		S	M	L		
Group 1 (1.48 m)		68	113	45	Plies	Quantity
	Marker 1	1	1	1	45	130
	Marker 2	1	2		23	50
	Marker 3		2		12	20
Group Total		68	113	45	80	200
Group 2 (1.45 m)		22	37	15		
	Marker 4	1	1	1	15	60
	Marker 5	1	2		8	30
	Marker 6		2		4	10
Group Total		22	37	15	27	100
Final Total		90	150	60	107	300

2. *Too many lays*: The number of markers to be made is doubled as well as the number of lays, which significantly increases the spreading time.

3. *Too many cuts*: By considering the same example above, the number of patterns or bodice cut is 16. The number of bodice cut will continue to increase with more groups, more markers and more number of lays.

4. *Dependency of fabric ratio*: The above explained example needs the fabric ratio to be preknown. This leads to a complex situation in the case of huge orders where the fabric is received in multiple drops and this leads to wastage of time as well as fabric.

15.3.5 Automated Grouping

This method compared to manual grouping should minimise the number of lays, markers, as well as bodice cuts. The same case when executed on automated software would give the solution as shown in Table 15.9.

15.3.5.1 Benefits in Automated Grouping Method

1. Less number of markers – Instead of 6 markers in the manual method, only 4 markers are adequate.

2. Less number of lays – Instead of 6 lays in the manual method, only 4 are sufficient.

3. Less number of bodices to cut – Only 12, as opposed to 16 in the manual case.

Suppose there is a shade-wise shipping requirement that could also be mentioned in the automated process and the system will ensure those criteria are followed.

TABLE 15.9

Automated Grouping of Cut Plan

		S	M	L		
Group 1 (1.46 m)		90	150	60	Plies	Quantity
	Marker 1	1	1	1	55	170
	Marker 2	1	2		25	30
Group Total		65	75	60	80	200
Group 2 (1.43 m)						
	Marker 3	1	2		15	60
	Marker 4		2		12	40
Group Total		25	75	0	27	100
Final Total		90	150	60	107	300

15.3.6 Characteristics of a Good Fabric Grouping

A good fabric grouping should

- Increase fabric utilisation, thus reducing overall consumption.
- Do not increase too many lays and markers as a lot of benefit gained by saving fabric will be lost in increased workload.
- Do not create too many groups as it will become difficult to manage and control by the factory workforce.
- Consider all scenarios for each case before deciding on the group range instead of following a standard set formula.
- Be easy and simple to understand and follow.

15.4 Performance Measurement Parameters in Cutting Section

Cutting is the vital process in garment manufacturing because it handles the costliest material resource, that is, the fabric and it is an irreversible process. Because of overemphasis on analysing the performance of the sewing section, there is a complete ignorance of analysing the performance of the cutting department including spreading and cut plan. This results in inefficiencies, leading to erosion of cost advantages (Ambastha 2013c). The few performance parameters to be analysed in the cutting section are discussed below.

15.4.1 Material Productivity

This provides value of output produced per unit of material used.

$$\text{Material Productivity} = \frac{\text{Output (value or unit or value added)}}{\text{Value of raw material used}}$$

This is an elementary re-examination of when, why and how raw materials are used. This evaluates how effectively or efficiently the material is utilised through the production system. Any material left in the fabric store is also a waste as it will be disposed off at a much cheaper rate.

Example: A garment industry bought 4500 m of fabric at a cost of Rs. 49/m for an order and produced 2500 garments and sold them at a cost of Rs. 310 each.

Therefore, material productivity is $(2500 \times 310)/(4500 \times 49) = 3.51$.

Therefore, the order generated Rs. 3.51 for every Rs.1 of material used.

15.4.2 Marker Efficiency

It is ratio between the fabric area used by the marker and the total fabric area. It is generally determined for each marker plan and should not be generalised for the entire garment order.

$$\text{Marker Efficiency} = \frac{\text{Area of marker used for garments}}{\text{Total area of marker}}$$

Marker efficiency around 80%–85% is considered good and varies based on the pattern shapes, constraints on pattern placements and fabric nature. This is a vital parameter to decide on the quality of the marker.

15.4.3 Marked Consumption

It is the consumption of a garment estimated as per the markers created by the design (CAD) section. In order to determine this parameter, the following procedures have to be followed.

- Cut order plan should be made stating the markers and number of fabric plies for each lay.
- All the markers should be made.
- Estimate the total length of fabric consumed in the lays.
- Divide this value by the total garments to be produced.

Example (Table 15.10):

Marker Lengths: S-1, M-1 = 3.8 m, M-2= 4.1 m
Total fabric needed = 3.8 × 100 + 4.1 × 50 = 585 m

Marked fabric consumption = 585/300 = 1.95 m. This value does not include wastages such as end loss or end bits.

15.4.4 Achieved Consumption

Actual fabric utilisation realised per garment after the completion of the entire garment production process is known as achieved consumption. This

TABLE 15.10

Example for Marked Consumption

Cut Order Plan				Order		
	Markers	S	M	S	M
Lay 1 = 100 plies	S-l, M-l	100	100	100	200
Lay 2 = 50 plies	M-2		100		

necessitates extensive estimation but the results will show a practical image of loss of material in the system.

$$\text{Achieved Consumption} = \frac{\text{Total fabric bought for the style}}{\text{Total garments shipped}}$$

The losses on raw material (fabric) incurred by the garment industry in terms of stock, end bits and cutting room wastages and rejection in stitching and finishing process as well as unshipped garments are incorporated in this calculation. If achieved consumption is only to be measured in the cutting section, then the formula should be slightly modified. Achieved consumption of the cutting department is determined by dividing the total fabric issued to the cutting department by the total cut garment panels issued to the sewing department.

Example: For the previous example, 640 m of fabric was finally bought (9.4% more than required) out of which 630 m was issued to the cutting department. The cutting room records show cutting of 315 garments. The factory finally shipped 290 garments. Twenty-five garments were rejected in the process.

Achieved consumption (factory level) = 640/290 = 2.21 m (11.8% wasted)

Achieved consumption (cutting room) = 630/315 = 2 m (2.5% wasted)

15.4.5 Fabric Utilisation

It is the ratio between the fabrics utilised on garments to fabric available to be used. This parameter gives information about the fabric utilisation status of the order.

$$\text{Fabric Utilization} = \frac{\text{Fabric used on garments}}{\text{Total available fabric}}$$

Total available fabric is nothing but the fabric allocated or procured for the particular order and the fabric utilised on the garment could be determined by the following methods.

a. By weight
  i. Weight of one garment in each size should be determined (the weight should be taken before sewing).
  ii. Multiply the weight of one garment with the number of garments cut in each size.
  iii. Divide the total weight by GSM and fabric width to obtain the total length of fabric (in meters) utilised in the garments.

TABLE 15.11

Marker Efficiency Calculation

	Markers	Marker Length (m)	Marker Efficiency (%)	S	M
Lay 1 = 100 plies	S-1, M-1	3.8	85	100	100
Lay 2 = 50 plies	M-2	4.1	80		100

b. By length

    i. The length of the marker should be multiplied with the marker efficiency of the particular marker plan and number of fabric plies spread in the marker.

    ii. The above estimation is carried out for each and every marker in the particular order and then the addition of all provides the total length of fabric (meters) used in the garments. The above will give fabric utilisation for the order. The formula can be extended to calculate overall fabric utilisation for the factory in a month.

Example: As per the previous example, total fabric available = 640 m. The fabric weight is 110 GSM and the fabric width is 1.07 m. The markers as mentioned in previous examples are shown in Table 15.11.

Fabric used in the garment:

a. By weight

    i. Weight of size 'S' garment = 187.1 g, size 'M' garment = 193 g

    ii. Total weight of order = $(100 \times 187.1) + (200 \times 193) = 57{,}310$ g

    iii. Total meters used on garments = $57{,}310/ (110 \times 1.07) = 487$ m

b. By length

    iv. Marker 1 = $3.8 \times 0.85 \times 100 = 323$ m; Marker 2 = $4.1 \times 0.80 \times 50 = 164$ m

    v. $323 + 164 = 487$ m

Fabric utilisation = $487/640 = 76.1\%$.

15.4.6 Cut Order Plan

A cut order plan is more efficient if it utilises the least number of plies and the least number of spreads while cutting an order. An inefficient cut order plan could lead to

a. Extra 4 to 6 cm fabric end loss on every increased number of fabric ply.

b. Smaller markers may have lower marker efficiencies.

c. More plies and lays could lead to increase in labour time for laying and cutting of fabrics.

d. More plies and lays could lead to higher fabric end bits and fabric wastages.

The least number of possible plies and lays for an order could be determined by the following formula.

$$\text{Least possible plies} = \frac{\text{Total order quantity}}{\text{Maximum pieces allowed in a marker}}$$

$$\text{Least possible lays} = \frac{\text{Total order quantity}}{\begin{array}{c}(\text{Maximum pieces in a marker}\\ \times \text{Maximum plies in a lay})\end{array}}$$

References

Aldrich, W. 2002. *Metric Pattern Cutting*. Blackwell Science, Oxford.

Ambastha, M. 2008. Cut order planning – The dot com way. *Stitch World* 17(7):22–4.

Ambastha, M. 2012. Fabric utilization – I cut order planning. *Stitch World* 22(6):12–5.

Ambastha, M. 2013a. 8 Fabric Losses Your Factory Faces Today. http://stitchdiary.com/8-fabric-losses-your-factory-faces-today (accessed on March 21, 2015).

Ambastha, M. 2013b. http://stitchdiary.com/8-ways-to-minimize-the-fabric-losses-in-your-factory (accessed on March 23, 2015).

Ambastha, M. 2013c. Are you wasting fabric in cut plan? http://stitchdiary.com/are-you-wasting-fabric-in-cutplan (accessed on March 21, 2015).

Anon. 1993. A system for made-to-measure garments by telmat informatique France. *Journal of SN International* 12(93):31–4.

Beazley, A. and T. Bond. 2003. *Computer-Aided Pattern Design and Product Development*. Blackwell Publishing, Oxford.

Burbidge, G.M. 1991. Production flow analysis for planning group technology. *Journal of Operation Management* 10(1):5–27.

Carr, H. and B. Latham. 2006. *The Technology of Clothing Manufacture*. Blackwell Science, Oxford, UK.

Chuter, A.J. 1995. *Introduction to Clothing Production Management*. Blackwell Scientific Publications, Oxford, UK.

Eberle, H., H. Hermeling., M. Hornberger, R. Kilgus, D. Menzer and W. Ring. 2002. *Clothing Technology*. Haan-Gruiten Verlag Europa-Lehrmittel Vollmer GmbH & Co, Nourney.

Fairhurst, C. 2008. *Advances in Apparel Production*. The Textile Institute, Woodhead Publication, Cambridge.

Glock, R.E. and G.I. Kunz. 2004. *Apparel Manufacturing – Sewn Product Analysis*. Prentice Hall, Englewood Cliffs, NJ.

Hui, C.L., S.F. Ng and C.C. Chan. 2000. A study of the roll planning of fabric spreading using genetic algorithms. *International Journal Clothing Science and Technology* 12(1):50–62.

Knez, B. 1994. *Construction Preparation in Garment Industry (in Croatian)*. Zagreb Faculty of Textile Technology, University of Zagreb, Croatia.

Mathews, M. 1986. *Practical Clothing Construction – Part 1 & 2*. Cosmic Press, Chennai.

Mehta, P.V. 1992. *An Introduction to Quality Control for Apparel Industry*. CRC Press, Boca Raton, FL.

Ng, S.F., C.L. Hui and G.A.V. Leaf. 1998. Fabric loss during spreading: A theoretical analysis and its implications. *Journal Textile Institute* 89(1):686–95.

Ng, S.F., C.L. Hui and G.A.V. Leaf. 1999. A mathematical model for predicting fabric loss during spreading. *International Journal Clothing Science and Technology* 11(2/3):76–83.

Rogale, D. and C.S. Polanovi. 1996. *Computerised System of Construction Preparation in Garment Industry (in Croatian)*. Zagreb Faculty of Textile Technology, University of Zagreb, Croatia.

Shaeffer, C. 2000. *Sewing for the Apparel Industry*. Woodhead Publication, Cambridge.

Solinger, J. 1988. *Apparel Manufacturing Handbook – Analysis Principles and Practice*. Columbia Boblin Media Corp, New York, USA.

Tyler, D.J. 1991. *Materials Management in Clothing Production*. BSP Professional Books, Oxford, UK.

16

Garment Production Systems

The garment production systems are a combination of production processes, materials handling, personnel and equipment that direct workflow and produce finished garments. It is a system that depicts how the two-dimensional fabric is transformed into a three-dimensional garment in a manufacturing system (Chuter 1995; Shaeffer 2000). The names of the production systems are based on the various factors like utilisation of a number of machines to assemble a garment, layout of machines, total number of operators involved to produce a garment and number of pieces moving in a production line during the production of a garment (Burbidge 1991).

Each garment production system needs a suitable management philosophy, materials handling procedures, plant layout for garments spreading and worker training. The garment industry could combine various production systems to achieve their specific garments' production needs like utilising only one production system or a combination of different systems for one product style (Babu 2006; Ramesh and Bahinipati 2011; Ahmad et al. 2012). The objectives of garment production systems are

- Examine the features of different kinds of garment production systems
- Compare and contrast the different production systems
- Assess and critically relate the merits and demerits of utilisation of different production systems in various circumstances

The most commonly used kinds of production systems in the garment industry are make through, modular production and assembly line production systems.

16.1 Make through System

It is the conventional method of production line where an operator assembles a single piece of garment at a time by carrying out all the sewing processes necessary to assemble a garment. After completion of assembling one garment, the operator will start assembling the next one and so on. In this system, an operator would be provided with a bundle of cut work pieces and would continue to assemble them based on his or her own method of work (Carr 1985; Burbidge 1991). This type of system is efficient when a huge category of garment styles has to be produced in very few quantities. The benefits of implementing the make through system are

- Quick throughput time
- Easy to supervise

The shortcomings of the make through system are

- Low productivity
- High labour cost
- It necessitates an experienced operator for assembling
- This system is limited to couture and sample making

16.1.1 Group System: Section or Process System

The group system is an improvement of the make through production system. In this system, an operator is specialised in one major component and assembles it from start to end. For example, if an operator is specialised in assembling the front panel, he or she would carry out the operations like assembling the front, setting the pockets, etc., and execute the entire operations essential to complete that particular component (Carr and Latham 2000; Ray 2014). With this type of garment production system, the sewing room should have a number of sections with flexible workers with sufficient skill to do all the required processes for the production of a specific style of garment (Kumar 2008). The sections are built according to the average garment produced, and include

- Preassembling (the preparation of small parts)
- Front making
- Back making
- Main assembling process like closing, setting collars and sleeves, etc.
- Lining making
- Setting linings
- Finishing operations like buttonholes, blind-stitching, etc.

16.1.1.1 Advantages

- The labour cost is lower and productivity is higher compared to the make through system as operators of various levels of skill and specialised machines are utilised in this system.

- This system is very efficient for producing a variety of styles in reasonable quantities.
- Automation and specialisation can be done.
- Breakdown of machines and absenteeism will not cause serious problems.

16.1.1.2 Disadvantages

- The garment quality should be strictly maintained as all levels of operators are involved in the work.
- Highly skilled operators are necessary to do simple operations within the section.
- The inventory cost is high due to high work-in-process (WIP), which is necessary in this system as a group of people are involved in each section.
- As the cut pieces are not bundled as in the case of a bundling system, there could be a chance for a lot of mix up, shade variation and sizes.

16.2 Whole Garment Production System

The two kinds of production systems that come under this class are complete and departmental production system. In the case of the whole garment production system, one operator assembles the entire garment from cutting the cloth to sewing as well as finishing the garment. The garment is ready for dispatch when the operator completes the final process. This type of production system is utilised in a very few circumstances, which are involved in custom wholesale. They are generally high cost and exclusively made for a specific customer (Schertel 1998). In the departmental whole garment production system, one operator does all the work with the particular machinery allocated to a department (Fairhurst 2008). For instance, one operator carries out all the cutting related jobs in the cutting section, and another person carries out the assembling jobs in the sewing department, and a third person does the pressing and packing work (Russell and Taylor 1999). Situations may arise where the operators could work on more than one piece of equipment to finish their respective job.

16.2.1 Advantages

- This kind of production system is more efficient while processing a huge variety of garment styles in small quantities.
- The operator could become an expert in his or her respective working area.

- As the wages or incentives are fixed based on the complexity of the job, the operators will try to complete the job without any problems.
- The inventory is minimised due to lesser WIP since one cut garment is given to one operator at a time.

16.2.2 Disadvantages

- Labour cost is higher due to the utilisation of highly skilled labourers for the particular job.
- As the wages are generally based on the number of garment pieces produced in a shift, the operator is more concerned with the number of pieces completed rather than the quality of the job.
- Lack of specialisation could lead to less productivity.
- This kind of production system is not suitable for processing bulk quantities of garment styles.

16.3 Assembly Line System

In this kind of garment production system, each operator is allocated to carry out only one job/operation repeatedly. Some of the characteristics of this system are

- The bundled cut garment pieces are moved successively from one job to another job.
- One bundle comprises all cut components that are necessary to finish the complete garment.
- Bundle tickets contain a master list of jobs for the particular garment style and corresponding coupons for each job.
- A ticket number will be allotted to each bundle which represents style, size and shade of the garment.

The two main types of assembly line production system followed in the industries are

- Progressive bundle systems
- Unit production system

16.3.1 Progressive Bundle System

In the progressive bundle production system (PBS), the bundles of cut garment pieces are moved from one process to another successively. This bundle production system is generally known as the conventional garments production system and is widely used in the apparel industry for many decades.

The PBS system of garment production comprises garment components necessary to complete a specific operation. For instance, pocket setting job in a bundle comprises a shirt front and pockets that are to be attached with garments. Bundle sizes could vary from 2 to 100 pieces. Some garment industries work with a standard bundle size of specific garment style, while most of the garment industries vary bundle sizes depending upon the cutting orders, fabric shade, size of the cut components in the particular bundle, and the specific job that is to be accomplished. The typical layout of the PBS is shown in Figure 16.1.

Bundles of cut pieces are carried to the sewing section and given to the operators scheduled to finish the garments' production operation. One operator will carry-out the same operation on all the cut garment components in the bundle and then retie the bundle and keep it separately until it is picked up and moved to the next job, which has to be carried out by another operator. This system may require a higher work in process (WIP) as the number of units in the bundles and the large buffer are required to ensure an uninterrupted work flow for all operators involved in assembling of garments (Solinger 1988; Tyler 1992).

This kind of production system could be used with a line layout based on the order that bundles are moved through the garment production line. Each garment style may have different processing requirements and thus different lines or routing. Routing identifies the basic operations, garment production sequence and the skill centres where garment operations are to be carried out (Mehta 1992; Periyasamy 2014). The main principles of this production system are

1. The various sections are kept based on the main process sequence and the layout of each section is dependent on the sequence of operations necessary to complete a particular component. For example,

FIGURE 16.1
Layout of PBS.

attachment of a sleeve could include a sequence of processes such as run stich of collar, collar ironing and top stitching of collar, etc.

2. A work store that is used to keep the completed work received from a previous job is located at the start and end of every section.

3. Because of these work stores, each section does not directly rely on the previous section. The PBS is somewhat cumbersome in operation and needs large quantities of WIP but is most likely one of the most stable systems as far as productivity is concerned (Mathews 1986).

4. Balancing and the changeover to new garment styles are also streamlined, owing to the amount of work held in reverse.

Advantages of PBS

- Labourers of all kinds of skill levels are involved in this system where the operations are split into small simple operations, which reduce the labour cost.
- The quality of assembled component will be better as every component is inspected during the assembling of each operation.
- The problem of lot mix-up, size and shade variation could be less as the cut garment pieces are moved in bundles from one operation to the next operation.
- Specialisation and rhythm of operation increase productivity.
- The higher WIP in this kind of system makes this a stable system where the production line will not be affected due to any breakdown of machines, absenteeism, etc.
- An efficient production and quality control system like time study, method study systems, operator training programmes and use of proper material handling equipment, etc., could be implemented.
- Tracking of bundles is possible.

Disadvantages of PBS

- Balancing the production line is tough and could be managed by an efficient production supervisor and IE engineer.
- Proper upholding of equipment and machinery is necessary.
- Proper planning is needed for every garment style and batch.
- Improper planning could lead to poor quality, lower productivity and labour turnover.
- Inventory cost will be high due to higher WIP in each.
- PBS is not effective for processing small quantities of orders and a variety of styles.

- Shuttle operators and utility operators are required in each batch for effective line balancing.

16.3.1.1 Straight Line or 'Synchro' Production System

This kind of production system is based on a harmonised or synchronised flow of work in each stage of the garment production process. Synchronisation of time is a critical factor of this system as synchronisation of workflow could not be done if more variations in standard time (SAM) of particular operations are present.

Assume if one operation has a SAM of 1.7 minutes, then all the other operations in the particular production line should have the same, or closer SAM value. Balancing of standard time for each operator could result in irrational combinations of whole or part operations which could minimise the efficiency of individual operators. The layout of a synchro-system for the manufacture of a full sleeve shirt is shown in Figure 16.2.

16.3.1.2 PBS Synchro Straight Line System

The PBS synchro system is not a flexible system and liable to frequent breakdown of machines and more absenteeism. The standby machines and operators should be made available to avoid bottleneck processes every time. Further, this system needs an adequate quantity of similar styles of garments for continuous operation of the line.

16.3.2 Unit Production Systems (UPS)

Though this production system has been in use for several years, major progress was made when computers were utilised for production planning, production controlling and regulating the work flow in the production line. The important features of this production system are

- It is mainly concerned with a single garment and not bundles.
- As per the predetermined sequence of processes, the cut garment pieces are transported automatically from one work station to another.
- The work stations are so constructed such that the cut components are accessible as near as possible to the operator to reduce the time taken for taking the component and positions the same for sewing.

The operational principles of a unit production system (UPS) are as follows:

- All the cut panels for one garment are loaded into a carrier at a workstation specially designed for this specific job. The carrier is divided into several sections, each having a quick-release clamp-like system to avoid falling out of cut panels during transportation through the system.

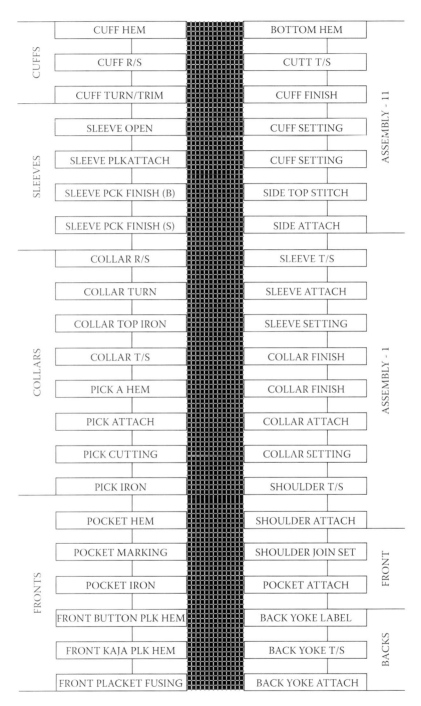

FIGURE 16.2
Layout for full sleeve shirt – batch system.

- When a particular batch of cut panels of the garments has been fixed into carriers, they are fed past an electronic device. This device counts and records the number of the carrier and addresses it to its first destination.

- The loaded carriers are then moved onto the main powered line, which is circulated between the rows of machines continually. Each workstation is connected to the mainline by means of junctions, which open automatically if the work on a particular carrier is addressed to that particular work station.

- The carrier is moving toward the left side of the operator and waits along with the other carriers in the work station. When the operator has completed the particular work on one carrier, he or she has to press a push button, which is positioned at the side of the sewing machine, to activate a mechanism that transports the carrier back to the main line so that another carrier will be fed automatically to take its place.

- A data collection system records when the carrier left the station and then it is addressed to its next destination.

The work station and carrier arrangement in the UPS system is shown in Figure 16.3. Normally, upright rails are interconnected to a computerised control system that routes and tracks production and provides the real time data related to the production. The automatic control of the work flow sorts work and balances the line.

Operation starts at a staging area in the sewing section of garments. Cut panels for one unit of a single garment style are grouped and loaded directly from the staging section to a hanging carrier. Loading is planned carefully; hence, minimum material handling is required to deliver garment panels accurately in the order and manner that they will be sewn. Various sizes and kinds of hanging carriers are available for different styles of garments. This production system avoids unnecessary handling of bundles as well as garment panels. The layout of machines in the UPS line is shown in Figure 16.4.

The integrated computer system present in the UPS monitors the work of each operator which eliminates the bundle tickets and operator coupons as

FIGURE 16.3
Work station and carrier system in UPS.

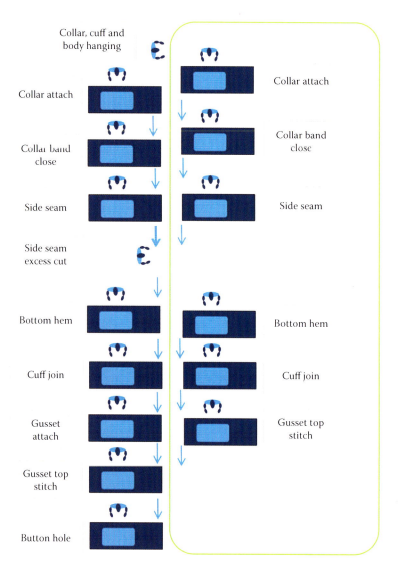

FIGURE 16.4
Machine layout for UPS line.

in the case of PBS. Individual bar codes are rooted in the carriers and read by a bar code scanner at each workstation and control points in garment units (Mok et al. 2013). Any additional information like style number, colour shade and lot which are required for sorting and processing could also be included.

These production systems have on-line terminals situated at every work station to gather data on each operation. Each operator could advance finished units, reroute units that require repair or processing to another station of garments. The terminals at each station facilitate a central control centre to track each unit at any given time and afford management with information to make instant decisions on routing and scheduling. The operators of the unit production system control centre can decide sequences of orders and colours to maintain operators supplied with work and to minimise change in machinery, operations and thread colours frequently (Sarkar 2011). It can control multiple routes and concurrent production of multiple styles of garments without streamlining production lines in garments.

The control centre may execute routing and balancing of work flow, which minimises the bottlenecks and stoppages. Information related to the working time of operators, time expended on each individual unit, number of units finished and the piece rate earned for each unit in garments could be collected (Gershenson 2007; Glock and Kunz 2004). It could estimate the earnings per hour, per day and the efficiency rate of each operator. This production system needs considerable investments. A comparison of the PBS and UPS production systems is given in Table 16.1 (Sarkar 2012).

Advantages
- Bundle handling completely eliminated.
- The time taken during the garment component pick-up and disposal is reduced to a minimum.
- Manual registering of work after the completion is eliminated as the output is automatically recorded in this system.
- The computerised systems automatically balance the work between stations.
- Up to 40 garment styles can be produced concurrently on one system.

Disadvantages
- UPS requires high investments.
- The payback period of the investment takes a long time.
- Effective production planning is required.

16.4 Modular Production System

The modular production system was first executed at Toyota in 1978 as part of Just-in-Time, which is called the Toyota Sewing System (TSS). It works on the principle of pull-type production systems, in which the job order arrives from the last step to previous steps (Subiron and Rosado 1995). Since the amount of work in the process is very low, smooth working when

TABLE 16.1

Comparison of PBS and UPS Production Systems

Parameters	Progressive Bundle System (PBS)	Unit Production System (UPS)
Movement of cut garment panels through the production line	• The bundled cut garment pieces have to be transported manually to the sewing department. • Continuity of the operator work is hindered as the operators have to break their work to get bundles. • Hence, it is less effective in production management aspect.	• The cut components are carried to each work station automatically. • It provides quick response time due to easy pick up and dispose at each work station without hindrance to operator workflow.
Production rate/throughput time	• Production rate is lesser/throughput time is longer in PBS compared to UPS. The through put time in PBS depends on the size of the bundle and the number of bundles kept inbetween two operators.	• Production rate is comparatively higher in UPS but not much higher since it has some WIP inbetween two operators.
Number of operators required	• The number of direct operators/labour involved in this system is high as typically the operators carryout tying and untying of bundles, positioning cut components in the bundle, drawing the bundle ticket and handling of work pieces.	• Here the number of labourers is less since the operator has only one job of assembling the garment panels. The secondary tasks that have to be carried out by the operator in case of PBS are avoided as the garment panels are kept in the overhead hanger and hence handling of assembled panels is avoided.
Level of WIP in production line	• The WIP in this system is quite high compared to UPS as the operators sew as many pieces as possible without considering the other operators. This leads to stacking up of unfinished garment components.	• UPS system involves less WIP inbetween operations since the automatic workstation has a limited number of hangers to hold the cut components.
Production requirements in the cutting section	• In the case of PBS, the productivity of the cutting department should be 60%–70% higher than the requirement of the sewing section due to the higher WIP required by the sewing section.	• In UPS, because of the need of WIP in the sewing section, a balanced flow of material between cutting and sewing section could be established without any load on the cutting department.
Inventory level	• Good amount of stock of fabrics and trims are needed due to high WIP in PBS.	• Minimum inventory is needed for fabric and trims in UPS due to low WIP.
Additional requirement of operators	• Necessitates more operators who can work overtime for repair work owing to some unfinished operations.	• UPS system requires less overtime workers as planning is easy in this system.

no inventory is possible. Some of the definitions of this production system are given below:

- A structured group of individuals are working together in a supportive way to complete a common goal.
- A group of people achieve their individual goals efficiently and effectively at the same time as achieving their team as well as organisational goals.
- A work team generally has a group of operators having corresponding skills who are devoted to achieve the set of performance goals.

This production system is a controlled, manageable work unit that comprises a powerful work team, machinery and work to be performed. The need of the industry, size of the industry and product line in garments decides the number of teams in a plant (Rogers 1990). As long as there is an order for the particular style of garment, the work teams could have a function but the achievement of this type of garments process is in the flexibility of being able to manufacture various kinds of garment styles in small quantities.

Several names are presently used to categorise modular garments production systems, like modular garments manufacturing, cellular garments manufacturing, flexible work groups and TSS in garments. The basic principle is alike among these production systems, although the industry and execution may differ. Based on the product mix, the number of operators on a team varies between 4 and 15. The basic principle is to find out the average number of operations required for production of a particular style being produced and divide by three. Team members are cross-trained and exchangeable among tasks within the group in this type of system.

16.4.1 Work Flow in a Modular System

It works on the principle of a pull system, with demands work coming from the next operator in a production line to stitch the particular garment. In this production system, work flow is continuous, produces minimum wastage and does not wait ahead of each process (Sudarshan and Rao 2013). This improves the flexibility of styles and quantities of products that can be produced in this system. The layout of equipment in a modular production system is shown in Figure 16.5.

Work teams in this system generally work as 'stand-up' or 'sit-down' units. Depending upon the sequence of operations involved in the production of a particular style of garment and the time required for completion of each operation, a module could be divided into many work zones. A work zone comprises a group of sequential garment operations. The operators are trained to execute the operations in their respective work zone and contiguous operations in adjoining work zones; thus, they can move freely from one operation to another operation as the garment progresses (Colovic 2013).

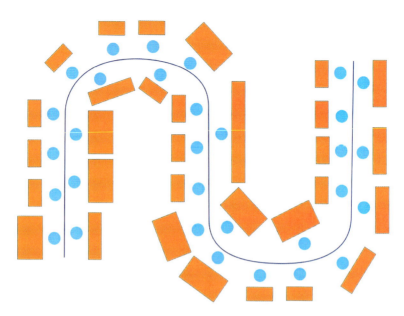

FIGURE 16.5
Layout of modular production system.

In the case of a module, the workflow could be based on a single-piece hand-off, Kanban or TSS. If a single-piece hand-off method is utilised, then the machines should be arranged in a very tight pattern. After completion of each operation, the part has to be handed to the next operator immediately for processing. Certain modules could operate with a small bundle of pieces of work or a buffer between the operators. If a small bundle is used, an operator should finish the operation on the whole bundle and take the bundle to the next operation.

A Kanban system utilises a selected work space between operations to balance supply with demand. The chosen space will preserve a limited number of completed components in the production line for the next process. If the chosen space is filled, there is no necessity to assemble the garment panels until it is needed and this confines the manufacture of product ahead of the next operation.

TSS (bump-back system) is an improvised section with flexible work regions and cross-trained operators, who have been trained in up to four different successive operations. This facilitates operators to move from one process to another process until the next operator is ready to start work on the garment. The operator who has been released from the garment production operation will then go back to the starting point of the work zone and start the work on another new garment. This system commonly uses a 4-to-1 ratio of machines to operators (Abernathy and Dunlop 1999).

An incentive reward for the team is based on group pay and additional benefits for meeting team goals for productivity as well as quality. Individual incentive rewards are not suitable for a team-based garment production system like this. Teams could perform all the operations or a certain segment of the sewing operations based on the organisation of the module and operations required (Tyler 1992).

With this team-based production system, the operators are given the responsibility for doing their module to meet the goals of their team with respect to productivity and quality. The team is accountable for keeping a smooth work flow, meeting production goals and maintaining a required quality.

16.4.2 Features of a Modular Production System

- Unconstrained to work with the operators.
- In one workstation, the workers/operators should be able to perform the operations in different sewing machines in a highly skilled manner.
- In this modular production system, in-line inspection locations are constructed into the production line so that the inspector could be able to return the defective garment panel to the concerned operator through the system.
- Productivity is high in this production system since the operator handles the garment only once for several operations, instead of handling it for every operation.
- Only a few garments will be processed in the particular production line as the throughput time is less in this system.
- A modular production system module could have up to eight work stations positioned around the transport system.

16.4.3 Advantages of a Modular Garment Production System

- High flexibility
- Fast throughput times
- Low wastages
- Reduced absenteeism
- Reduced repetitive motion ailments
- Operator ownership of the production process is high
- Empowered employees
- Improved quality of product

16.4.4 Disadvantages of a Modular Garments Production System

- A high capital investment in equipment
- High investment in initial training
- High cost incurred in continued training

16.5 Evaluation of Garment Production Systems

The evaluation of garment production systems can be done by taking into consideration four primary factors such as

1. *Processing time:* It is the total working time of all the processes involved in assembling a garment.
2. *Transportation time:* It is the total time consumed for movement of semifinished or finished garments from one workstation or department to another.
3. *Waiting time of unfinished garments:* It is the idle time of a work bundle when it waits for the next operation.
4. *Inspection time:* It is time taken for in-process inspection of semifinished garments or final inspection of finished garments before packing.

The main goal of all the production systems is to decrease the total production time which leads to reduction in inventory cost. The appropriate selection of a suitable garment production system for an industry is influenced by the product style and policies of the industry and on the labour capacity. The cost of inventory decides the choice of a production system in most circumstances in an apparel industry (Hanthiringe and Liyanage 2009). When material, labour, space and interest costs are high, a synchronised subassembly system, which gives the minimum possible in-process inventory, is more suitable.

References

Abernathy, F.H. and J.T. Dunlop. 1999. *A Stitch in Time – Apparel Industry.* Blackwell Scientific Publications, Oxford, UK.

Ahmad, S., A.A.B. Khalil and C.A.A. Rashed. 2012. Impact efficiency in apparel supply chain. *Asian Journal Natural Applied Science* 1(4):36–45.

Babu, V.R. 2006. Garment production systems: An overview. *Indian Textile Journal* 117(1):71–8.

Burbidge, G.M. 1991. Production flow analysis for planning group technology. *Journal of Operation Management* 10(1):5–27.

Carr, H. and B. Latham. 2000. *The Technology of Clothing Manufacturing.* Blackwell Scientific Publications, Oxford.

Carr, H.C. 1985. *Production Planning and Organization in Apparel Manufacture.* Clothing Institute, London.

Chuter, A.J. 1995. *Introduction to Clothing Production Management.* Blackwell Scientific Publications, Oxford, UK.

Colovic, G. 2013. Management in the garment industry – Modular Production System. http://gordanacolovic.blogspot.in/2013/10/modular-production-system.html (accessed on March 23 2015).

Fairhurst, C. 2008. *Advances in Apparel Production*. The Textile Institute, Woodhead Publication, Cambridge.

Gershenson, C. 2007. *Design and Control of Self-Organizing Systems*. Coplt ArXives, New York, USA.

Glock, R.E. and G.I. Kunz. 2004. *Apparel Manufacturing – Sewn Product Analysis*. Prentice Hall, Englewood Cliffs, NJ.

Hanthiringe, G. and K. Liyanage. 2009. Simulation based approach to evaluate modular manufacturing system in the apparel industry. *Pakistan Textile Journal* 23:22–5.

Kumar, A. 2008. Production Planning and Control: Lesson 8 Course material Delhi University. www.du.ac.in/fileadmin/DU/Academics/course_material/EP_08. pdf (accessed on March 14, 2015).

Mathews, M. 1986. *Practical Clothing Construction – Part 1 & 2*. Cosmic Press, Chennai.

Mehta, P.V. 1992. *An Introduction to Quality Control for Apparel Industry*. CRC Press, Boca Raton, FL.

Mok, P.Y., T.Y. Cheung, W.K. Wong, S.Y.S. Leung and J.T. Fan. 2013. Intelligent production planning for complex garment manufacturing. *Journal of Intel Manufacturing* 24(1):133–45.

Periyasamy, A.P. 2014. Garment production systems: An overview. http://textiletutors. blogspot.in/2014/09/garment-production-systems-overview.html (accessed on October 27, 2014).

Ramesh, A. and B.K. Bahinipati. 2011. The Indian apparel industry: A critical review of supply chains In: *International Conference on Operations and Quantitative Management (ICOQM) Nashik*, India.

Ray, L. 2014. Production Planning for Garment Manufacturing. http:// smallbusiness.chron.com/production-planning-garment-manufacturing-80975. html (accessed on October 5, 2014).

Rogers, G.G. 1990. Modular Production Systems: A Control Scheme for Actuators. PhD Dissertation Loughborough University Loughborough, UK.

Romero-Subiron, F. and P. Rosado. 1995. The design of a line control system for the modular furniture industry. *International Journal of Production Research* 33(7):1953–72.

Russell, R.S. and B.W. Taylor. 1999. *Operations Management*. Prentice Hall, Upper Saddle River, NJ.

Sarkar, P. 2011. Functions of Production Planning and Control (PPC) Department in Apparel Manufacturing. http://www.onlineclothingstudy.com/2011/12/ functions-of-production-planning-and.html (accessed October 5, 2014).

Sarkar, P. 2012. Comparison between Progressive Bundle System and Unit Production System (UPS). http://www.onlineclothingstudy.com/2011/02/comparison-between-progressive-bundle.html (accessed on April 22 2015).

Schertel, S. 1998. New Product Development: Planning and Scheduling of the Merchandising Calendar (Master Dissertation). North Carolina State University, Raleigh, NC.

Shaeffer, C. 2000. *Sewing for the Apparel Industry*. Woodhead Publication, Cambridge.

Solinger, J. 1988. *Apparel Manufacturing Handbook – Analysis Principles and Practice*. Columbia Boblin Media Corp, New York, USA.

Sudarshan, B. and N.D. Rao. 2013. Application of modular manufacturing system garment industries. *International Journal Science Engineering Research* 4(2):2083–9.

Tyler, D.J. 1992. *Materials Management in Clothing Production*. Blackwell Scientific Publications, Oxford, UK.

17

Flow Process Grid

The production department has three main groups, namely, cutting, sewing and packing. These departments come under the manufacturing section, each section having section in-charges. For example, a pattern master is an in-charge for the cutting unit. They monitor placement of patterns on the fabric lots and cutting the garment parts in an efficient manner. Line supervisors are the in-charges for the sewing section. General maintenance also comes under the production department. Service engineers are the in-charges in this department. They are servicing or repairing sewing machines and also cutting machines.

17.1 Flow Process Grids and Charts

Most production managers, engineers and manufacturers are familiar with the use of the flow process chart as a tool for designing production systems and plant layouts. However, one's ability to use any tool efficiently will vary with the design principles on which the tool is built. A study of flow process charts in various textbooks and technical magazine articles showed that the flow process chart, used in most places, is actually an inadequate tool for production planning purposes (Chuter 1995; Glock and Kunz 2004). These flow process charts are inadequate because they are diagrams without any time or space scales. Any production blueprint, diagram, or chart must be based on these principles – time and space relationships – if it is to be a worthy engineering tool for calculating production systems and plant layouts. The core of production system efficiency is time, whereas the core of plant layout efficiency is time value based on space relationship plus space values (McBride 2003).

If a flow process chart is to be an effective planning device, it should be created with mathematical graph concepts using the grid formation of a y-axis, the ordinate and an x-axis, the abscissa. The y-axis could be the timeline of the apparel production system and plant layout (Solinger 1988; Mannan and Ferdousi 2007). This timeline represents the time relationships that exist between the work and temporary storage stations in the production flow. The x-axis represents the lateral space relationships among the work and temporary storage stations. The work flows from the base of the graph, the first time level, to the top of the graph, the final time level.

The total production time is equal to the sum of the y time levels. Each time level is equal to the sum of the y time levels. Each time level is equal to the time required to produce a required amount of product units. The production equipment and workers per work station on the graph will be equal to that required to yield the required amount per unit time level. Such a flow process chart, one with a graph grid structure containing ordinate and abscissa values, will be referred to here as a 'flow process grid' (Solinger 1988; Russell and Taylor 1999).

17.1.1 Differences between a Flow Process Grid and a Flow Process Chart

The flow process grid is a dimensional graph of the whole production process of a garment, which measures and depicts the distinct time and space interaction of all factors in the production process such as process stations, inspection stations, temporary storage stations and transport activities, necessary to dispatch a given amount of garments in an intended time and space. The flow process chart is merely a diagram of production sequence without regard to the time and space relationships in the sequence (Kumar 2008; Nayak and Padhye 2015).

The format of the flow process grid must possess the following factors: (1) the spatial relationship necessary between work stations for the best plant layout and (2) the time relationships required among work stations necessary to yield minimum total production time. In order to illustrate these relationships, it is imperative that the grid contains a phrase, word, or symbol for every process inspection station and transportation and storage stations in the production sequence.

Symbols are used on various flow process charts to prefix reports identifying each stage of the apparel production system. This allows one to sort quickly the category of each stage of production. If flow process charts were built on the grid concept, then one could easily identify high rations of transportation and storage in relation to processing. This would enable one to assess a production system very quickly and accurately (Ramesh and Bahinipati 2011). It could also highlight the measures necessary to change the system in order to improve the production efficiency. The common process flow symbols used are given in Table 17.1.

17.2 Construction of Flow Process Grids

Since the largest percentage of production labor in apparel manufacturing is most often engaged in the sewing department and since this department usually has the greatest allover production problems of coordinating

TABLE 17.1

Process Flow Symbols

Symbol	Description
	Operation/Process: Any operation for making, altering or changing the job is said to be an operation.
	Decision: Represents a decision making point.
	Transport: Process/material flow or movement. Movement or travel of the job.
	Storage: Keeping, holding and storing the job and other things.
	Breakdown, interface or time required for some adjustments. A temporary halt in the process.
	Inspection: Checking of the quality and quantity.

production between work stations, the principles for making a flow process grid will be developed and illustrated by make of a flow process grid for the sewing production of a simple garment, a men's T-shirt (Solinger 1988).

Step 1: List all the sewing operations necessary to produce the garment.

Step 2: Group the operations in levels according to the numerical order in which these operations may be performed to give the quality specifications for the garment. For example, assume that the T-shirt is going to be produced (Table 17.2 seam grids) and the quality specification for seam sequence for the sewing operations of this particular T-shirt is given in Table 17.3 (operations are listed alphabetically).

From these quality specifications, it could be noted that the side seam and underarm seams may be combined into one operation if this is quantitatively better than making it as listed, two separate operations. This permits us to list the operations on the following possible levels of operational sequence:

- Side seams, shoulder seam, collar seam, underarm seam
- Neck seam, sleeve hem, hip hem
- Covering stitch
- Armhole seam

TABLE 17.2

Seam Sequence for T-Shirt

Sewing Grid for Level Listing I			Sewing Grid for Level Listing II		
Level	Line 1	Line 2	Level	Line 1	Line 2
Y7	Sleeve hems		Y7		
Y6	Hip hems ↑		Y6	Arm hole seam	
Y5	Side – underarm seam ↑ ↑		Y5	Hip seam ↑ ↖	Sleeve hem
Y4	Arm hole seam ↑		Y4	Side seam ↑ ↑	Sleeve seam
Y3	Covering stitch ↑ ↑ ↖		Y3	Covering seam ↑ ↑	
Y2	Neck seam		Y2	Neck seam ↑ ↖	
Y1	Shoulder seam	Collar seam	Y1	Shoulder seam	Collar seam

On examining this arrangement of four levels, it could be observed that the side and shoulder seams cannot be made simultaneously at two different work stations because the same parts are needed for both operations. Further, both hems (sleeves and hip) may be made at any time after the second level, third, fourth, fifth, etc. The shoulder seam and collar seams must be made on the first level because the neck seam cannot be made until these two operations are sewed (Glock and Kunz 2004). This means that the side seams will be made at the fourth level or one of the successive levels if the armhole seam remains on the fourth level. The crucial question that must be considered is whether the armhole seam should be made before or after the side and underarm seams. If the armhole seam is made at the fourth level, the side and underarm seams will become one operation, side-underarm seams and it will be on the fifth level. The hems will then follow on the sixth and seventh levels.

TABLE 17.3

Seam Sequence for Sewing Operations

Sl. No.	Operation	Federal Specification	Quality Sequence
1	Armhole (2)	SSa-I-504	Catch the covering switch
2	Back neck and shoulder covering stitch (1)	SS-h02 -406	
3	Collar seam (1)	SSa-I-504	
4	Hip seam (1)	Efc-I-503	Cross the side seam
5	Neck seam (I)	SSa- 1-5-4	Catch both ends of collar seam
6	Shoulder seam (2)	SSa-I-504	
7	Side seam (2)	SSa-1-5-4	
8	Sleeve hem	EFc-1-503	Cross the underarm seam
9	Underarm seam (2)	SSa-I-504	

If the armhole seams are to be made after the side and underarm seams, the side and the underarm seams will usually be made on the fifth level, the armhole seam must be placed on the sixth level, and vice versa (Oliver et al. 1994; Mok et al. 2013). Whether it is best to make the armhole seam on the fourth, sixth, seventh or eighth level (it is possible to make the hip hem any time after the side seam is made) will depend on the following factors:

- The contour and size of (a) the sleeve cap and (b) the armhole
- The size of the whole sleeve and the body
- The working characteristics of the fabric
- The available sewing equipment and auxiliaries
- The ability and capacity of the operator

If the available information does not permit one to make a definite decision with a high degree of confidence, the next step is to make methods studies of both sequences in order to come to a definite decision as to the best operational sequence (Ray 2014). The factors that must be evaluated in order to determine the best sequence of operations for any product are

- The length of the seam or stitching
- The contours of the sewing line and the edges of the part or parts being sewed
- The area dimensions or size of the parts being sewed
- The bulk or space dimensions of the parts being sewed
- The working characteristics of the fabric(s)
- The available equipment: the sewing machine bed type and the auxiliary tabling
- The ability and capacity of the operator

For example, two different shirts may have armhole sleeves whose lengths are both equal, 22″, but sleeve A may have a 7″ high sleeve cap and sleeve B may have a 3″ cap. Sleeve cap A will have more curvature (greater contour) than cap B. The size area or dimensions of the T-shirt sleeve are the same regardless of whether the armhole seam is made before or after the side and underarm seams, but the bulk or space factors change with the operational sequence. The space dimensions of the open sleeve without the underarm seam are different from the space dimensions of the closed sleeve with the underarm seam (Schertel 1998). The same is true for the body of the shirt before and after the side seams are made. The bulk factor (the space dimensions) of the operation just as much as the size of the area dimension of an operation, or the contour of the sewing edges, will help or hinder. What may be an advantageous bulk shape for one operation may be a decidedly disadvantageous bulk shape for a different operation, although the surface area and weight of shape is alike for both situations. If the evaluation of the factors shows that it is best to place the armhole on the fourth level, the final level sequence for the T-shirt is as follows:

Level listing I:

1. Shoulder seam, collar seam
2. Neck seam
3. Covering stitch
4. Armhole seam
5. Side-underarm seam
6. Sleeve hems
7. Hip hems or vice versa

Should the evaluation show that it is more economical to make the underarm and side seams before the armhole seams, then the level breakdown may be either one of the variations of the following sequence depending on which variation the evaluation favours:

Level listing II:

1. Shoulder seam, collar seam
2. Neck seams
3. Covering stitch
4. Side seams (or, if quality specifications permit, it may also be made after the hip hem), sleeve seams
5. Hip hem, sleeve hem
6. Armhole seam

17.3 Operation Breakdown

The work in each style is broken down into operations. An operation is one of the processes that must be completed in converting materials into finished garments. An operation breakdown is a sequential list of all operations involved in assembling a garment, component or style (Solinger 1988; Sarkar 2011).

17.3.1 Benefit of Breakdown

1. Could notice all operations of the garment at a time.
2. Could expect the difficulties of doing a crucial operation.
3. Could make layout in an easy, simple and less time-consuming way.

4. Could calculate the SMV for target setting and equal time distribution to the operator during layout.

5. Easy to select appropriate operator for the specific process.

6. Could know the quantity and kind of machine required to produce the required garment.

7. Could achieve the production target within a very short period.

8. Could be aware about quality to meet the buyers standard.

9. Could know about additional guide, folder and attachment.

17.3.2 Calculation of Operation Breakdown

- Target per hour $= \dfrac{\text{Worker} \times \text{Working hour} \times 60}{\text{SMV}} \times \text{Effciency } \%$

- Basic Pitch Time (BPT) $= \dfrac{\text{SMV}}{\text{Total Manpower}}$

- Upper Control Limit (UCL) $= \dfrac{\text{BPT}}{\text{Wanted Organizational Efficiency (0.85)}}$

- Lower Control Limit (LCL) $= \text{BPT} \times 2 - \text{UCL}$

17.3.3 Operation Breakdown and SAM of the Full Sleeve Formal Men's Shirt

The operation breakdown for the men's shirt is shown in Table 17.4 (Sarkar 2012a,b).

17.3.4 Operation Breakdown and SMV of a Trouser

The operation breakdown for the men's trouser is shown in Table 17.5.

17.3.5 Operation Breakdown and SMVs of a Jacket

The operation breakdown for the men's jacket is shown in Table 17.6.

17.4 Control Forms in Production Department

In the process of quality control, the control at various levels of production of garments is monitored by various control forms. This helps in maintaining continuity in the quality control. Further monitoring of the production process will be controlled by documentation (Mehta 1992). The stage-by-stage documentation helps in not only achieving the expected quality but also in completion of production within the target time.

TABLE 17.4

Operation Breakdown of Men's Shirt

Sl. No.	Operations	SMV	Machine Type
	Collar		
1	Collar pieces joint 2	0.60	SNLS
2	Collar run stitch	0.42	SNLS
3	Collar trim and turn	0.25	SNEC
4	Collar top stitch	0.47	SNLS
5	Band hem	0.34	SNLS
6	Band set	0.65	SNLS
7	Collar label attachment	0.24	SNLS
8	Band top	0.45	SNLS
9	Band and collar ready trim	0.41	SNEC
	Cuff preparation		
10	Cuff pieces joint 2	0.80	SNLS
11	Cuff hem	0.53	SNLS
12	Cuff run stitch	0.73	SNLS
13	Cuff trim and turn	0.57	SNEC
14	Cuff top and sleeve pleat making	0.66	SNLS
	Sleeve preparation		
15	Sleeve panel attach	0.70	OL5
16	Sleeve panel top	0.65	SNLS
17	Under placket attach and tacking	0.65	SNLS
18	Big placket attachment	1.18	SNLS
	Front		
19	Button hole placket trim and attachment	0.56	KANSAI
20	Button placket sew and wash care label	0.38	SNLS
21	Button placket piping attach	0.40	SNLS
	Back		
22	Label attachment	0.35	SNLS
23	Yoke attach and top	0.43	SNLS
24	Dart mark and sew	0.56	SNLS
25	Front and back trim and pair	0.70	MANUAL
	Assembly		
26	Shoulder attachment and top	0.46	SNLS
27	Collar attach	0.54	SNLS
28	Collar finishing with content label	0.85	SNLS
29	Sleeve no set and attach	0.85	OL5
30	Top stitch at armhole	0.71	SNLS
31	Side seam attach and sleeve trim	0.70	OL5
32	Cuff no set and attaching	0.88	SNLS
33	Bottom hem	0.67	SNLS
	B\S and B\H		
34	Buttonhole front-7, cuff-2, sleeve placket 2	1.08	BH

(Continued)

TABLE 17.4 (*Continued*)

Operation Breakdown of Men's Shirt

Sl. No.	Operations	SMV	Machine Type
35	Button mark	0.27	MANUAL
36	Button front -7, spare-1, cuff 2, sleeve placket 2	1.17	BTN
37	Button melting	0.23	IRON
	TOTAL	22.10	

TABLE 17.5

Operation Breakdown of Trouser

Sl. No.	Operation Description	M/C Type	SAM
	Front		
1	Facing OL	OL3	0.350
2	Facing attachment to pocket bag	SN	0.450
3	Pocket bag close	SN	0.550
4	Turn and top stitch on pocket bag	SN	0.500
5	Pocket bag mark and attachment front	SN	0.600
6	Trim and turn	HLP	0.400
7	Top stitch	DN	0.550
8	Dummy stitch	SN	0.500
9	Inside fly run and ready	OL3	0.350
10	Zip attachment to fly	SN	0.450
11	Inside fly attachment	SN	0.400
12	Zip finish	SN	0.450
13	J fly attachment and top stitch	SN	0.550
14	J stitch	DN	0.450
15	Front rise attachment	OL5	0.400
16	Bar tack fly	BARTACK	0.100
17	FR trim and inspection	HLP	0.450
	W/B		
18	W/B joint	SN	0.350
19	WB iron	HLP	0.450
20	WB run stitch	SN	0.600
21	Trim and turn	HLP	0.400
22	Loop ready	FL	0.350
23	Mark and trim	HLP	0.450
	Back		
24	Dart stitch	SN	0.450
25	Welt attachment	DN	0.500
26	Pocket bag attachment	SN	0.450
27	Notch and turn	HLP	0.60

(*Continued*)

TABLE 17.5 (*Continued*)

Operation Breakdown of Trouser

Sl. No.	Operation Description	M/C Type	SAM
28	Side tacking	SN	0.65
29	Welt close	SN	0.45
30	Bottom edge stitch	SN	0.65
31	Pocket bag close	SN	0.60
32	Bag top stitch	SN	0.55
33	Welt top edge stitch	SN	0.650
34	BK rise	OL5	0.45
35	Trim and inspect	HLP	0.45
	Assembly		
36	FR and BK attachment (side)	OL5	0.750
37	Inseam	OL5	0.700
38	Loop attachment	SN	0.650
39	WB and body no set	HLP	0.450
40	WB attachment	SN	0.600
41	WB finish	SN	1.000
42	Loop finish	SN	0.700
43	Bar tack on loop	BARTACK	1.10
44	Bottom hem	SN	0.800
45	Hook and eye attachment	Special	0.500
46	Back pocket button attachment	BS	0.260
47	Button hole sew	BH	0.260
48	Trimming and checking	HLP	2.00

TABLE 17.6

Operation Breakdown for Jacket

Sl. No.	Operation	SAM	Machine Type
	Cuff preparation		
1	Inner mark for run	0.3	Manual
2	Mark for run	0.3	Manual
3	Run stitch with lining	0.7	SNLS
4	Trim and Turn	0.45	Manual
5	Edge stitch	0.65	SNLS
6	Topstitch	0.6	SNLS
	Sleeve preparation		
7	Sleeve panel attach	0.7	SNLS
8	Sleeve panel top	0.75	DNLS
9	Inner sleeve decostitch	1.6	SNLS
10	Inner sleeve with lining ready stitch	1	SNLS

(Continued)

TABLE 17.6 (*Continued*)

Operation Breakdown for Jacket

Sl. No.	Operation	SAM	Machine Type
	Front top welt pocket		
11	Bone ready with lining	0.25	SNLS
12	Welt pocket mark	0.3	Manual
13	Bone dummy stitch	0.3	SNLS
14	Bone ready	0.3	SNLS
15	Bone attach	0.35	SNLS
16	Inside patch mark	0.3	SNLS
17	Inside patch attachment	0.5	SNLS
18	Welt pocket attach	0.3	SNEC
19	Trim, notch and turn	0.45	SNEC
20	Tacking	0.25	SNLS
21	Welt edge stitch	0.3	SNLS
22	Pocket bag close	0.5	SNLS
23	Welt finish	0.3	SNLS
24	Welt pocket fold stitch	0.4	SNLS
25	Welt pocket bartack 2	0.4	Bartack
26	Inspection and pairing	0.4	Manual
	Front inner welt pocket		
27	Bone ready with lining	0.25	SNLS
28	Welt pocket mark	0.3	Manual
29	Bone dummy stitch	0.3	SNLS
30	Bone ready	0.3	SNLS
31	Bone attach	0.35	SNLS
32	Inside patch mark	0.3	SNLS
33	Inside patch attachment	0.5	SNLS
34	Welt pocket attach	0.3	SNEC
35	Trim, notch and turn	0.45	SNEC
36	Tacking	0.25	SNLS
37	Welt edge stitch	0.3	SNLS
38	Pocket bag close	0.5	SNLS
39	Welt finish	0.3	SNLS
40	Welt pocket fold stitch	0.4	SNLS
41	Welt pocket bartack 2	0.4	Bartack
42	Inspection and pairing	0.4	Manual
	Front piping pocket preparation 2		
43	Pocket pleat making and tacking	0.5	SNLS
44	Pocket pleat iron	0.4	SNLS
45	Piping ready	0.35	SNLS
46	Pocket hem	0.6	SNLS
47	Pocket mark for piping	0.35	Manual
48	Pocket piping no set	0.4	SNLS

(*Continued*)

TABLE 17.6 (*Continued*)

Operation Breakdown for Jacket

Sl. No.	Operation	SAM	Machine Type
49	Piping attachment	3	SNLS
50	Pocket inner pcs attach	4	SNLS
51	Pocket ol3	0.6	Ol3
52	Trim and turn	0.4	SNEC
53	Pocket side edge stitch mark	0.3	Manual
54	Pocket side edge stitch sew	0.6	SNLS
55	Pocket topstitch	2	SNLS
56	Mark for pocket attach	0.35	SNLS
57	Pocket attach	2	SNLS
58	Flap mark	0.35	Manual
59	Piping ready	0.35	SNLS
60	Flap mark for piping	0.5	Manual
61	Piping no set	0.4	SNLS
62	Piping attach with lining	2	SNLS
63	Ready trim	0.25	SNEC
64	Inner flap pc attach	2.4	SNLS
65	Trim and turn	0.45	SNEC
66	Flap edge stitch	1.5	SNLS
67	Flap top stitch	1.2	SNLS
68	Mark for attach	0.35	Manual
69	Flap attach	0.5	SNLS
70	Flap edge and top stitch	0.65	SNLS
71	Bartack pocket 6, flap 4	0.8	Bartack
72	Inspection and pairing	0.4	Manual
	Front facing preparation (overlap panel)		
73	Front facing lining attach	1.2	SNLS
74	Facing pcs joint 2	0.4	SNLS
75	Inner facing pcs joint 2	0.4	SNLS
76	Rib mark and trim	0.45	SNLS
77	Rib ready	0.45	SNLS
78	Rib attach at neck	0.5	SNLS
79	Rib finish at neck	0.45	SNLS
80	Inner rib attach at top facing	0.6	SNLS
81	Top rib attach at inner facing	0.6	SNLS
82	Rib tacking at joint	0.4	SNLS
83	Facing patch iron	0.5	SNLS
84	Facing patch mark	0.3	SNLS
85	Facing patch dummy stitch	0.4	SNLS
86	Facing neck stay stitch	1.2	SNLS
87	Facing bottom run stitch 2	0.5	SNLS
88	Notch and turn the bottom 2	0.4	Manual

(Continued)

TABLE 17.6 (*Continued*)

Operation Breakdown for Jacket

Sl. No.	Operation	SAM	Machine Type
89	Zipper attach 2	0.8	SNLS
90	Zipper finish 2	1	SNLS
91	Facing top stitch 2	1.2	SNLS
92	Facing stay stitch 2	0.5	SNLS
	Front facing panel and inner body assembly attach		
93	Mark for attach	0.5	Manual
94	Facing no set	0.4	Manual
95	Facing panel to inner body attach	2	SNLS
96	Inner facing lining attach	1.2	SNLS
97	Inner facing attach 2	1	SNLS
98	Inner facing and facing panel top stitch	1.5	DNLS
	Back		
99	Back panel attach	0.7	SNLS
100	Back panel top	0.75	DNLS
101	Back yoke attach	0.5	SNLS
102	Back yoke top	0.45	DNLS
103	Inner back with lining ready stitch	0.6	SNLS
104	Iron and trim	0.3	Manual
105	Front and back pairing	0.4	Manual
	Top assembly ready		
106	Shoulder joint	0.6	SNLS
107	Shoulder topstitch	0.5	DNLS
108	Patch iron	0.8	Iron
109	Patch ready mark	0.6	Manual
110	Patch mark for attach	0.3	Manual
111	Patch attach	0.8	SNLS
112	Patch deco stitch	1.5	SNLS
113	Sleeve no setting	0.4	Manual
114	Sleeve dummy stitch	0.45	SNLS
115	Sleeve attaching	0.85	SNLS
116	Top stitch at sleeve	0.75	DNLS
117	Side seam	0.65	SNLS
118	Inspection	0.4	Manual
119	Collar lining attach	0.5	SNLS
120	Collar mark for deco stitch	0.3	Manual
121	Collar deco stitch DN	0.35	DNLS
122	Collar deco stitch SN	0.3	SNLS
123	Collar mark for run stitch	0.3	Manual
124	Collar no set	0.3	SNLS
125	Collar run stitch	0.45	SNLS

(*Continued*)

TABLE 17.6 (*Continued*)

Operation Breakdown for Jacket

Sl. No.	Operation	SAM	Machine Type
126	Turn and iron	0.35	Iron
127	Collar attach to body	0.5	SNLS
128	Collar closing at neck	0.4	SNLS
129	Bottom lining attach	0.6	SNLS
130	Bottom mark for run stitch	0.35	Manual
131	Bottom run stitch	0.6	SNLS
132	Turn and iron	0.4	Iron
133	Bottom patch no set	0.4	SNLS
134	Bottom attach to body	0.5	SNLS
135	Bottom top mark	0.4	SNLS
136	Bottom top dn	0.6	SNLS
137	Zipper attach 2	0.8	SNLS
138	Zip closing tacking at top	0.25	SNLS
139	Cuff no set	0.4	Manual
140	Cuff attach	1	SNLS
	Inner lining assembly ready		
141	Front and back pairing	0.4	Manual
142	Label attach	0.5	SNLS
143	Shoulder joint	0.6	SNLS
144	Sleeve no setting	0.4	Manual
145	Sleeve attaching	0.85	SNLS
146	Side seam	0.7	SNLS
147	Side seam ol3	0.6	Ol3
	Both panels assembly		
148	Assembly no set	0.4	Manual
149	Attaching mark	0.35	Manual
150	Top and lining neck attach at front	0.6	SNLS
151	Top and lining neck attach at back	0.4	SNLS
152	Top and lining bottom attach	0.8	SNLS
153	Top and lining left front attach	0.6	SNLS
154	Top and lining right front attach	0.6	SNLS
155	Cuff attach to lining assembly	1	SNLS
156	Tacking at armhole	0.4	SNLS
157	Garment turning	0.5	Manual
158	Inner lining assembly side seam close	0.6	SNLS
159	Inner welt pocket close	0.5	SNLS
160	Front zipper panel top stitch	2	SNLS
161	Bottom top stitch	1	DNLS
162	Bottom patch top stitch	0.6	DNLS
163	Bottom patch top bartack 4	0.35	Bartack

(*Continued*)

TABLE 17.6 (*Continued*)

Operation Breakdown for Jacket

Sl. No.	Operation	SAM	Machine Type
	Button hole and button		
164	Button hole pocket 2, cuff 2	0.4	Button hole
165	Button hole and button mark	0.3	Manual
166	Button pocket 2, cuff 2	0.45	BTN
167	Button wrapping	0.4	SPL
168	Trimming and checking	3	Manual

Note: SNLS: 1 Needle lock stitch; SNCS: 1 Needle chain stitch; SNEC: 1 Needle edge cutter; BS: Blind stitch; DNLS: 2 Needle lock stitch; DNCS: 2 Needle chain stitch; OL3: 3 thread over lock; OL4: 4 thread over lock; OL5: 5 thread over lock; APW: Auto pocket welting; BT: Bartack machine; BH: Button holing; BA: Button attach; CHK: Checking.

17.4.1 Sales Tally Form

This is made out in the production control department. It collects all sales orders to enable production control to determine the total amount needed per style and colour. It gives production control the information needed for ordering raw materials and scheduling the cutting orders with respect to size and colour distribution and time. It contains the information such as style, colour and style listing in chart form, the date, time span (daily or weekly) and the serial number.

17.4.2 Purchase Order

This gives a merchant the authority to ship raw materials to the industry. It is usually made out by purchasing, production control, or the departments using the materials or supplies. Some industries may use purchase requisition forms, which are made out by production control or the department using the items. The purchasing department then issues purchase orders according to the needs listed on the collations. The purchase order for raw materials contains the following information:

- The date the order is given
- The authority for the order and the order number
- The delivery date or dates and amounts per date
- Where and how to be shipped: also any pertinent packing instructions
- The firm's name
- The vendor's style and colour names and/or numbers for the raw material
- The prices and terms, and the width, finish of fabric and other pertinent quality specifications for the fabric, such as pick count and tensile strength that are needed for this fabric

17.4.3 Receiving Memo

This is made out by the receiving department and lists the specification and amounts of all raw materials and supplies that are received.

It contains the following information:

- Date received
- Item received: specifications and amount (the shipper's style number or name, also the firm's style number or name)
- From whom received (carrier)
- The shipper's name
- Firm's purchase order number for the shipment
- The shipper's sales order number for the shipment
- The signature of the one receiving it signifying the purchase order authorising the shipment has been checked

17.4.4 Cutting Order

This form is the initial work order made by production control. Large industries may have systems which comprises both cutting orders and cutting tickets. Nevertheless, there is a difference between a spreading ticket and a cutting ticket or a cutting order although in certain situations the three may be synonymous (Solinger 1988; Rogale and Polanovi 1996). The precise definitions are given below: A cutting order is an authorisation to cut a number of garments made from one or more types of fabric and which may be cut in one or more spreads. A cutting ticket is an authorisation to cut a number of garments that must be cut in two or more spreads depending on the different types of materials used. A spreading ticket is an authorisation to cut one of the fabric requirements for a given number of garments on one cutting table upon which one can spread. When a garment is made of only one type of fabric and the cut ordered is made in one spread, the spreading ticket and the cutting order become synonymous. Cutting orders should carry the following information:

- The date the order is issued and the serial number of the order
- Style number and/or name of the garment
- General description of the garment
- Listing of the types of fabrics to be cut for the garment
- The mill number or name of specific fabrics to be cut
- The colour and size distribution (and totals per size and colour)
- The date these garments are required for shipment
- Special remarks with reference to items such as zippers, buttons, etc.

Cutting tickets should contain the following information:

- The cutting order number against which the cutting ticket is made and the cutting order date
- The cutting ticket number and its date
- The code numbers of the markers to be used
- The colour and size distribution of the amount to be cut
- The number of bundles made for the sewing department
- The work or pay control ticket numbers issued against the cutting ticket
- The date work began on the ticket; the date the work was completed
- The calculated yardage to be used
- The actual yardage used
- Remarks: any discrepancy between the above two yardages
- The names of those who worked on the lot

If spreading tickets are used, it may be advisable to enter a piece goods listing against each spreading ticket. Spreading tickets would have essentially the same information as cutting tickets plus the cutting table number and/ or cutting table area. Spreading tickets should list the spreading, cutting and bundling times for the spread lot.

17.4.5 The Cutting Production Control Chart

This could be made in the cutting department for the purpose of controlling cutting department activities. Besides the usual delineations for scheduled production, actual production and time, this control chart should also have provision for each cutting table, each cutting table area if long tables are used, each fixed band knife and each activity such as spread, mark, cut or bundle that takes place at each table.

17.4.6 Cutting Projection Tally

This is a device that may be used to update the sewing department as to the exact time the sewing department will receive each style every week. This form should normally carry the following information:

- The day, date and hour each style cut is ready for the sewing department
- The amount, number of colours and number of work bundles in each cut

- The cutting order number, cutting ticket number and move ticket number of the cut (or job order numbers)
- The date each cut is required for shipment
- The date of the projection

This form would usually have two copies: one for the cutting department and one for the sewing department; in certain situations, it may be advisable to give a third copy to production control.

17.4.7 Recut or Swatch Ticket

This would initiate in the cutting apartment whenever garment panels that are damaged and cannot be used have to be cut again. The same ticket may be used for cutting sample swatches for sales. Each ticket should contain the date of the recut, the style number and/or name of the fabric, the colour, the inventory piece number, the yardage used, the garment section(s) recut, the bundle number of the garment and the cutting ticket number (Ambastha 2012).

17.4.8 Bundle Ticket

This control form originates in either the cutting department or the payroll department. It is generally used for pay control as well as production control purpose. The bundle ticket can be used for unit flow as well as bundle flow production systems. The exact form of the ticket will depend on the production system used (Solinger 1988; Sudarshan and Rao 2013). In sectionalised production systems, the bundle ticket should be perforated in sections equal in number to the total number of subassembly and assembly lines used to produce the product. Each section should have subsections perforated, equal in number to the number of jobs in the subassembly line covered by the ticket section. Each of these individual subsections will be detached from the main body of the subsection by the operator after he or she completes the job on the bundle. The main body of the subsection is the division that is returned to pay control by the production supervisor after all the jobs in the section have been completed. Each of these divisions should contain the following information:

- The serial number of the entire bun-die ticket
- The name of the subsection (such as sleeve, collar, front, etc.)
- The style name or number
- The cutting ticket number (or spreading ticket or more ticket number)
- The date of the compilation of the bundle in the cutting section
- The size, amount and colour of the bundle

- The name of each process in the section and the number of the operator who has carried out the job against the respective job name. Also, the date the operation was completed
- The signature of the supervisor of the particular production section covered by the ticket section
- The subsection that each operator takes. After job completion, the following information should be listed:
  - The name of the job
  - The bundle ticket number
  - The amount size and colour of the bundle
  - The price of job (if a piece works usage system is used)
  - The style name or number
  - The move ticket number controlling the bundle (or the cutting ticket number)

17.4.9 Move Ticket

This ticket would originate in the cutting department. It can be dispensed with under certain production systems. It controls all the bundle tickets issued against a cutting or spreading ticket. For example, assume a lot of 150 dozen in 6 colours, 3 sizes, has been cut on a given cutting ticket which controlled three different spreads: one of body fabric, one of lining, the other of trimming fabric (Vijayalakshmi 2009). The move ticket lists the bundles made and the bundle ticket numbers assigned to this cutting ticket. The move ticket should contain the following information:

- The cutting ticket number (and spreading ticket numbers)
- The listing of bundle ticket numbers assigned
- The move ticket number
- The amount, colour and size of each bundle
- The date the move ticket was compiled
- The cutting order number
- The style name or number
- The completion date for the move ticket

A move ticket should actually be a style or job order control device. If it is a job order control device, it may list more than one style when the production system and sequence are alike for the style listed on the mover ticket (Babu 2006). Job order forms should contain the following information:

- The job order number
- The date the job order was made

- The date the job order is required for delivery
- The listing of move tickets against this job order

Each of these move ticket listings should give the style name or number of each move ticket, the move ticket number, the cutting number of the move ticket, the total number of bundles on each move ticket, the total amount per size and colour, the grand total per move ticket, customer's name (or code name or number) for when this order was made and the sales order number(s) against which this job order was made.

17.4.10 Sewing Department Project Tally

This could initiate in the sewing department after a cutting department projection has been received. The form informs the pressing department as to the exact time the pressing department will receive each style (and the amount) during the next week. This form should generally contain the following information:

- The date and hour each style or job order to be sewed will be ready for off-pressing (or finishing)
- The amount, colours, size and bundle distribution of each style
- The cutting order number, cutting ticket number and move ticket number (or job order number) of each bundle
- The date each bundle, style, move ticket or job order is required for shipment
- The date the projection is made and the time span the projection covers; a 3-copy distribution would usually send one copy to production control, one to the sewing department and one to the pressing department

17.4.11 Pressing Projection Tally

This originates in the pressing projection and would usually duplicate most of the information listed in the sewing projection. A 3-copy distribution of this form would send one copy to production control, one to pressing and one to the packing and shipping department.

17.4.12 Packing and Shipping Projection

This would initiate in the shipping department after the pressing projection is received and it would list the following information:

- The day and date of each shipment scheduled to be made in the coming week, the customer's name and the shipment destination
- The style, colour and size distribution of each shipment

- The sales order number against which the shipment is made
- The cutting order (or cutting ticket number) against which the shipment is made
- The date and time projection is made within the time span the projection covers

17.4.13 Shipping Memo

This is the form made out for each shipment when the shipment is made. It lists the contents of each shipment and gives the distribution of each shipping container (cartons, cases, etc.). A packing memo would be a form used to list the contents shipped in one container. As to whether one or both of these two forms are used will depend on the shipping system used. Each form, however, will contain the name, initials, or number of the clerk who packs the shipping container(s) as well as the style, colour and size distribution in each container, the date the shipment was made, the carrier, the sales order number of the shipment and the customer's purchase order number for the shipment if there is such number.

17.4.14 Invoice or Bill

This initiates in the accounting department after the charge or shipping memo is received. This usually contains all the information on the shipping memo plus the price of each item and the total amount due for the shipment. A 2-copy distribution would send one copy to the customer and the other copy to the accounting department.

17.4.15 Production Control Ledger Cards

These forms are used by production to control purchasing, production and inventory activities. Two basic forms may be used to control all these activities. Production Control Card I may be called as planning or purchasing control card for each style and colour, if a style is made in six colours, six cards would be used to control the planning for this style (Fairhurst 2008). Each card or form would contain the following information:

- Style number or name, colour, fabric description, fabric width, yards per dozen or unit.
- The purchase record section – this section of the record contains the date of each purchase of fabric in this colour, the purchase order number, the mill to which the purchase was issued, the yardage ordered, the price, the terms, the sales value of the fabric ordered and the amount of garments that can be cut from this purchase order.

- The sales record section – this section of the card lists the size distribution sold daily of this style and colour. It also lists the total amount sold and the amount to be sold for which fabric is purchased and available.
- Receiving record – this section lists the date cloth for this style and colour has been received. It also lists the receiving memo number, purchase order number, mill name of each cloth shipment received, the yardage received, the yardage cut, the cutting ticket number of each cut, and the cutting order number of each cut.
- The cutting order record – this section lists the date and number of each cutting order, the size distribution of each cutting order and the receiving memo number of the cloth to be used for each cutting order. This Production Control Card I acts as a raw material inventory record as well as a purchase control.

Production Control Card II is the finished garment inventory control for each style and colour. Each card or form would contain the following information:

- Style number or name, colour, fabric description, fabric width, yards per dozen or unit.
- The cutting ticket record – this section contains the date each cutting ticket was completed, the cutting order number, the amount cut per size, time total amount cut and the yardage used.
- The finished garment record – this section consists of three subsections: 'shipped garments', 'on call or waiting garments' and 'stock sell garments'. 'On call or waiting garments' are garments made against specific sales orders.

These garments are waiting to be shipped on specific dates listed on the sales order or they are waiting to be shipped as called for by the customer. Each subsection has daily listings for time size distribution for the style, and the total amount shipped, on call, or in stock. A common date line is used for all three subsections (Ambastha 2012). These two production control forms, raw material inventory and finished garment inventory, may be incorporated into one control form for controlling the complete inventory on the basic raw material used in the garment (Tyler 1991; Shaeffer 2000). This may be done to save form space; duplicate date or total amount columns may be eliminated. Posting time is also saved in such situations.

17.4.16 Equipment Maintenance Record

This is a form on which a record is kept of the maintenance and repair work done on production equipment such as cutting, spreading, sewing and pressing machines (Ahmad et al. 2012). This record enables one to determine when it has become economically feasible to replace the machine. The record should contain the following information:

- Maker, model number and serial number
- Date of acquisition and cost
- Date and man hours spent on each repair and down-time adjustment
- The parts replaced in each repair and the cost
- Sum of operating hours between repairs and adjustment
  - Summary of operating speeds and condition
  - Summary of types of operation
  - The operator (or operators); remarks section for cause of breakdown

17.4.17 Equipment Inventory Record

This record lists each machine by maker, model number and serial number, date of acquisition, cost and relative performance value. In many firms, forms 17 and 18 can be combined as one for some of the representative forms listed in this section.

17.4.18 Receiving Quality Control Sheet

This lists the quality specifications to be measured when fabric or other raw material is received. It lists those specifications that must be measured as soon as the fabric is received and checked off against the supplier's shipping memo sent with the fabric.

17.4.19 Laboratory Quality Control Sheet

This lists the quality specifications of raw material which must be measured in a testing laboratory.

17.4.20 Rejection Memo

This is another quality control document forwarded by the receiving department to purchasing, listing materials rejected with reasons for the same.

References

Ahmad, S., A.A.B. Khalil and C.A.A. Rashed. 2012. Impact efficiency in apparel supply chain. *Asian Journal Natural Applied Science* 1(4):36–45.

Ambastha, M. 2012. Performance measurement Tools-5. *Inventory Managements Stitch World* 21:34–37.

Babu, V.R. 2006. *Garment Production Systems: An Overview*. http://www.indiantextile journal.com/articles/FAdetails.asp (accessed on March 12, 2015).

Chuter, A.J. 1995. *Introduction to Clothing Production Management*. Blackwell Scientific Publications, Oxford, UK.

Fairhurst, C. 2008. *Advances in Apparel Production*. The Textile Institute, Woodhead Publication, Cambridge.

Glock, R.E. and G.I. Kunz. 2004. *Apparel Manufacturing – Sewn Product Analysis*. Prentice Hall, Englewood Cliffs, NJ.

Kumar, A. 2008. *Production Planning and Control: Lesson 8 Course Material*. Delhi University. www.du.ac.in/fileadmin/DU/Academics/course_material/EP_08. pdf (accessed on March 14, 2015).

Mannan, M.A. and F. Ferdousi. 2007. *Essentials of Total Quality Management*. The University Grants Commission of Bangladesh, Dhaka.

McBride, D. 2003. The 7 Manufacturing Wastes. http://www.emsstrategies.com/ dm090203article2.htm (accessed on March 25, 2015).

Mehta, P.V. 1992. *An Introduction to Quality Control for Apparel Industry*. CRC Press, Boca Raton, FL.

Mok, P.Y., T.Y. Cheung, W.K. Wong, S.Y.S. Leung and J.T. Fan. 2013. Intelligent production planning for complex garment manufacturing. *Journal of Intel Manufacturing* 24(1):133–145.

Nayak, R. and R. Padhye. 2015. *Garment Manufacturing Technology*. Woodhead Publication, Cambridge.

Oliver, B.A., D.H. Kincade and D. Albrecht. 1994. Comparison of apparel production systems: A simulation clothing. *Textile Research Journal* 12(4):45–50.

Ramesh, A. and B.K. Bahinipati. 2011. The Indian apparel industry: A critical review of supply chains In: *International Conference on Operations and Quantitative Management (ICOQM)*. Nashik, India.

Ray, L. 2014. Production Planning for Garment Manufacturing. http://smallbusi-nesschron.com/production-planning-garment-manufacturing-80975.html (accessed on October 5, 2014).

Rogale, D. and C.S. Polanovi. 1996. *Computerised System of Construction Preparation in Garment Industry (in Croatian)*. University of Zagreb, Faculty of Textile Technology, Croatia.

Russell, R.S. and B.W. Taylor. 1999. *Operations Management*. Prentice-Hall, Upper Saddle River, NJ.

Sarkar, P. 2011. *Functions of Production Planning and Control (PPC) Department in Apparel Manufacturing*. http://www.onlineclothingstudy.com/2011/12/functions-of-productionplanning-and.html (accessed October 5, 2014).

Sarkar, P. 2012a. Operation Breakdown and SMV of a Trouser. http://www. onlineclothingstudy.com/2012/01/operation-breakdown-and-smv-of-trouser. html (accessed October 22, 2014).

Sarkar, P. 2012b. Operation Breakdown and SMVs of a Basic Jeans. http://www. onlineclothingstudy.com/2012/07/operation-breakdown-and-smvs-of-basic. html (accessed October 22, 2014).

Schertel, S. 1998. New Product Development: Planning and Scheduling of the Merchandising Calendar (Master Dissertation). North Carolina State University, Raleigh, NC.

Shaeffer, C. 2000. *Sewing for the Apparel Industry*. Woodhead Publication, Cambridge.

Solinger, J. 1988. *Apparel Manufacturing Handbook – Analysis Principles and Practice.* Columbia Boblin Media Corp, New York, USA.

Sudarshan, B. and N.D. Rao. 2013. Application of modular manufacturing system garment industries. *International Journal of Science and Engineering Research* 4(2):2083–2089.

Tyler, D.J. 1991. *Materials Management in Clothing Production*. BSP Professional Books, Oxford, UK.

Vijayalakshmi, D. 2009. Production Strategies and Systems for Apparel Manufacturing. http://www.indiantextilejournal.com/articles/FAdetails.asp (accessed on March 12, 2015).

18

Plant Loading and Capacity Planning

18.1 Setting Up of a Garment Industry

The factors to be considered while starting a new garment unit are discussed below.

18.1.1 Selecting Appropriate Product Category

Deciding product categorisation to be focussed on during set-up of a garment industry could play a crucial role. At the initial face of starting a garment unit, the various kinds of garments such as T-shirts, polo and woven products should not be considered at the same time and only one or two product profiles should be considered (Shaeffer 2000).

18.1.2 Estimation of Production Requirement

It would be helpful to have an idea about quantity of garments that can be produced per day so that it would be helpful in future planning based on the budget and customer demand. This necessitates the process of determination of the production capability of an industry.

18.1.2.1 Plant Loading

Plant loading is defined as the allotment of workers or machines for future processing of an order by considering the sequence of processes as in a route sheet and the priority sequencing and utilisation of work centres (Tyler 1991; Chuter 1995). Loading establishes the volume of load every work centre should have in a forthcoming period which results in load schedules indicating the evaluation of labour and machine hours necessary to get the master production schedules with the available labour and machine hours in every planning schedule in the short term.

18.1.2.2 Capacity Study

A capacity study is the evaluation of a garment industry, manufacturing process, machine, or operator to estimate the maximum rate of production. The objective of the capacity study is

- To find-out the deviation between the actual rate of production to its capacity
- To evaluate the causes for lagging in the actual production
- To achieve the actual production closer to its actual capacity using proper methods and reducing the idle time

There are various types of capacity available for a factory.

- *Maximum capacity* – Number of hours available in a given time under normal conditions.
- *Potential capacity* – Maximum capacity adjusted for expected efficiency.
- *Committed capacity* – Total hours formerly allocated for production during a certain time period.
- *Available capacity* – The difference between committed and potential capacity is known as available capacity.
- *Required capacity* – It is garment SAM necessary to manufacture a specified volume in a certain period of time.

Calculation of Capacity

Consider the following cutting plan example:

Size	10	12	14	16	18
Qty	40	90	80	25	25

The limitations on lay sizes are

- Maximum height of lay = 15 plies
- Maximum length of lay = 4 garments marked
- Time for laying one fabric ply = 1 minute
- Marking time = 5 minutes
- Cutting time = 10 minutes
- Working hours = 8

Solution

Plan the cutting lay out.

Lay I – 25 Plies (Sizes – 16, 18, 12, 12)
Lay II – 40 Plies (Sizes – 10, 14, 14, 12)

Lay I

- Maximum number of garments in Lay I = $25 \times 4 = 100$
- Laying time for 25 plies = 25 minutes
- Laying time for one garment = $25/100 = 0.25$ minute
- Marking time for one garment = $5/100 = 0.05$ minute
- Cutting time for one garment = $10/100 = 0.10$ minute
- Total processing time for Lay I = $25 + 5 + 10 = 40$ minutes
- Total processing time per garment = $0.25 + 0.05 + 0.10 = 0.40$ minute

Lay II

- Maximum number of garments in Lay I = $40 \times 4 = 160$
- Laying time for 40 plies = 40 minutes
- Laying time for one garment = $40/160 = 0.25$ minute
- Marking time for one garment = $5/160 = 0.03$ minute
- Cutting time for one garment = $10/160 = 0.06$ minute
- Total processing time for Lay II = $40 + 5 + 10 = 55$ minutes
- Total processing time per garment = $0.25 + 0.03 + 0.06 = 0.34$ minute

Capacity

- Capacity/hour for Lay I = $60/0.40 = 150$ garments
- Capacity/day for Lay I = $480/0.40 = 1200$ garments
- Capacity/hour for Lay II = $60/0.34 = 176$ garments
- Capacity/day for Lay II = $480/0.34 = 1412$ garments
- Total time essential to complete the order = $40 + 55 = 95$ minutes

18.1.3 Number of Machines

After deciding on the type of product and production capacity, the number of sewing machines and other machinery requirements could be calculated. Otherwise, it can be carried out conversely, that is, after deciding to set-up a factory for a specific number of machines as well as type of product, projected production per day can be determined.

18.1.4 Type of Machines

The succeeding process is to select the proper kinds of machines suitable for the production of garments as well as the number of machines to be purchased in each kind of machine. This step would be useful for estimating the capital investment in machines. Apart from the sewing machines, list other essential equipment such as pressing tables, spreading tables, boiler, generator, furnishings etc.

18.1.5 Raw Materials Requirement

After selection of product category and machines, raw materials such as fabric and other accessories and trims to make the garment with their average consumption have to be listed (Ramesh and Bahinipati 2011). This would be helpful for preparing the budget on material sourcing.

18.1.6 Factory Space Requirement

The space needed for setting up of machines, equipment and administrative centre has to be estimated. According to the estimation the factory layout could be planned.

18.1.7 Manpower Requirement

After setting up the machine and materials, the labour, the primary resources for a garment industry could be planned. The manpower calculation includes number of office staff, supervisors and workers. Further, an estimation has to be done for their salaries.

18.1.8 Project Cost

To determine the budget for setting up an apparel industry, one could prepare the cost of the project. For doing that, the assessment of total capital investment, EMI amount, salary for staff, workers' wages and running costs have to be taken into consideration.

18.1.9 Internal Process Flow

Plan out the detailed process flow for execution of an order. This will facilitate deciding what all the departments need to set up and plan to employ the people accordingly.

18.1.10 Supplier Listing

Finding out the good and reliable suppliers for fabrics, trims and other necessary items required to manufacture the garments is crucial for completion and dispatch of the orders in time.

18.2 Plant Layout

It is a floor plan for deciding and orchestrating the chosen equipment and machinery of an industry in the best suitable location to permit the quicker flow of materials at a minimum cost and with the least amount of material handling during the manufacturing process from the receipt of raw materials to the shipment of the finished garments (Fairhurst 2008; Ahmad et al. 2012).

18.2.1 Principles of Plant Layout

The following principles have to be followed to have an ideal plant layout. The understanding of these principles would help in learning the aspects that are influencing the plant layout.

18.2.1.1 Principle of Minimum Travel

Workers and materials must pass through the shortest distance between the processes to avoid wastage of labour and time and reduce the cost of materials handling. This is mainly important for garment industries where each department is interconnected and the movement of the labour from one department to another must be minimised for increased productivity.

18.2.1.2 Principle of Sequence

Machineries as well as processes should be arranged in a sequential order which is achieved in the product layout. It contains the arrangement of the working area for each operation in the same order. For a proper flow of materials, the plant layout must offer easy movement of raw materials to the production department and to the packing department (Nahmias 1997; Ramesh Babu 2012). The plant layout, following the principle of sequence, needs to consider the frequency of movement between the different departments, volume of production in each department, total working area available in each department and the nature of operations in each department.

18.2.1.3 Principle of Usage

Every foot of existing space should be effectively utilised. It includes the proper usage of space both horizontally and vertically. Apart from using the floor space of a room, if the ceiling height is also utilised, more material can be stored in the same room. Use of overhead space saves a lot of floor space.

18.2.1.4 Principle of Compactness

There should be harmonious fusion of all the related factors so that the final layout looks well integrated and compact.

18.2.1.5 Principle of Safety and Satisfaction

This layout has built in options for workers to ensure they are safeguarded from the occurrence of fire. The comfort and convenience of the worker has been considered more important while planning this layout. In an apparel unit, factors such as proper lighting, ventilation and prevention of hazardous conditions are very important (Nahmias 1997). Employees must be protected from excessive heat, dust from the raw materials such as fabrics and the trimmings of the threads in sewing, glare and fumes. The safety of workers both during operation, maintenance and transportation of materials should be taken care of.

18.2.1.6 Principle of Flexibility

The layout must allow modifications with minimum complications and at minimum cost.

18.2.2 Influencing Factors of Plant Layout

The plant layout changes from industry to industry, location to location and plant to plant. The plant layout is influenced by the 3M's, namely materials, machinery and men (Oliver 1994; Kumar 2008).

18.2.2.1 Materials

It is the important aspect that influences the plant layout. For any industry there is a need to offer a proper storage and movement of raw materials, which are necessary for the production of a product, until they are transformed into finished products. It is a common principle that every industry procures the raw materials economically when they are available. This creates the need for appropriate storage so that the goods are moved according to the requirement through production departments.

18.2.2.2 Worker

While outlining the design it is imperative to consider the type, position and prerequisites of workers. Worker facilities, for example, wellbeing and related services, locker rooms and public facilities influence the design. Employee safety ought to additionally be considered.

18.2.2.3 Machinery

The machinery required is reliant on the type of product, quantity of production, the type of process and management policy. These decide the size and type of the machinery to be installed which, in turn, influences the plant layout.

Production is the combination of men, materials and machines. The ratio in which these elements are used depends on their costs and on the production processes selected. Before laying out a plant, it is necessary to determine which of these elements are to be stationary and which will be moving during the selection process. The plant layout must offer the space for storage of fuel, be it coal, oil or gas.

18.2.2.4 Product

A layout is generally designed with the objective of manufacturing a product. Whether the product is light or heavy, small or big, its arrangement related to the plant location affects the plant layout. The quantity of production, quality of product, size of machinery and space requirement for a machine and other facilities are based on the sales demand and plant layout. A product with relatively inelastic demand should be produced on a mass scale with less specialised equipment.

18.2.2.5 Management Policies

Management policies also influence plant layout. Some of the managerial policies are

- The volume of production and provision for expansion
- The extent of automation
- Making or buying a particular component
- Desire of rapid delivery of goods to the customers
- Purchase policy
- Personnel policies

18.2.3 Types of Layout

A layout alludes to the organising and grouping of machines which are intended for production of materials. Grouping is done on diverse lines. The factors influencing the selection of a proper layout of machines for a particular style of garment relies on several factors as given below.

18.2.3.1 Process Layout

It includes grouping of similar machines in a particular department. The process arrangement is meant by the grouping together of similar machines based upon their uniqueness. A volume of raw material is allotted to a particular machine which accomplishes the first operation. This machine could be arranged at anyplace in the industry. For performing the next operation, a different machine could be necessary, which may be situated in another place of the industry (Solinger 1988; Russell and Taylor 1999; Vijayalakshmi 2009). For carrying out the production process, the material must be transported to the other equipments. This kind of layout is appropriate for the intermittent kind rather than continuous type of production. While grouping machines based on the type of process, the following points must be considered.

- The distance between the departments must be as small as possible to minimise the material movement.
- The machines that are similar are grouped in one section/ department.
- It must be convenient for supervision and inspection.

Advantages:
- Investments on machines are reduced as they are general purpose machines.
- There is greater flexibility in the production.
- Better supervision is achievable through specialisation.
- This layout provides better use of men and machines.
- It is easier to handle any breakdown of machines through taking the machine to another machine station.
- The investment costs on machines are comparatively lower.

Disadvantages:
- Movement of materials is difficult.
- Requires more floor space.
- Since the work-in-progress has to move from one place to another to look for a machine, the production time is generally high.
- The WIP accumulates at different places.

18.2.3.2 Product Layout

In this kind of layout, the machines are generally arranged in a series based on the process sequence required for manufacturing the garment. In this layout, the process starts at one side of the line and the assembled product is delivered at another side of the line. In between, partly finished goods move automatically or manually from one machine to another. The output of one machine becomes the input of the next machine (Russell and Taylor 1999; Ray 2014).

Advantages:
- Materials handling is automated, hence reduction in materials handling cost.
- Bottlenecks in production line could be avoided.
- Lesser manufacturing time.
- The layout helps in better production control.

- It necessitates less floor space per unit of production.
- WIP is reduced and investment thereon is minimised.

Disadvantages:
- Expensive and inflexible layout.
- Supervision is difficult.
- Expansion is difficult.
- Breakdown of any machinery in a line could disturb the whole system.

18.2.3.3 Fixed Position Layout

In this kind of layout, the product remains stationary in a fixed location, where men and machine have to move toward it which is desirable as the cost of moving them is lower than the cost of moving the product (Solinger 1988; Glock and Kunz 2004).

Advantages:
- Men and machines can be utilised for numerous kinds of operations manufacturing different products.
- The investment on layout is less.
- The costs of transportation for a bulky product are avoided.

18.2.3.4 Cellular Manufacturing (CM) Layout

In this kind of layout, the machines are generally assembled into cells which function fairly like a product layout within a process layout. Every cell in this design is shaped to produce single parts, all with common attributes, which typically means they necessitate the same machines and have similar machine settings (Babu 2006).

Advantages:
- Lower WIP inventories.
- Reduced material handling costs.
- Flow time of materials is less in production planning.
- Improved visual control of process which enables quicker set ups.

Disadvantages:
- Manufacturing flexibility.
- Reduced machine stoppage time.
- Spare equipment could be necessary so that parts need not be transported between cells.

18.2.3.5 Combined Layout

It is a mixture of the product and process layouts, which could be observed in many of the apparel units. Each process is situated as a single unit and a number of such units is arranged in a product layout. It is feasible to have both types of layout in a capably combined form if the products manufactured are fairly similar and not complex (McKelvey and Munslow 2003).

18.2.3.6 Service Facility Layout

The major distinction between the service facility and manufacturing facility layouts is that many service facilities exist to bring collectively customers and services together. Some of the requirements of service facility layouts are large, well organised and amply lighted parking areas and well-designed walkways to and from parking areas.

18.2.3.7 Classification of Layout Based on Flow of Material

The layout can also be classified based on the flow of the materials (Figure 18.1) as follows:

1. *Linear:* The sewing area is in the middle of the floor with cutting and finishing areas on either end of the sewing line (Figure 18.1).
2. *U-shaped:* This layout is suited where supply of materials and reciept of finished goods are done through the same place. Parts production stations may be placed inside the U. The same workers can therefore handle both supplying materials and taking the finished goods away from the line. It is therefore easier to supply materials at the same rate finished goods come off the line, and thus maintain a constant number of goods in progress (Figure 18.1).
3. *Comb-shaped:* Achieved by combining plural linear lines, each of the part's lines is also linear and the parts lines are connected to the main line at the point where the parts are needed (Figure 18.1).

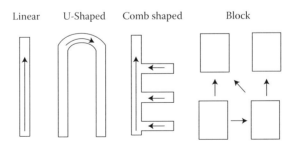

FIGURE 18.1
Types of layout based on material flow.

4. *Block:* Plural units are combined to form individual blocks, each of which comprises the required sewing machines. This format is suited to organisation by production groups or to semipermanent layouts for small lot production lines where the goods being produced change frequently (Figure 18.1).

18.3 Line Balancing

A line is defined as a group of operators under the control of one production supervisor. It is a function of the work study office to provide management with information to help the efficient and productive running of the factory, and part of this information is the process known as line balancing (Solinger 1988).

Line balancing is crucial in the efficient running of a production line and the objective of line balancing is to balance the workload of each operation so that the flow of work is smooth, no bottleneck processes are created and the operators could be able to work at higher performance throughout the day. It is intended to cut down the waiting time to a minimum or with the use of work in progress (WIP) to get rid of waiting time completely. Line balancing is defined as 'the engagement of sequential work activities into production line to achieve a high utilization of labour and equipment and hence minimizes idle time'. Balancing may be accomplished by adjustment of the work stations or by including machines and/or workers at some of the production lines so that all operations take about the same quantity of time (Sarkar 2011).

18.3.1 Need for Balancing

- Keeping inventory cost low
- To enable the operator to work at an optimal pace
- To enable the supervisor to attend other problems
- To enable better production planning
- Balancing production line results in on time shipments, low cost and ensures reorders

18.3.2 Goals for Balancing

- Meet production schedule
- Avoid the waiting time
- Minimise overtime
- Protect operator earnings

18.3.3 Production Line Balancing

The line balancing approach is to create the production lines flexible enough to take up external and internal abnormalities in production. There are two types of line balancing:

- *Static balance* – It is a long-term difference in capacity over a period of several hours or more. Static imbalance could lead to underutilisation of machines, men and production lines.
- *Dynamic balance* – It is nothing but short-term changes in capacity like for a minute or an hour maximum. Dynamic imbalance occurs from product mix changes and difference in work time unrelated to product mix.

18.3.4 Points to Be Noted When Balancing

- Meet production target by usage of
  - Regular operators
  - Utility operators
  - Shuttle operators
- Work flow should be constant throughout all operations
- Avoid overtime
- Determine human resources
- Check absences daily
- Assign utility shuttle operators based on need
- Update daily production every two hours

18.3.5 Micro-Steps in Line Balancing

The steps to a properly balanced line are

- Calculation of the labour requirements
- Operation breakdown
- Theoretical operation balance
- Initial balance
- Balance control

18.3.5.1 Calculation of Labour Requirements

With good work measurement records, the work content of a new garment can be calculated. The number of people required will depend upon the probable efficiency of the line selected and the percentage of the time that they are at work and doing their own specialist jobs.

18.3.5.2 Sectionalisation

This is the extent to which the manufacture of the garment is split among different operations, in the interest of greater specialisation and thus efficiency.

18.3.5.3 Operation Breakdown

This usually takes the form of the element descriptions from the method study, together with the appropriate standard times and a note of the type of machinery required. Special work aids and attachments should also be mentioned on it.

18.3.5.4 Theoretical Operation Balance

The elements are grouped together to match the number of people selected, in the calculation of labour requirements. No allowance is made for the varying ability of the people who will man the workstations.

18.3.5.5 Skills Inventory

This consists of a list of the people in the section or factory, which shows their 'expected performance' at various types of work. It provides both a talent list for section/team manning and also a means of planning the growth of the skills of the workforce.

18.3.5.6 Initial Balance

The expected performance of the people available must be taken from the skills inventory, in order to manage the line in a way that smoothes out the potential variations in output between the stations shown in the theoretical balance. It is usual to select 'floaters' at this stage, who will help to cope with absenteeism and imbalance.

18.3.5.7 Balance Control

Balance control is perhaps the most vital skill in a supervisor, with its objective to maintain the highest output and not just to keep people busy. For simplicity, the worked examples in the text and in three of the questions in the next chapter are taken from the same case study.

18.3.6 Macro-Steps in Line Balancing

The method of line balancing can vary from factory to factory and depend on the type of garments manufactured, but in any instance, line balancing concerns itself with two distinct applications. They are 'setting up' a line and 'running' a line (Bubonia 2012).

18.3.6.1 Setting Up a Line

Before a new style is introduced to a production line, it is necessary to establish the operation sequence, the time, the type of equipment and the attachments required to manufacture the order. Management must have this information before the commencement of the order, so that the line can be balanced and laid out in such a way as to maximise productivity. Two methods can be used to set up a line:

Method 1: Calculating the number of operators necessary to achieve a given production rate per hour.

Method 2: Calculating the number of garments to be produced by a given number of operators.

Using either technique, certain information is required before commencing the calculations, which are given below:

1. The number of operators in the line
2. A list of operations involved in making the garment
3. The standard minute values for each operation
4. Output required from a given group of operators

Further, the following information is required for balancing a line:

- The size of the group
- An operation sequence
- The standard time for each operation
- The total standard time for the garment

The method of calculating the line balance is as follows:

1. Add up the operation times for the whole of the style.
2. Establish the percentage of each operation of the total time.
3. Work out the theoretical balance using each operation's percentage of the total number of operators on the line.
4. Round off the theoretical balance to the nearest half an operator, either up or down.
5. List the type of equipment required for each operation at the side of the rounded figure.

6. The equipment that has half operators could be combined with similar equipment to get 'full' operators.

7. If odd half operators are there, it should be rounded up.

8. The number of garments that would be produced per hour on each operation should be calculated by multiplying the number of operators by 60 (minutes) and dividing by the total minutes for the style.

Line Imbalance

A series of operations is involved in producing a garment. In bulk garment production, generally a group of people works in a particular assembly line and every operator is capable of doing only one specific operation and then hands over the product to the next operator to carry out the next operation. Under some circumstances, in the assembly line it could be observed that work is started to pile up in a particular production line and a few operators are idle. When this situation arises in the production line, it is known as an imbalanced line (Solinger 1988; Mok 2013) and it happens due to two main reasons: difference in work content in dissimilar operations and variation in performance level of an operator.

The main important aspect to be considered for imbalance in a line is the identification of the bottleneck area in the production process. Each individual operator's capacity should be compared with the target capacity. The operators whose ability is less than the target output are bottleneck operations for the production process. Without improving the bottleneck operation in the production line, it is practically not possible to increase the output of imbalanced line (Sudarshan and Rao 2013). Therefore, to remove the bottleneck operation, the following methods could be used depending upon the situation.

- Group the operations wherever possible – An operator could be given another operation with less work content in case of availability of higher capacity than the target output.

- Shuffle operators – For the operations that have low work content, a low performing operator can be allotted and consequently for the operations having higher work content efficient operators could be allotted.

- Reduce cycle time – Working aids such as guides or attachment could be used to aid the operator in handling parts during sewing, positioning, cutting and finishing.

- Improve production layout – The most significant zone for recuperating output from a particular process is by means of the best production layout and the best working method.

- More operators in bottleneck operations – Include one or two extra machines in tougher tasks. Before doing this, evaluate the cost benefits of putting additional machines on the line.

18.3.6.2 *Running a Line*

1. There should be a reasonable level of work in progress. A recommended level is between 30 minutes to 1 hour between operations. Anything below 30 minutes will not give the supervisor sufficient time to react to a breakdown. Anything above 1 hour's supply is unnecessary.
2. Work in progress should always be kept in good order and full view.
3. Have a number of additional machinists trained on many operations so that they can be used where necessary to cover for absenteeism. Therefore, if absenteeism is 5%, a squad of skilled operators would be required to cover this amount.
4. Space should be made available within the line for spare machines in case of a breakdown.
5. Ensure that the mechanics keep the machines regularly serviced.
6. If a bottleneck keeps occurring at a particular place in the line, improve the method to eliminate the bottleneck. It is most important to establish where this point is on the line.
7. Supervisors must know the capabilities and skills of the operators under their control.
8. Supervisors must learn that the amount of work waiting for each operation will increase or decrease over a period of time, and must plan when to take appropriate action.
9. Supervisors could carry out balancing duty regularly at 2-hour intervals, checking every operation on the line to ensure that the WIP level is within the correct limits.
10. Balancing duties should be carried out on time irrespective of what else the supervisor is doing.
11. The supervisor should be able to make up his or her mind about what to do if the levels are not correct, and not have to wait for a manager to make the decision.

18.3.7 Important Aspects in Line Balancing

There are certain aspects that have to be determined for line balancing as given below:

- Determination of the cycle time
- Determination of the ideal number of work required in the line
- Balancing efficiency

18.3.7.1 *Determination of Cycle Time (CT)*

Cycle time is the time interval at which completed garments leave the production line. When the quantity of output units required per period is specified and the available time per period is given, then

$$\text{Cycle time (CT)} = \frac{\text{Available time per period}}{\text{Output units required per period}}$$

18.3.7.2 Determination of the Ideal Number of Workers Required in the Line

Ideal number of workers required in the assembly line and production line

$$= \frac{\text{(Total operation or task time)} \times \text{(Output units required per period)}}{\text{Available time per period per worker}}$$

$$\text{i.e. } N = \Sigma t \times \frac{1}{(CT)} = \frac{\Sigma t}{CT}$$

18.3.7.3 Balancing Efficiency

A well-organised line balancing system could reduce the idle time and could be determined as

$$\text{Balancing efficiency (\%)} = \frac{\text{Output of task time}}{\text{Input by workstation times}} = \frac{\Sigma t}{CT \times N}$$

where
  Σt = Sum of the actual worker times or task times to complete one unit
  CT = Cycle time
  N = Number of workers or work stations

$$\text{Effb} = \frac{\text{Theoretical number of workers}}{\text{Actual number of workers}}$$

Example: The preference diagram for assembly activities A to G is shown in Figure 18.2. The element times required for the activities are also shown in the diagram in minutes. The garment line operates for 7 hours per day and an output of 550 units per day is desired. The determination of idle time of activities is shown in Figure 18.3.

(i)

$$\text{Cycle time} = \frac{\text{Available time per period}}{\text{Output units required per period}}$$

$$= \frac{(7 \times 60) \text{ min}}{550}$$

$$CT = 0.76$$

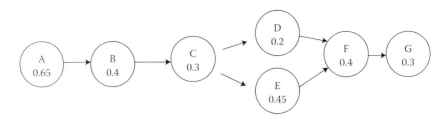

FIGURE 18.2
Example of an assembling activity with SAM values.

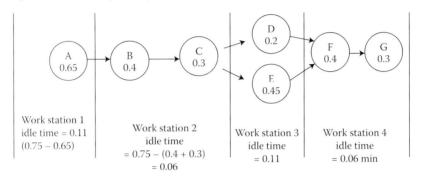

FIGURE 18.3
Determination of idle time of activities.

(ii)

$$\text{Theoretical minimum no. of workers} - \frac{\sum t}{CT}$$

$$= \frac{0.65 + 0.4 + 0.3 + 0.2 + 0.4 + 0.45 + 0.3}{0.76}$$

$$= 3.552$$

Grouping of work stations is done arbitrarily in such a way that the number of work stations is equal to the number of actual workers and the idle time in each work station is minimal. The sequence of operation should not be disturbed while grouping.

Total operation time = 2.7 minutes

Total idle time = 0.34 minutes

Total cycle time = 2.7 + 0.34 = 3.04 minutes (operation time + idle time)

$$\text{Balancing efficiency} = \frac{\sum t}{CT \times N} = \frac{2.7}{0.76 \times 4} = 0.8881$$

TABLE 18.1

Line Balance Matrix

		Output		
		Low	Medium	High
WIP	Low	Put extra operator in previous operation	Analyse the operator in previous operation	No change
	Medium	Put extra operator in current operation	Analyse the operator in current operation	No change
	High	Remove the operator from previous operation and balance in current operation	Remove the operator from previous operation	Remove the operator from previous operation and balance where production is low

Or $\text{Eff}_B = 0.8881 \; 100\% = 88.81\%$
Alternatively,

$$\text{Eff}_B = \frac{\text{Theoretical minimum no. of workers}}{\text{Actual number of workers}} \times 100$$

$$= \frac{3.552}{4} \times 100 = 88.81\%$$

18.3.8 Line Balance Matrix

The line balance matrix at different situations is given in Table 18.1.

18.4 Determination of Machinery Requirements for a New Factory

While executing an order for bulk production, a variety of machines is required to complete the assembling process and production. The machines should be selected based on the product category. The factors influencing the machinery requirement are given below.

18.4.1 Selection of Product Type

The requirement of machine type depends on product types and product styling. If multiple product styles are planned, then the number of units to be produced per day for each product should be decided.

18.4.2 Daily Production Target

The next step is to decide the number of garment pieces to be produced monthly or daily.

18.4.3 Estimation of Line Efficiency

Though one could not be aware of the line efficiency at the starting stage of set-up of a garment industry, we have to consider a line efficiency level the garment factory could perform 6 months ahead to calculate the machine requirement. As most of the garment factories run at 40% efficiency (India), this value could be taken for the calculations.

18.4.4 Preparation of Operation Bulletin

The industrial engineer will practise the operation bulletin of the product. The industrial engineer has to study the product and ensure the types of operations needed to assemble the garment, select the correct kind of sewing machines for each operations and determine the SAM of the product, number of sewing machines required for each process/operation and total machines needed per line to accomplish the production target.

18.4.5 Calculation of Number of Lines

The number of lines required for the production of a particular style is given by a ratio between daily total production target and estimated production per line.

18.4.6 Preparation of a Matrix of Machine Mix

Once production line number of machines for each product is considered, then preparation of a table with a list of machines for each product style has to be done as shown in Table 18.2.

18.5 Estimation of Production Capacity of a Garment Factory

In the garment industry, production capacity is one of the significant criteria used for merchant selection by the buyers. So it is crucial that the marketing and planning department should be conscious about the production capacity of the production lines (Solinger 1988; Sarkar 2011). Production capability of a garment factory is mainly stated by means of total machines in the factory or number of pieces the factory produces on a daily basis for the specific

TABLE 18.2

Machine Mix Matrix

Sl. No.	Machines Sewing Machines	Product Wise No. of m/c Requirement				
		Trouser (2 Lines)	Shirt (4 Lines)	Tee (8 Lines)	Polo (2 Lines)	Total
1	SNLS (with thread trimmer) or	50	112	14	26	202
	SNLS (without thread trimmer)					0
2	SNLS (with edge cutter)		12			12
3	3THO/L	5			12	17
4	4TH O/L		32			32
5	SNCS	4				4
6	FOA (Feed of the arm)		6			6
7	Key hole making (for trousers)	2				2
8	Button holing (computer controlled)	2	6			8
9	Button sewing (computer controlled)		6			6
10	Bartack machine	2			1	3
11	SNAP button attaching machine	1			1	2
12	Flat lock (flat bed)			24	6	30
13	Flat lock (cylinder bed)					0
	Total no. of machines	66	142	46	46	300

product style. Normally, total numbers of machines in an industry remain the same for a particular period of time. But an industry could manufacture different styles of product during the particular season. According to the style, the machinery requirement could change and the average production in each style may differ. To determine the daily production capacity in terms of number of pieces, the information such as factory capacity in terms of hours, SAM of the product and line efficiency is needed.

18.5.1 Calculation of Factory Capacity (in Hours)

For the calculation of this we need the number of machines in the garment unit and the number of running hours per day. For example, if

Total number of machines = 300

Shift hours per day = 8 hours

Then total factory capacity (in hours) = 300 × 8 hours = 2400 hours.

18.5.2 Calculation of Product SAM

The SAM of different product styles has to be obtained from the industrial engineer in the factory through proper study.

18.5.3 Factory Average Efficiency

This information is collected from the industrial engineer or could be calculated from past data. Assume the average line efficiency is around 50%.

18.5.4 Calculation of Production Capacity (in Pieces)

With the above data, the following formula could be used to determine the production capacity:

$$\text{Production capacity (in pieces)} = \frac{\text{Capacity in hours*60}}{\text{SAM of product}} \times \text{line efficiency}$$

For example, a garment industry has 8 sewing lines and each line has 30 machines for a total of 300 machines and a working shift is 8 hours per day. Total factory capacity per day is 2400 hours (300 machines × 8 hours). If a garment industry is making a formal shirt having a SAM value of 28 minutes and has utilised daily production capacity of all 320 machines at 55%, then

$$= (2400 \times 60/28) \times 55\%$$

$$= 2828 \text{ Pieces}$$

18.6 Sewing Room Capacity

Capacity planning or production planning is generally done based on sewing capacity. Apart from production planning, the production planner should also have knowledge on the capacity in other processes to meet the deadline (Solinger 1988; Schertel 1998). The sewing room capacity can be calculated by the given formula.

$$\text{Sewing room capacity per day (min)} = \{(\text{Number of sewing machine}$$
$$\times \text{work hours in a day} \times 60)$$
$$- \text{worker absenteeism \%}\}$$
$$\times \text{efficiency \%}$$

$$\text{Monthly capacity} = \text{Daily capacity} \times \text{number of working days in a month}$$

TABLE 18.3

Example of a Sewing Capacity

Line No.	No. of Operator	Minutes/Day (Daily Working Hours × 60)	Line Efficiency (%)	Absenteeism (%)	Capacity Available
Line 1	20	480	50	8	4416.0
Line 2	26	480	55	10	6177.6
Line 3	30	480	45	9	5896.8
Line 4	32	480	50	10	6912.0
		Total sewing floor capacity per day (in minutes)			23,402.4

For example, a garment industry has 4 lines and it works for 8 hours day. The number of total operators, line efficiency and absenteeism percentages are as given in Table 18.3.

Available capacity of the line will vary on factors such as

- Number of operators working in the line
- Line's existing efficiency
- Operator absenteeism percentage

Capacity could also be expressed in number of garment pieces by dividing the total capacity (in minutes) by SAM of the garment. Assume that a garment industry produces a full sleeve shirt of SAM 21. Shirt production capacity of the floor will be 1394 pieces per day (29,268/21).

18.7 Determination of Operator Efficiency

In a garment manufacturing system, skills and ability of a sewing operator are stated as 'operator efficiency'. An operator having higher efficiency produces more garments than an operator having lower efficiency for the same period of time, which could minimise the cost of manufacturing a garment. In addition, the capacity of the industry is determined according to the operator efficiency (Solinger 1988; Sarkar 2011). Thus, efficiency is one of the predominantly used performance assessment tools. SAM of the garment and the list of operations performed by the operators are required to calculate the efficiency of the operator using the formula given below.

$$\text{Operator Efficiency (\%)} = \frac{\text{Total minute produced by an operator}}{\text{Total minute attended by him}} \times 100$$

where

Total minutes produced = Total pieces made by an operator

$$\times \text{SAM of the operation [minutes]}$$

Total minutes attended

$$= \text{Total hours worked on the machine} \times 60 (\text{min})$$

For instance, an operator was carrying out an operation with a SAM of 0.65 minute. In a shift of 8 hours, he produces 420 pieces. Then the operator's overall efficiency is given by

$$= (420 \times 0.65)/(8 \times 60) * 100\%$$

$$= 56.87\%$$

18.7.1 On-Standard Operator Efficiency

Operator efficiency could be articulated in a more precise manner as 'on-standard efficiency'. An operator may be attending all the hours in a shift but if he or she has not been allotted any on-standard job to carry out in the particular shift, then he or she will not be in a position to achieve the SAM as per his or her capacity and skill level. The operator's on-standard efficiency could be determined using the following formula (McBride 2003; Sarkar 2011):

$$\text{Operator on-standard efficiency (\%)} = \frac{\text{Total minute produced}}{\text{Total on standard minute attended}} \times 100$$

where

$$\text{Total minutes produced} = \frac{\text{Total pieces made by an operator}}{\text{SAM of the operation (min)}}$$

Total on-standard minute attended

$$= (\text{Total hours worked} - \text{loss time}) \times 60 \, [\text{minutes}]$$

For example, an operator produces 450 pieces per shift of 8 hours with an operation SAM of 0.70 minutes. He was 'waiting for work' for 30 minutes and his machine broke down in a particular shift for 40 minutes. Then, the operator's on-standard efficiency is given by

$$= (450 \times 0.70)/\{480 - (30 + 40)\} * 100\%$$

$$= 76.8\%$$

18.8 Determination of Efficiency of a Production Line

Similar to the calculation of individual operator efficiency, the efficiency of a production line could also be equally vital for an apparel industry (Mannan and Ferdousi 2007). For the determination of efficiency of a production line for a day, the following information is required.

1. Number of operators working in the line in a day
2. Working hours in a day
3. Production (number of pieces) per day
4. Expected garment SAM for the particular style

From the above information, the following factors need to be determined:

1. Total minutes produced by the line
2. Total minutes attended by all workers in the particular production line
3. Line efficiency (%)

An example for the calculation of line efficiency is given in Table 18.4.

18.9 Line Loading Plan for Garment Production

In a line loading plan, the person from the production department decides on a date a particular style is to be loaded in the line and the number of lines to

TABLE 18.4

Calculation of Line Efficiency

Number of Operator (A)	Working Time in Hours (B)	Production (Line Output) (C)	SAM of Garment (D)	Total Attended Minutes (E = A × B × 60)	Total Minute Produced (F = C × D)	Line Efficiency (%) (F/E ×100)
42	8	160	42.25	20,160	6760	33.53
45	10	220	41.25	27,000	9075	33.61
32	8	310	22	15,360	6820	44.40
35	11	420	25	23,100	10,500	45.45
34	10	339	24	20,400	8136	39.88
37	8	230	24	17,760	5520	31.08
35	9	210	34	18,900	7140	37.78
34	11	331	35	22,440	11,585	51.63
34	10	350	34	20,400	11,900	58.33

TABLE 18.5

Example of Order List in an Industry

Order No.	Description of Product	Order Quantity (in pieces)	Production Completion Date
ASS101	Dress	2000	10th May
ASS102	Blouse	3000	12th May
ASS103	Trouser	5000	15th May
JKY104	Long sleeve Tee	3000	17th May
JKY105	Skirt	3000	21st May
JKY106	Dress	1500	21st May
PEN107	Long sleeve Tee	10,000	10th May
PEN108	Skirt	1200	04th May
	Total	28,000	

be allocated for the particular style to meet the production target date (Jang et al. 2005; Sarkar 2011). This is an essential job for a production planner. He or she has to do backward as well as forward planning based on lead time available. The stepwise procedure for the line loading plan is given below:

Step 1: Construct a list of ongoing orders with detailed information like order number, description of style, quantity and production target date as shown in Table 18.5.

Step 2: If a garment industry has five production lines, then the available capacity of the line could be determined based on the capacity calculation formula taking into account absenteeism (10%) and line efficiency as mentioned previously and as shown below.

Available capacity in hours =

$$\left\{ \left(\text{Number of} \frac{\text{operators}}{\text{machines}} \times \text{no. of working days in a month} \right. \right.$$

$$\times \text{daily work hours} \times 60 \Big) - \text{absenteeism } \% \right\} \times \text{Efficiency } \%$$

For example, available capacity in each line has been given in Table 18.6. It is considered that each line is equipped with 20–32 operators (machines), a factory's normal shift time is 8 hours (480 minutes) and line efficiency is in the range of 45%–55%.

Step 3: Subsequently, the required capacity for each order in minutes and in days has to be determined. Consider that all lines are empty and there is no concern with the starting date, then the decision has to be made on which line to be chosen for the processing of the

TABLE 18.6

Example of Available Capacity Calculation

Line No.	No. of Operator	Minutes/Day (Daily Working Hours × 60)	Line Efficiency (%)	Absenteeism (%)	Capacity Available
Line 1	20	480	50	8	4416.0
Line 2	26	480	55	10	6177.6
Line 3	30	480	45	9	5896.8
Line 4	32	480	50	10	6912.0

TABLE 18.7

Example of Capacity Calculation

Order No.	Loaded to Line No.	Order Quantity	Style SMV	Capacity Required (minutes)	Capacity Available Per Day	Capacity Required (Days)
ASS101	Line 1	1800	26	46,800	4416.0	11
ASS102	Line 2	2000	24	48,000	6177.6	8
ASS103	Line 3	2500	22	55,000	5896.8	9
JKY104	Line 4	2700	13	35,100	6912.0	5
JKY106	Line 1	1800	24	43,200	4416.0	10
PEN107	Line 4	9000	12	108,000	6912.0	16

particular style. Assign the order to the line as per the product category and line set up. For example, in Table 18.7 the line number has been revealed against the order number. The following formula has to be used for the calculation of capacity.

$$\text{Capacity required in minutes} = \text{Order quantity} \times \text{SMV of a style}$$

$$\text{Capacity required in days} = \frac{\text{Capacity required in minutes}}{\text{Capacity available per day}}$$

Step 4: After that, backward calculation has to be made to find out the date for style loading and number of days (excluding Sundays and holidays) required for completing the production on target date. One to two days could be added initially for setting up a line as per the style requirement. If needed, one or two days of buffer could be added. To make it easy in determining the loading dates taking into account the above points, the spreadsheet-based planning board (Figure 18.4) could be used. Finally, another final table (Table 18.8) has to be prepared demonstrating the loading date against the respective orders.

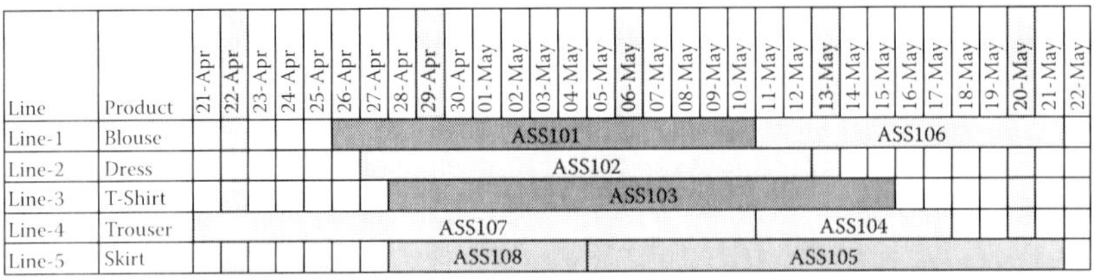

Planning board (Line / Product / dates 21-Apr through 22-May):

Line	Product	Schedule
Line-1	Blouse	ASS101 ... ASS106
Line-2	Dress	ASS102
Line-3	T-Shirt	ASS103
Line-4	Trouser	ASS107 ... ASS104
Line-5	Skirt	ASS108 ... ASS105

FIGURE 18.4

Example of a planning board.

TABLE 18.8

Example of Order Loading Date

Order No.	Garment Description	Loaded to Line #	Production Completion Date	Capacity Required (Days)	Loading Date
TKK101	Dress	Line 1	10th May	11	26th April
TKK102	Blouse	Line 2	12th May	12	27th April
TKK103	Trouser	Line 3	15th May	13	28th April
PGG104	Long sleeve Tee	Line 4	17th May	5	11th May
PGG105	Skirt	Line 5	21st May	13	5th May
PGG106	Dress	Line 1	21st May	8	11th May
PGG107	Long sleeve Tee	Line 4	10th May	16	21st April
PGG108	Skirt	Line 5	04th May	5	28th April

References

Ahmad, S., A.A.B. Khalil and C.A.A. Rashed. 2012. Impact efficiency in apparel supply chain. *Asian Journal of Natural Applied Science* 1(4):36–45.

Babu, V.R. 2006. Garment Production Systems: An Overview. http://www.indiantextilejournal.com/articles/FAdetails.asp (accessed on March 12, 2015).

Bubonia, J.E. 2012. *Apparel Production Terms and Processes.* Fairchild Books, New York.

Chuter, A.J. 1995. *Introduction to Clothing Production Management.* Blackwell Scientific Publications, Oxford, UK.

Fairhurst, C. 2008. *Advances in Apparel Production.* The Textile Institute, Woodhead Publication, Cambridge.

Glock, R.E. and G.I. Kunz. 2004. *Apparel Manufacturing – Sewn Product Analysis.* Prentice Hall, Englewood Cliffs, NJ.

Jang, N., K.G. Dickerson and J.M. Hawley. 2005. Apparel product development: Measures of apparel product success and failure. *Journal of Fashion Marketing Management* 9(2):195–206.

Kumar, A. 2008. *Production Planning and Control: Lesson 8 Course material.* Delhi University. www.du.ac.in/fileadmin/DU/Academics/course_material/EP_08.pdf (accessed on March 14, 2015).

Mannan, M.A. and F. Ferdousi. 2007. *Essentials of Total Quality Management.* The University Grants Commission of Bangladesh Dhaka.

McBride, D. 2003. *The 7 Manufacturing Wastes*. http://www.emsstrategies.com/dm090203article2.htm (accessed on March 25, 2015).

McKelvey, K. and J. Munslow. 2003. *Fashion Design: Process Innovation and Practice*. Blackwell Publishing, Oxford, UK.

Mok, P.Y., T.Y. Cheung, W.K. Wong., S.Y.S. Leung and J.T. Fan. 2013. Intelligent production planning for complex garment manufacturing. *Journal of Intel Manufacturing* 24(1):133–45.

Nahmias, S. 1997. *Production and Operations Analysis*. Irwin, Chicago, IL.

Oliver, B.A., D.H. Kincade and D. Albrecht. 1994. Comparison of apparel production systems: A simulation. *Clothing Textile Research Journal* 12(4):45–50.

Ramesh, A. and B.K. Bahinipati. 2011. The Indian apparel industry: A critical review of supply chains. In: *International Conference on Operations and Quantitative Management (ICOQM)* Nashik, India.

Ramesh Babu, V. 2012. *Industrial Engineering in Apparel Production*. Woodhead Publishing India Pvt Ltd, New Delhi, India.

Ray, L. 2014. *Production Planning for Garment Manufacturing*. http://smallbusinesschroncom/production-planning-garment-manufacturing-80975.html (accessed on October 5, 2014).

Russell, R.S. and B.W. Taylor. 1999. *Operations Management*. Prentice-Hall, Upper Saddle River, NJ.

Sarkar, P. 2011. Functions of Production Planning and Control (PPC) Department in Apparel Manufacturing. http://www.onlineclothingstudy.com/2011/12/functions-of-productionplanning-and.html (accessed October 05, 2014).

Schertel, S. 1998. New Product Development: Planning and Scheduling of the Merchandising Calendar. Master dissertation, North Carolina State University, Raleigh, NC.

Shaeffer, C. 2000. *Sewing for the Apparel Industry*. Woodhead Publication, Cambridge.

Solinger, J. 1988. *Apparel Manufacturing Handbook – Analysis Principles and Practice*. Columbia Boblin Media Corp, New York, USA.

Sudarshan, B. and N.D. Rao. 2013. Application of modular manufacturing system garment industries. *International Journal of Science and Engineering Research* 4(2):2083–9.

Tyler, D.J. 1991. *Materials Management in Clothing Production*. BSP Professional Books, Blackwell Scientific Publications, Oxford, UK.

Vijayalakshmi, D. 2009. Production Strategies and Systems for Apparel Manufacturing. http://www.indiantextilejournal.com/articles/FAdetails.asp?id1/41988 (accessed on March 12, 2015).

19

Garment Merchandising

In the textile industry, merchandisers have a predominantly significant role owing to the exhaustive nature of product range. The practice of buying and selling materials and services is called merchandising. Merchandising activity coordinates different departments in the garment industry. It develops a valuable relationship with the buyers. It builds an excellent relationship with the buying houses and the merchandiser concentrates on queries, order processing and assessment of apparel products. All these aspects make the merchandising activity an important role in the garment industry (Kunz 2005).

The function of merchandising differs relying on whether it is performed in retail or manufacturing. It involves the conceptualisation, development, obtainment of raw materials, sourcing of production and dispatch of product to buyers.

19.1 Types of Merchandising

Two kinds of merchandising are practiced in the export of garment units

1. Marketing merchandising
2. Product merchandising

The main purpose of marketing merchandising is product development and costing of the same. Product merchandising comprises all the responsibilities from sourcing of materials to dispatching of finished goods and is done in the garment unit itself.

19.2 Evolution of Merchandising in Garment Unit

The conception of the word 'merchandising' can be followed back to the early historic period with the materialisation of exchange or trade between nations. There are various data available about the presence and progress of trade between the civilisations in those periods like Greek, Roman, Indian, Chinese, and Egyptian. But amid those days, the importance of the word merchandising was limited merely to just exchange of commodities which were profited from nature. They were not produced for a particular purpose or customers (Fan and Hunter 2004).

The word merchandising got its importance after the industrial revolution that appeared after World War II. During this time, there was enormous demand for product development. Hence, merchandising grew as a connection between the design and marketing and sales to fulfil the needs of the public. The inevitability of merchandising is essential due to various reasons like intense growth of the garment industry, intricate raw material and processes, arrival of fresh garment styles, shorter product life cycle (PLC), innovations in textiles, rapid growth of application of computers in textiles etc.

19.3 Merchandiser

An individual who is associated with merchandising activity is called a merchandiser. The merchandiser synchronises with the design team to successfully exhibit the product (Tyler 1992; Kunz 2005). He or she creates colours and specifications and carries out the market research to decide the most effectual ways to sell and promote the product. Excellent communication, ability to negotiate and analytical competences are essential qualities required for a merchandiser. Further, he or she also desires to be a creative and innovative thinker. The qualities required for the merchandiser are shown in Figure 19.1. He or she should be able to plan meticulously and control the operations involved in production of products, sourcing them and dispatching them to the customer on time.

A merchandiser should be partially a designer able to think creatively, partly an engineer able to develop the product, partly a computer expert

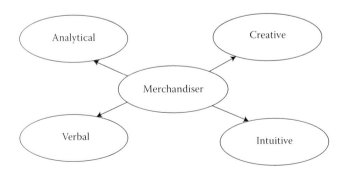

FIGURE 19.1
Myers theory on merchandising.

able to communicate online, partly a marketer able to market and sell the product, and partly an entrepreneur. According to *Theory on Merchandising*, Myer defines merchandising as 'Careful planning, capable styling and production or selecting and buying, and effective selling' (Diamond 2000).

19.3.1 Qualities of Merchandiser

- *Planning capability:* Merchandiser must be competent enough to plan the activities based on the order that is to be followed. Otherwise, it will directly affect the delivery time of the order.
- *Decision making:* It is a very important quality required for the merchandiser to deliver the product on time to buyers.
- *Communication skills:* Oral as well as written communication are important to endorse the business activity as well as to have a good relationship with the buyers.
- *Loyalty:* It is a crucial character of human beings, particularly for business persons.
- *Technical knowledge about the field:* The merchandiser must have ample knowledge about the garment production activities, and technical knowledge to communicate with different levels of persons in the apparel industry.
- *Coordinate and cooperate:* The merchandiser is the person who coordinates with the various departments in an apparel industry to get the job done.
- *Monitoring ability:* He or she must supervise the various activities in different departments to speed up the orders to dispatch it on time to the buyers.

19.3.2 Function of Merchandisers

1. Development of new garment styles and samples and execution of the same
2. Garment costing based on the order
3. Arrangement of raw materials, accessories and trims for execution of an order
4. Production scheduling
5. Approval of patterns and various samples
6. Follow up of preproduction activities
7. Coordinating with inspection agencies
8. Production controlling
9. Identification of bottlenecks in the process and materials and resolve the same

10. Monitoring of in-house production activities as well as follow-up of subcontract work given outside
11. Reporting the progress of orders to the buyer as well as top management
12. Maintenance of proper records for individual garment styles
13. Ensuring constant production rate by taking preventive as well as corrective actions
14. Attending meetings with superiors and furnishing the required details about merchandising

19.3.3 Types of Garment Merchandising

19.3.3.1 Fashion Merchandising

It includes all the activities beginning from fashion forecasting, design and development of product to retail activity and this also comprises production as well as retail merchandising. Fashion merchandise consists of items of retail merchandise that have ornamental value either with or without having any functional value. It includes predominantly items of apparel because they can all be ornamental as well as functional (Chuter 1995; Evelyn 1999; Stone 2001; Glock and Kunz 2004). The meadow of fashion merchandising exists to service the designer and customer relationship. The process flow of a fashion merchandiser is shown in Figure 19.2.

19.3.3.1.1 Fashion Forecasting

It demonstrates/directs the apparel industry for new fashion presentation for the forthcoming season, thus it can recognise new fashion concepts, and the retail store's merchandising approach.

19.3.3.1.2 Design Development

The design is created in light of themes utilising the components of design, namely, colour, texture, shape and implementing the principles of fashion such as proportion, balance and harmony. The design development must be practical which collects the current fashion trends and also viable to convert them into a finished garment.

19.3.3.1.3 Sample Development

The collection of designs developed by the fashion designers and the designs that have excellent prospects should be selected and taken for sample development process.

19.3.3.1.4 Product Specification

The specifications with regard to the particular product or style could be helpful in setting the product specification for the particular style of garment.

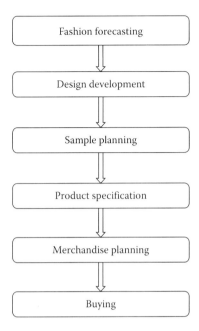

FIGURE 19.2
Process flow of a fashion merchandiser.

It will direct the production department in the industry to develop and plan their merchandising activities and production planning and organise the thing for effective and efficient production of the merchandise.

19.3.3.1.5 Merchandising Plan or Policy

It is a long range criterion for fashion buying and selling also for associated activities such as sales promotion. Merchandising plans are planned quite a few months ahead of the selling season.

19.3.3.1.6 Buying

It is a very vital task of fashion merchandising because it is the real process of manufacturing the product and displaying it for the sales at retail stores. A buyer's task consists of both the buying and selling features of retailing.

19.3.3.2 Apparel Export Merchandising

It could be defined as all the planning as well as activities involved from the buyer communication and order receiving to the dispatch of the product by fulfilling the subsequent factors:

- *Right merchandise:* Retailers should fill their shelves with the merchandise that the customer needs.

- *Right place:* The place/location of the merchandise is of significance as it decides the ease of access.
- *Right time:* Since the majority of the merchandise is based on seasons and seasonal based changes in fashion and the related requirements should be on hand when it is mainly needed.
- *Right quantity:* A lucrative balance between volume of sales and quantity of inventory is the required objective.
- *Right price:* Merchandiser could arrive at a cost that is adequate to provide the retail store profit and yet low enough to meet the competition and customer's expectations.
- *Right promotion:* Correct balance between the investment and the demand created for the customers.

19.4 Merchandising Workflow

The work activities of a merchandiser will include the following:

- Examining the buyer requirement, understanding and communicating them to the specific departments and exhibiting the product to the buyers they need.
- Confirming the quality during production as well as ensuring timely delivery of an order.
- Developing a time and action (TNA) calendar for completing the schedules of various activities like cutting, sewing, finishing, dispatch etc. The WIP (work in progress) and the status of the order have to be monitored by the merchandisers regularly.
- Coordinating and tracking the sourcing activities and confirming that all the raw materials and accessories are delivered on time.
- Follow-up of postshipment activities to keep a long-term relationship with the buyer.
- Accompanying the buyers on visits to manufacturers to understand production processes.
- Meeting with suppliers for negotiating the cost and handling of stocks.
- Ascertaining the difficulties related to production and supply of an order and dealing with it when they occur.
- A preproduction meeting (PPM) is held among staff in the garment industry to discuss the style, trims, construction etc. if there are further clarifications, an external PPM is held with the QC, merchandiser, and buyer.

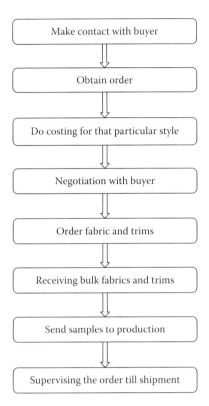

FIGURE 19.3
Merchandising workflow.

The general merchandising workflow is shown in Figure 19.3 (Davis and Nancy 2002).

19.5 Merchandising Process Flow

The merchandising process flow is shown in Figure 19.4.

19.6 General Merchandising Process

19.6.1 Order Enquiry

This is the first step where the buyers have an enquiry with the merchandiser about a new order.

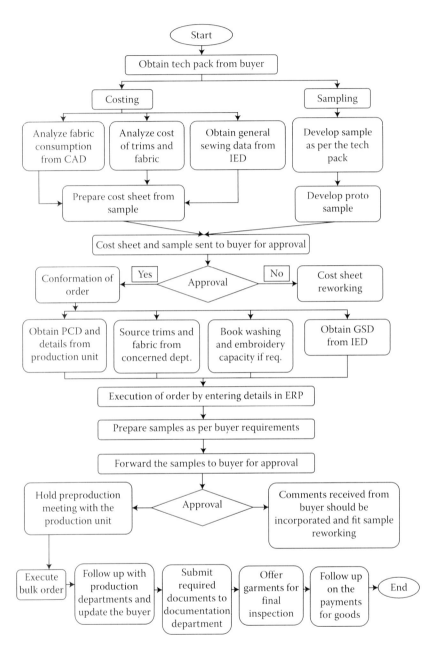

FIGURE 19.4
Merchandising process flow.

19.6.2 Forwarding Tech Pack

After the order enquiry has been completed, the buyer will send a 'specification sheet or tech pack' to the merchant. It covers all the details of a particular product style such as product style design, measurement details of garment, type of fabric and construction, style code of a product, surface ornamentation details if any etc.

19.6.3 Product Development

After the receipt of specification sheet (tech pack), the merchandisers have to organise the information provided in the specification sheet in a single format by categorising different product styles and their details. Consequently, the merchandiser should provide proper instructions to the junior merchandiser about the product style and hence he or she could assist the sample coordinators to prepare the development samples from the sampling department.

19.6.4 Approval of Development Samples

When the development samples are prepared, two or three samples have to be sent to the buyer for its approval. The main objective of a development sample is to realise how the particular style of garment looks with the specific details. These samples are prepared with the available fabric in the industry similar to the exact requirement. In the development sample, surface embellishments and fit analysis are followed as per the specification sheet.

19.6.5 Costing

Once the development sample is approved by the buyer, then the costing has to be done taking into account various costs incurred to produce a garment. It contains various factors such as

- Fabric cost
- Trims and accessories cost
- CMT (cut-make-trim)
- Finishing or washing cost
- Bank charges
- Buffer value
- Miscellaneous costs such as rejection cost, wastage etc.

19.6.6 Order Placement

After the determination of a garment cost which is also approved by the buyer, then the buyer will place the order with the necessary quantity of order and other main details to the merchant.

19.6.7 Order of Fabric and Trims

After the conformation of the order by the buyer, the merchandiser can place the order for requisite fabric by considering various parameters such as colour, GSM, weave structure etc., which is necessary for the specific garment style. The requirements are forwarded by the merchandiser to the purchase department and they will place the orders.

19.6.8 Lab Dip

The lab dips for a particular garment style, containing many shades of the fabric colour which the buyer is asked have to be sent to the buyer for the approval before going for further production.

19.6.9 Fit Sample

The fit garment sample is made after the development sample is approved by the buyer. The fit sample is generally produced in a medium size and with original fabric to check the fit. All the measurements should be verified as per the specification sheet. After checking the fit sample, the buyer returns the fit approvals sheet which comprises all the actual measurements and deviation in the garment has occurred for the purpose of correction. The order is confirmed only after the approval of the fit sample by the buyer.

19.6.10 Preproduction Samples

After the approval of fit samples, the preproduction (PP) samples (otherwise known as red seal samples) have to be produced. The red seal sample, which has to be produced as per the buyer's requirement, should have all the specifications of the particular style with the original or exact fabric, trims, colour, surface ornamentation etc. Two or three garment samples in each size (S, M, L, XL) have to be sent to the buyer for approval and the buyer can advise on any corrections if required.

19.6.11 Size Set Samples

These size set samples are prepared for the intension of inspection of various sizes of the same style with respect to measurements, fit, styling etc. Further, the size set samples are produced to verify whether the assigned unit is capable of producing the specific garment style as per the requirements and specifications in all the sizes (Fairhurst 2008). The work flow of sampling during the merchandising process is shown in Figure 19.5.

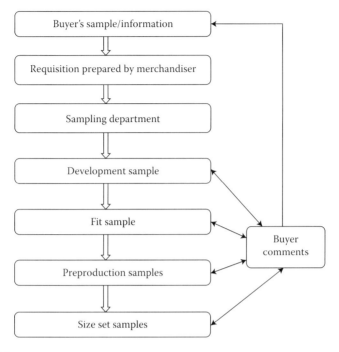

FIGURE 19.5
Workflow of sampling.

19.6.12 Preproduction Meeting (PPM)

After all the size set samples are approved by the buyers and all raw materials are organised in the stores, then the bulk production of the garment style can be started. A preproduction meeting should be organised by the merchandiser prior to bulk production with the production manager and other department heads to plan the production process to evade any delays in target time.

19.6.13 Hand-Over the Production File to Production Planning and Control

The production file comprising all the information of a particular garment style has to be prepared by the merchandiser. It is then forwarded to CPC along with the approved preproduction samples which are further forwarded to PPC after examination of the file by the CPC. Some of the details in the production file include the following items:

- Measurements for the specific garment style
- Export order sheet
- Colour details of the fabric and trims

- Brief description of style
- Type of packing required
- Instructions related to print/embroidery
- Material requirement sheet
- Job details for CAD and fabric order
- Marker plan
- TNA order sheet
- 2D style diagram and trims
- Packing information

19.6.14 Ensuring the Availability of Fabrics and Trims

After the receipt of the production file, the PPC have to study and check every detail in the file related to the particular garment style and simultaneously have to check the availability of the particular fabric and trims in the store.

19.6.15 Inspecting the Surface Ornamentation of the Particular Style

Surface embellishment may consist of embroidery, printing or appliqué and these are made as per the requirements of the buyer; hence, the PPC department should check the ornamentation details and plans accordingly to complete the same.

19.6.16 Checking the Status of Stitching Materials In-House

Sewing accessories are materials that are used for sewing such as sewing threads and accessories that assist production. Hence, simultaneously the merchandiser has to make arrangements for the stitching materials in-house.

19.6.17 Checking the Patterns with Master

The production file includes all the information about the patterns and the merchandiser has to forward the original patterns alongside with the production file to the production planning and control department. Once they receive the patterns, they will forward the patterns to the pattern master in the department and he or she will check the pattern and confirm the same.

19.6.18 Grading and Final Cross Check of Patterns

After inspection of all the patterns of different product styles by the pattern master, it is forwarded to the automatic grading section or manual grading section in the industry. Grading is a process of proportionately increasing or decreasing the sizes from the basic one. Finally, the graded pattern should be inspected and confirmed by the pattern master before going for further process.

19.6.19 Spreading and Cutting

It is the process of arranging fabrics on the spreading table as per length and width of the marker in stack form which has to be carried out carefully without any wrinkles or tension in the fabric. After spreading is completed, the patterns are placed over the top layer of the spread as per the marker plan and the plies are cut using a straight knife and band knife cutting machines. After cutting is completed, the cut components of the garment are sorted and bundled.

19.6.20 Garment Wash

In some specific cases, the cut components will go for washing as per the specification sheet or buyer requirement.

19.6.21 Fabric Printing/Embroidery

After the completion of garment washing, the bundles could be sent for printing/embroidery if required for the specific style. Surface ornamentations should be carried out on cut garment panels as it minimises the risk of destroying the entire garment if some defects occur during printing.

19.6.22 Loading the Order in the Production Line

After the surface embellishment is completed on cut components, the bundles are moved on to the sewing department as per the production plan. In the sewing section, the bundles are allotted to each worker as per his or her work and the sewing process moves from one end to the other end where the complete garment is assembled in a line.

19.6.23 Finishing

After the complete garment is assembled and inspected at the end of the each line, it is forwarded to the finishing section where the following operations will be carried out:

- Inspection: For any defects and stains
- Trimming: Protruding threads are removed to provide a neat appearance to the garment
- Ironing: To remove or introduce crease marks in the garment
- Packing: The finished garments are folded and packed in the polythene covers

19.6.24 Dispatch

Dispatch is the final process in which the garments are generally packed in wooden cartons with the dimensions specified by the buyer and shipped to the buyer.

19.7 Documents to be Maintained by the Merchandiser

19.7.1 Production Order (PO)

A production order comprises all the information needed by the PPC department to generate a line loading plan and it should be available for each style of every buyer. The production order consists of the description of style and style number, sizes, order quantity, quantity for each size, fabric consumption, specifications of interlining and trims, packing instructions, label specifications, etc.

19.7.2 Bill of Materials (BOM)

From the PO, the requisite quantity of fabric and trims could be determined for a single product which is then multiplied by the number of shirts being produced. The required quantities are given as a bill of materials for various trims such as sewing threads, buttons, zippers and cuff links. The BOM is issued to the store to get the required amount of trims.

19.7.3 Specification Sheet/Tech Pack

This form is vital for the execution of any order. It consists of all the technical information regarding the specific garment style such as fabric, tolerances, interlining details etc. for the style processing (Jarnow and Dickerson 1996). It provides necessary information required for various departments. For example, for the cutting department documents such as marker planning, marker consumption etc. and for the sewing department, details such as construction details, measurements etc. are provided by the tech pack. The specification sheet along with the PO form is issued to all the sections in the industry while the style is moving from one department to another.

19.7.4 Order Status Report

In this order status report, all the styles and their diverse activities are updated in an Excel work sheet and is retained by the senior merchandisers. Therefore, he or she could easily track the current progress of a particular style.

19.8 Apparel Retail Merchandiser

The retail business entails dividing a smaller part from a large good or a product and selling it to the consumers. Retail merchandising consists of all the operations associated with direct selling of products or even services to the consumers of a particular product. The retail merchandiser sells the products in small quantities and coordinates as an intermediary between the wholesaler and consumers (Kiell and Maynard 2001; Kunz 2005). The process flow of retail merchandising is shown in Figure 19.6.

19.8.1 Functions of a Retail Merchandiser

- Providing personal services to all required consumers.
- Giving two-way information such as from producer to consumer and vice versa.
- Assisting in standardisation and grading of products.

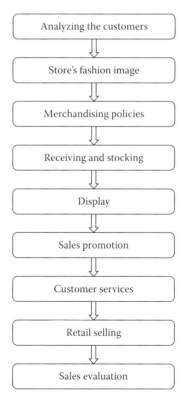

FIGURE 19.6
Apparel retail merchandising process.

- Undertaking transportation and storage of products.
- Assembling various products from different suppliers and wholesalers.
- Keeping adequate stock of various products to supply to consumers.
- Providing credit facilities to the consumers.
- Drawing the consumer's attention by bestowing a window display of products, conducting fashion events etc.
- Carrying out marketing activities.
- Assuming risk by stocking and providing goods to consumers.

19.8.2 Tasks of a Retail Merchandiser

19.8.2.1 Analysing the Local Customers

Sales in a particular retail store could be influenced by various factors such as

- The geographical location of the store
- The population content of the area
- The social activities that the area offers
- The economical conditions and level of the local population
- The fashion influences of the particular population

19.8.2.2 Selection of a Fashion Image

Each retail store has to create a retail image in the mind set of customers irrespective of whether the retail store wants it. A retail merchandiser should create his or her own store's image which could draw the attention of customers.

19.8.2.3 Buying the Fashion Merchandise

It is challenging work as it demands a huge amount of activities such as budget planning, merchandise selection, supplier selection etc. Further, it is also important to decide on the quality of the products to be ordered.

19.8.2.4 Receiving and Stocking the Merchandise

After the arrival of ordered goods at the retail stores, it should be inspected vigilantly for the quantities against the packing list or shipping invoice sent by the supplier. The quantity of the products/goods received at the retail store should be inspected prior to display or selling.

19.8.2.5 Display

The manner in which the merchandise/product is displayed in the retail store is significant for sale of goods. Better showcasing of goods always boosts the sales of the retail store and vice versa. The store's image is based on the principle of use of space for display.

19.8.2.6 Sales Promotion

It is a vital process for successful marketing and sale of a product in the store. It refers to promotion of sale of merchandise, ideas or services. It contains activities such as advertising, publicity, fashion shows, personal selling, visual displays, special events etc.

19.8.2.7 Sales Evaluation

After the completion of the selling season, the retail store has to analyse the sales of the goods in the particular season. This consists of analysing the sales by style, product, size and colour, fast sold goods, unsold items, price line, etc.

19.9 Performance Measurement Tools for Merchandising

Performance measurement is a technique of gathering and reporting information concerning the performance of an individual or organisations. Contrasting to the manufacturing process where task activities are small enough to have repeatability; hence, it follows the work measurement principle. Merchandising activities have a long duration and follow a project management principle (Stone 2001; Sumathi 2002; Tate 2004).

19.9.1 Enquiry Response Time

It is the time interval between the enquiries to the buyer for the order to the confirmation of the order. A response must be sent within 24 hours or the next working day. A costing request should be replied to within 48 hours with possible alternatives.

$$\text{Enquiry response time} = \frac{\text{Enquiries replied within time frame}}{\text{Total enquiries replied}} \times 100$$

19.9.2 Sample Acceptance Percentage

This represents the capacity of the design team in a garment industry in realising the buyer's tastes, costs and the trends of the current season.

A higher percentage of sample acceptance level facilitates a long-term relationship with the buyer and a better chance of receiving production orders for the particular styles.

$$\text{Sample adoption \%} = \frac{\text{Samples accepted by buyer}}{\text{Samples presented to the buyer}} \times 100$$

19.9.3 Order Conversion Rate

This shows the percentage of successful conversion of sampling to actual buyer orders. Lower percentage adds significant cost to the manufacturer and time delays.

$$\text{Order conversion rate (or) Sample hit rate} = \frac{\text{Number of styles ordered}}{\text{Number of styles sampled}}$$

19.9.4 On-Time Sample Delivery Percentage

It represents the total time taken from the date of order enquiry from the buyer to the dispatch of the order. A proto sample should be sent within four days and a fit sample within five days to the buyer to ensure a quick response from the buyer. This information can be valuable for evaluating the performance of the merchant and for acquiring further orders by showing a good delivery record to the buyers.

$$\text{Sample delivery} = \frac{\text{Number of samples sent on or before time}}{\text{Total number of samples sent}} \times 100$$

19.9.5 Sample Rejection Percentage

It is ratio between the number of garments rejected and the number of garments dispatched. The rejected garment includes proto sample, fit sample or any other type of samples depending on industry requirement. The rejection percentage evaluates the ability of pattern makers as well as merchandisers in understanding the tech pack. It is vital for the merchandisers to have a good impression with the buyer.

19.9.6 On-Time File Handover to Production Department

This measurement index represents the ability of the merchandising department in handing over the order related files to the production department in a stipulated time to avoid any bottleneck in the delivery of the order

(Stone 2001). Any change in the schedule disrupts the planning schedule. This index can be determined on a weekly or a monthly basis.

$$\text{File handover achievement} = \frac{\text{Number of on time deliveries}}{\text{Number of planned deliveries}} \times 10$$

19.9.7 Number of Orders Handled per Unit Time

It represents the number of orders successfully carried out by a merchandiser in a particular month or season. This indirectly evaluates the quantum of work carried out by the merchandiser in a specific period of time.

Orders handled per unit time = Total orders handled in a month or year

19.9.8 Value Handled per Unit Time

It signifies the value that the merchandiser is bringing into the company.

$$\text{Value handled per unit time} = \sum \text{order values in a month or year}$$

References

Chuter, A.J. 1995. *Introduction to Clothing Production Management*. Blackwell Scientific Publications, Oxford, UK.

Davis, L. and B. Nancy. 2002. *The Business of Fashion – Designing Manufacturing and Marketing*. Fairchild Publication, New York.

Diamond, J. 2000. *Fashion Retailing – A Multi-Channel Approach*. Prentice Hall, Englewood Cliffs, NJ.

Evelyn, C. 1999. *Moore Math for Merchandising*. Wiley Eastern Inc., India.

Fairhurst, C. 2008. *Advances in Apparel Production*. The Textile Institute, Woodhead Publication, Cambridge.

Fan, J., W. Yu and L. Hunter. 2004. *Clothing Appearance and Fit*. Textile Institute, Woodhead Publishing Limited, UK.

Glock, R.E. and G.I. Kunz. 2004. *Apparel Manufacturing – Sewn Product Analysis*. Prentice Hall, Englewood Cliffs, NJ.

Jarnow, J and K. Dickerson. 1996. *Inside the Fashion Business*. Prentice Hall Publication, Upper Saddle River, NJ.

Kiell, B. and Z. Maynard. 2001. *Industrial Engineering Handbook*. McGraw Hill Inc., New York.

Kunz, G. 2005. *Merchandising Theory Principles and Practice*. Fair Child Books, New York.

Stone, E. 2001. *Samples Fashion Merchandising – An Introduction.* McGraw Hill Book Co., New York.

Sumathi, G.J. 2002. *Elements of Fashion and Apparel Designing.* New Age International Publication, New Delhi, India.

Tate, S.L. 2004. *Inside Fashion Business.* Prentice Hall, New Delhi.

Tyler, D.J. 1992. *Materials Management in Clothing Production.* Blackwell Scientific Publications, Oxford, UK.

20

Garment Costing

Cost accounting is a structure of determination of costs of products or services. It was fundamentally created to address the issues of administration. It gives exhaustive information about the cost to various levels of administration for proficient execution of operations. Finance accounting provides data about profit, loss etc., of the combined activities of the business. It doesn't give the information with respect to expenses by departments, products and processes. The losses due to idle plant capacity and time, labour inefficiency, poor raw materials etc. are not taken into account completely in financial accounting. Cost accounting deals with the determination of past, present and future expenses of products produced (Barbee 1993). It gives elaborative cost information to various levels of management for proficient execution of their operations.

Costing is the procedure of determination of production and marketing cost of each product in the line. Costing decisions include every functional division of an industry. Pricing is the process of determination of selling price of the products that are manufactured (Koshy 2006). It is based on information given in the costing process, the value customers will place on the product, and the competition in the retail market.

Under the circumstances of untainted contest, the supply and demand of the particular product decides the cost/price of the product. In the apparel business sector, product pricing is the responsibility of the manufacturer. An industry's success is mostly decided by the top management's perception of the company's cost structure, the market, pricing options and source of profit.

20.1 Purpose of Ascertaining Cost

For the determination of cost, the entire industry should be segregated into small elements of sections and every small section should be taken as a cost centre, to which costing has to be done. A cost centre may be a locality or machinery for which the assessment of costing should be carried out and which is utilised for cost control. The main purpose of determination of cost of a cost centre is cost control.

Cost estimation is concerned with the calculation of actual costs. Ascertainment of actual costs uncovers nonprofitable exercises or activities and losses. Cost evaluation is the process of foreordaining expenses of products or services. The costs are estimated in advance of manufacture of the product. Generally, the estimated costs are arrived based on the earlier actual cost which is adjusted for expected changes in the future. These are used in the preparation of the budgets and evaluating performance (Cooper and Kaplan 1988; Barbee 1993).

20.2 Manufacturing Costs

The cost accounting structure is normally planned by cost centres; hence, the unit costs for every operation can be estimated. Manufacturing costs comprise all the expenditures that are involved in the production of a final product. These costs are called 'cost of goods manufactured' on the income statement. Manufacturing costs are split up into three parts such as raw material cost, direct labour cost and factory overhead.

Raw materials like fabric, sewing thread and trim are called direct variable costs. Direct labour costs in most of the garment units comprise wages of supervisors and employees who work on an incentive, piece rate or hourly wage basis. Factory overhead includes both variable and nonvariable indirect manufacturing costs. Factory overhead costs are exclusive to each industry; however, they are normally subdivided into (1) indirect labour, (2) factory occupancy costs and (3) other overhead.

Indirect labour includes quality control, service personnel, material handlers, maintenance workers, industrial engineers and security. The job of these persons is vital to efficient production of a product line. Nonvariable factory tenancy costs comprise rent, depreciation, insurance, property taxes and security. Machine parts and repairs and needles are examples of variable factory costs and other overhead costs include machinery and equipment costs, materials management and cost of compliance with regulations.

General working cost/expenses or administrative overhead are indirect costs that incorporate the costs of working the offices and all departments that are not directly concerned with the working of the industry but are important to the operation of the firm. Cost centres such as the accounting department, computer programming, management information systems, secretarial and clerical staff, personnel office, design and merchandising, marketing and sales and management could be considered as part of administrative overhead (Cooper and Kaplan 1988; Anonymous 2015a).

20.3 Methods of Costing

Costing is the operation for ascertaining the total resource investment essential to manufacture and market a product. Costing includes

1. Determination of variable and nonvariable material cost and labour cost necessary to manufacture a product
2. Overhead required to operate the industry
3. General operating cost

For efficient costing, information related to cost must be specific and accurate. Inaccurate costing could lead to cancellation of an order having good profit potential. Management in an organisation utilises costing to estimate

1. The producibility of a style within an established price range
2. The profit potential in a style
3. To decide whether a style could be added to the line

Costing could also be used to defend the procurement of new machinery or the extension of production facilities. The method of costing indicates the systems and processes involved in the estimation of costs and it depends on the type and nature of manufacturing activity (Chuter 1995). The two basic methods of costing are

1. *Job costing:* This is the cost estimation for a particular work order where the estimation of cost was carried out separately.
2. *Process costing:* This method is practiced in bulk manufacturing units where cost is accumulated for each department.

Product costing needs in-depth knowledge in materials, product development, production processes and plant operations. The costs involved in manufacturing a product are only taken into account in product costs. Two kinds of product costing are generally used in the garment industry, such as absorption costing and direct costing. Manufacturing costs are separated between variable manufacturing costs and nonvariable manufacturing costs. Direct costing and absorption costing are included in variable manufacturing costs, but only absorption costing is included in nonvariable manufacturing costs (Anonymous 2015b).

20.3.1 Absorption Costing

Absorption costing comprises every single manufacturing cost, both variable and nonvariable, to be product costs that should be allocated to products. Overhead is generally calculated as a percentage of direct labour. The estimation of a sensible overhead application rate is a major drawback in absorption costing. Further, it is also difficult to concentrate on the actual variable costs and profit prospective related to a particular product (Solinger 1998; Hergeth 2002). Factory overhead costs that do not differ with deviations in volume are included as part of cost of products produced with absorption costing. Industries repeatedly project the anticipated total overhead for the particular period based on the past year's costs and expected changes.

The risks connected with using absorption costing are the dependency of the costing system on the accuracy of the estimation of direct labour and the determination of the overhead application, which is arbitrary. For these two reasons, direct costing is mostly recommended for cost ascertainment in place of absorption costing (Koshy 2006).

20.3.2 Direct Costing

Marginal cost is the increase in the total production cost that results from manufacturing one more unit of output and variable costs. Direct costing is a theory that takes into account only the variable costs like labour, material costs and sales commission to be product costs. Nonvariable costs, namely, manufacturing and nonmanufacturing, are treated as time period costs (Solinger 1998).

A direct costing system gives information about costing in a way that can be easily understood and used by management. Since the individual product costs are obviously identified, direct costing makes it easier to evaluate the cost of production and the contribution of product to nonvariable sewing and administrative costs and profit (Bheda 2002). Direct costing makes it easier to categorise the product styles with the highest contribution rate and the most profit potential.

20.4 Stages of Costing

Costing could be carried out at various stages of production, like

- Preliminary or precosting, which is carried out during product development before samples are made.
- Final costing, which is done before the production and price fixing.
- Recosting is done where there is a change in machinery, production processes, materials or garment components.
- Actual costs are determined during production.

20.4.1 Preliminary Costing

The preliminary costing could be useful for the fashion product manufacturers, who can use it in the development stage to come to a conclusion of whether the fashion design developed by the designers is reproducible and merchantable within the established cost range. Generally, it provides only a rough assessment of costs of manufacturing a specific garment style based on determination of raw materials cost as well as labour costs of previously produced similar styles. Costing at this early phase of development of product is especially crucial for the manufacturer because of the wider range of ideas the designer could use.

20.4.2 Cost Estimating

Cost estimating, which is done just prior to price setting and production, requires a detailed analysis of garment components and the specific assembly procedure for each style. Cost estimating determines the expected investment in materials, direct labour and overhead required to produce a single unit of a style. It requires more detail and greater accuracy than preliminary costing. Costing at this stage is based on production samples and standard data (Solinger 1998; Bheda 2002).

20.4.3 Materials Costing

Direct costs of fabric, trim and materials for a particular product are based on estimates arrived in the process of sample manufacturing. The initial step in materials costing is to estimate the yardage and materials required for the production of one garment (Bheda 2002; Clayton 2008). Industries with computerised design systems use the data entered for each product to estimate the required fabric yardage for a single garment. Other direct materials costs like inspecting and shading of fabrics are figured on a per yard basis.

Materials costs are influenced by the rate of utilisation and it relies on quantity of material, which is used compared to the total purchased. Poor use can originate from inadequately engineered designs, inconsistent widths, imprudent cutting etc. Many industries have setup benchmarks for fabric utilisation.

20.4.4 Labour Costing

The time is the origin of production standards and labour costing and hence it should be determined beforehand if it can be controlled and managed. A production standard reveals the normal time necessary to finish one operation using a particular method that will give predictable quality. Production standards are set up as a measure of productivity of labour and operators under standard conditions. Production standards aid to develop consistency of an operator and to discover the most cost-effective method of production.

Production standards used for estimation of labour costing are generally based on work measurement techniques such as predetermined time (PMTS), standard data and time studies. The time values are normally expressed in terms of standard allowed minutes (SAM). An operation breakdown represents the complete list of all the sequence of operations involved in sewing a specific garment style (Lowson 2002; Hogan 2011). Each operation is recorded in the sequence in which it will be performed along with SAM of every operation. The costing of each operation has to be done independently and could be then converted to dollars per unit. In the garment industry which gives hourly wages to the operators, production could be based on production standards representing what an operator is anticipated to finish in a definite period of time.

While estimating the direct labour cost, the production standard stipulates the SAM to finish one cycle. The direct labour cost could be estimated by multiplying the quantity or volume that could be produced in one hour and base rate and divided by the actual quantity produced in an hour. The certain percentage of benefits like insurance, sick leave and vacations should be added to the above cost (Kothari and Joshi 2012). Machine time is regularly determined separately from handling time which is almost the same when an operation is done. But, the total of time required to complete a line of stitching differs with the seam length, stitches per inch (SPI) and the machine speed (Brown and Rice 2001). The stitching time may be calculated by taking into account the time variations that may occur with the stitching process.

$$\text{SAM for stitching} = \frac{\text{seam length} \times \text{stitches per inch}}{\text{machine speed (rpm)}}$$

20.4.5 Recosting

Recosting is determined after garments are put in the production line and the working patterns are developed. At this stage, alterations could be done to cut down the fabric and sewing time. In a few circumstances, the pattern maker could normally advise an increase in costs in order to improve the quality level.

20.4.6 Actual Costs

Actual costs are estimated by the collection of information from the production department. After a particular style has reached the assembling section, an industrial engineer could face some rates that are too tight and that more time is needed to complete specific operations. If a rate adjustment is required, it will certainly influence the costs.

20.5 Components of Cost of Garment

Normally, the costing is prepared by considering the raw material cost, market demand, operating cost of the industry and forecasted profit of the firm and also considering the expectations of the buyer (Carr and Latham 2006; Hogan 2011). The various elements in garment costing are

- Fabric
- Trims and accessories
- CMT (cut, make and trim) charges
- Embroidery, appliqué, printing, washing and other value added processes
- Garment testing
- Logistics and transportation cost
- Profit of the industry

20.5.1 Fabric

Fabric, the raw material for garment manufacturing, itself accounts for 65%–75% of the garment cost, hence it is the most vital parameter in garment costing. In many circumstances, analysing the quantity of fabric as well as quality of it in the garment provides a better indication of cost of production (Fairhurst 2008; Easterling and Ellen 2012). The type of fabric and fibre content of the same, value added finishes applied on the fabric and fabric GSM determine the cost of the fabric.

20.5.1.1 Influencing Parameters for Fabric Cost

20.5.1.1.1 Unit of Measurement

It is basically a number used as a basic criterion for evaluating the fabric cost. It is expressed in meters or yards in case of woven fabric and in kilograms (kg) for knitted fabrics.

20.5.1.1.2 Minimum Order Quantity

It represents the minimum quantity of fabric that the fabric manufacturer could supply to the garment manufacturer. Minimum order quantity (MOQ) is based on the fabric type and construction and on capacity of the merchant. It plays a vital role while ordering the fabric because it directly influences garment cost. If the ordered quantity of fabric is less than the determined MOQ, then the merchant could claim higher price as compared to regular charges.

20.5.1.1.3 Order Quantity

The fabric cost could differ with the order quantity. The larger the order quantity, the more costly the fabric; fabric cost could be optimised up to a certain level. However, this relies on the fabric type and construction and capacity of the fabric manufacturer in addition to the intercession between supplier and fabric buyer.

20.5.1.1.4 Incoterm Used

When importing the fabric from another country, the merchandiser must deal with the supplier for transportation or shipment of the fabric based on incoterms, namely, EXW, FOB, CIF, DDP, etc. based on these, who can bear the transportation cost can be decided. Whatever type of incoterm used, all the cost should be claimed from the buyer. For instance, if the fabric is purchased under EXW incoterm, the merchandiser should add the cost of transportation in addition to the custom clearance charges and fabric cost while determining the cost of the garment. The fabric cost can be determined by

$$\text{Fabric cost} = \text{Yarn Cos} + \text{Fabric manufacturing cost} \\ + \text{Dyeing cost} + \text{Finishing cost}$$

20.5.1.2 Cost Calculations of Fabric in Garment

For example, the fabric consumption of a knitted T-shirt can be determined as

$$\text{Fabric consumption (kgs)} = \frac{\begin{array}{c}(\text{Body length} + \text{Sleeve length} + \text{Allowance}) \\ \times (\text{Chest} + \text{Allowance}) \times 2 \times \text{GSM}\end{array}}{10000}$$

Similarly, for woven shirt fabric, the fabric consumption can be calculated as

$$\text{Fabric consumption (meters)} = \frac{\begin{array}{c}(\text{Full length} + \text{Sleeve length} + \text{Allowance}) \\ \times (\text{Chest} + \text{Allowance}) \times 2 \times \text{Fabric width}\end{array}}{39.37}$$

These types of methods are used to estimate the fabric consumption at the sampling stage by the merchandiser. Normally, fabric wastage and the buffer value of 0.03%–0.08% in the fabric consumption will be included while calculating the fabric consumption.

20.5.2 Trims

Trims comprise all materials other than fabric utilised in the garment such as sewing threads, zippers, buttons, elastics, labels, etc. Quality and quantity of trim and labour necessary to apply it on a garment depend on the cost of the garment (Cooklin et al. 2006). MOQ, quality of raw material utilised for making the trims and lead time are the parameters to be taken into account while calculating trim cost.

20.5.2.1 Thread

After fabric, which is a main component, thread is another item that needs to be taken into account for estimating the cost of garments. The consumption of sewing thread is determined by the industrial engineering (IE) department. It is based on the type of seam and stitch density (Solinger 1998; Somasekhar and Rajmogili 2002). While purchasing the sewing thread, the operation breakdown for the particular style and total number of sewing machines necessary to complete the particular style of garment should also be considered. For the determination of thread consumption software is also available which could give the precise thread consumption. The sewing thread wastage of around 10%–15% should be considered while ordering it.

20.5.2.2 Labels

Various kinds of labels are used in garments like the main label, content label and care label. The cost of it depends on its manufacturing process, for instance, based on the fibre content, printed labels, size of labels, colours etc.

20.5.2.3 Zippers

The types of zippers, such as plastic zippers, moulded zippers, metallic zippers, invisible zippers etc. play a significant part in the cost of the zipper. The merchandiser must be aware of the various parameters of the zipper for negotiation and accurate costing (Tyler 1991; Solinger 1998). Minimum order of quantity is the parameter that influences the cost of the zipper.

20.5.2.4 Buttons

Another kind of closure, buttons, could be made up of different types such as nylon, plastic, wood, shell, or metal. Each kind of button has its own minimum order of quantity decided by the manufacturer of it. Buttons are purchased on a bulk basis with the lignes specified.

$$1 \, gross = 1 \, packet = 144 \, buttons = 12 \, dozens$$

20.5.2.5 Polybags

The cost of polybags is mainly based on thickness, dimension and raw material and is procured in terms of number of pieces. The cost of polybags is also vital because it makes a difference while considering the entire order quantity.

20.5.2.6 Cartons

The cost of cartons varies based on the material used and their dimensions. The cartons are procured based on their dimensions, number of plies and

GSM of the paper that is used to make the carton box. In general, 3, 7 and 9 plies are utilised in a carton box.

20.5.2.7 Hand Tags

These are normally used as packing material and the cost of it depends on the raw material used, printing over it and the minimum order quantity.

20.5.2.8 Shanks and Rivets

Generally these types of trims are made up of metal and the cost of these trims is dependent on MOQ and the raw material used to make them.

20.5.2.9 Hangers

Hangers are generally made up of hard plastics, seldom with wood material. The hanger cost depends on the raw material used to make it, size of the hanger, colour of the hanger and any printing on it.

20.5.2.10 Tapes and Velcro

Generally, tapes are purchased based on the width, hence, the width of the tape as well as MOQ influences the cost of the tape (Solinger 1998; Cooklin et al. 2006; Pareek 2013).

20.5.2.11 Other Charges

Trims charges are normally determined based on the way of transportation, for air transportation the cost will increase by 15%–25% and for transportation though sea, it will increase by 10%–15%. If it is domestic, then the local taxes are added. Supplementary charges involved in the garment costing are

Rejection and wastage charges	–2%–5%
Inspection charges	–1%–2%
Buying house commission	–1%–1.5%
Transportation charges	–$1–2/piece
Profit margin	–10%–15%

20.5.3 Cut-Make-Trim (CMT) Cost

The cost of making completed 'in-house' is given by

$$= \frac{\text{Total(cost/hour)} \times \text{total hours required for a style}}{\text{Number of units produced}}$$

1. $\text{Labour cost/min (@ 100\% efficiency)} = \dfrac{\text{Operator salary/month}}{\text{Available minutes in a month}}$

2. $\text{CM Cost} = \dfrac{\text{SAM of garmet} \times \text{Labour cost per minute}}{\text{Line Efficiency (\%)}}$

20.5.4 Value Added Processes

This denotes the cost of value added processes such as embroidery, printing and washing used to impart the type of finish the buyers need. Cost of these kinds of value added services varies depending on different styles.

Hence, by considering all these aspects, CMT charges can be determined by the following manner.

1. *Available capacity per month (in minutes)* $= 26$ working days/month
$$\times 8 \text{ hours/day} \times 60$$
$$= 12,480 \text{ minutes}$$

2. *Labour cost per minute* $\big(@100\% \ efficiency\big) = $ (Salary of an operators/
month/Available
capacity/month
$$= 10000/12480$$
$$= \text{Rs. } 0.8$$

3. *Sewing cost* $=$ (Garment sewing SAM $\times$ Labour cost/min)/
Line efficiency (%)
$$= (14 \times 0.80)/55$$
$$= \text{Rs. } 20.36$$

4. *Cutting cost* $= \big($SAM of cutting $\times$ Labour cost/min$\big)/$
Cutting efficiency (%)
$$= 8 \times 0.8/55$$
$$= \text{Rs. } 11.6$$

5. Trimming cost is considered as Rs. 3 as it depends upon how many operators are there for trimming.

Production cost of garment (CMT) $=$ sewing cost $+$ cutting cost
$$+ \text{ trimming cost}$$
$$= 20.36 + 11.6 + 3$$
$$= \text{Rs. } 34.96$$

20.6 Costing for Men's Shirts (Long Sleeve)

The measurement chart for the men's log sleeve shirt is given below:

1. Collar length – 17", allowance – 4.5"
2. Collar width – 2.5", allowance – 1"
3. Chest – 50", allowance – 2"
4. Centre back length (CBL) – 33", allowance – 1.5"
5. Sleeve length – 36"
6. Yoke length – 22" (drop shoulder), allowance – 3"
7. Arm hole depth (1/2) – 0.6"
8. Cuff length – 9.5", Allowance – 2.5"
9. Cuff width – 2", allowance – 0.5"
10. Pocket – 6.5" (allowance – 1.5") × 5.5" (allowance – 0.7")
11. Yoke width – 4", allowance – 0.8"

1. Fabric requirement for back panel $= \dfrac{((\text{CBL} + \text{allowance}) \times ((1/2)\ \text{Chest} + \text{allowance}))/36}{44}$

$= \dfrac{((33 + 1.5") \times (25 + 2))/36}{44}$

$= 0.588\ \text{yds}$

2. Fabric requirement for Front panel $= \dfrac{((\text{Body length} + \text{allowance}) \times ((1/4)\ \text{Chest} + \text{allowance})) \times 2/36}{44}$

$= \dfrac{((33 - 1.25 + 1") \times (12.5" + 2")) \times 2/36}{44}$

$= 0.608\ \text{yds}$

3. Fabric requirement for Yoke $= \dfrac{((\text{Yoke length} + \text{allowance}) \times (\text{Yoke width} + \text{allowance}))/36}{44}$

$= \dfrac{(22" + 3") \times (4 + 0.8)/36}{44} = 0.075\ \text{yds}$

4. Fabric requirement for sleeve $= \dfrac{((\text{Sleeve length} + ((1/2)\text{of drop shoulder} + 0.5")) \times (\text{arm hole depth} + \text{allowance})) \times 2}{36 \times 44}$

$= \dfrac{(36" + 11.5") \times (22 + 1.2) \times 2}{36 \times 44} = 1.39\ \text{yds}$

5. Fabric requirement for cuff $= \dfrac{\begin{array}{c}((\text{Cuff length} + \text{allowance})\\ \times (\text{Cuff width} + \text{allowance})) \times 2\end{array}}{36 \times 44}$

$= \dfrac{((9'' + 2.5'') \times (2 + 0.5'')) \times 2}{36 \times 44} = 0.036\,\text{yds}$

6. Fabric requirement for Pocket $= \dfrac{\begin{array}{c}((\text{Pocket length} + \text{allowance})\\ \times (\text{Pocket width} + \text{allowance})) \times 2\end{array}}{36 \times 44}$

$= \dfrac{((6.5'' + 1.5'') \times (5.5 + 0.7'')) \times 2}{36 \times 44}$

$= 0.062\,\text{yds}$

7. Fabric requirement for Collar $= \dfrac{\begin{array}{c}((\text{Collar length} + \text{allowance})\\ \times (\text{Collar width} + \text{allowance})) \times 4\end{array}}{36 \times 44}$

$= \dfrac{((17'' + 2.5'') \times (2.5 + 1'')) \times 4}{36 \times 44}$

$= 0.172\,\text{yds}$

8. Total Consumption for one Garment $= 0.588 + 0.608 + 0.075 + 1.39$
$+ 0.036 + 0.062 + 0.172$
$= 2.931\,\text{yds/garment}$

9. Total fabric consumption per dozen garments $= (2.932 \times 12)$
$+ 5\%\,\text{wastage}$
$= 36.94\,\text{yds}$

Consider the cost of the fabric is $1 per yard, and then the calculation of the garment cost is given in Table 20.1.

20.7 Costing for Men's Basic T Shirts

Consider the 100% cotton single jersey knitted fabric having GSM of 150, with the order quantity of 15,000 pieces. The measurements for the production of garment are given below.

1. 1/2 Chest – 73 cm, Allowance – 5 cm
2. Body length – 86 cm, Allowance – 5 cm
3. Sleeve length – 36 cm, Allowance – 5 cm

TABLE 20.1

Garment Costing of Long Sleeve Shirt (Woven)

Fabric cost/dozen garments	$36.94 \times 1 = \$36.94$
Accessories cost/dozen garments	$7 (approx.)
CM cost/dozen of garments	$10 (approx.)
Sub total	$53.94
Transportation cost (0.5%)	$0.27
Clearing and loading cost (2%)	$1.08
Overhead cost (0.5%)	$0.27
Net cost	$55.56
Profit (15%)	$8.33
Net Free On Board price	$63.89
Freight (4%)	$2.55
Net C and F price	$66.44
Insurance (1%)	$0.66
Net CIF price	US$67.10
Net CIF price/piece	US$5.59

Assumptions:

1. Yarn cost/kg: $3.40
2. Knitting and washing cost/kg: $1.10
3. Dyeing cost/kg: $1.90
4. Printing cost/dozen: $4.50
5. Accessories cost per dozen: $1.50

1. Fabric Consumption =

$$\frac{(\text{Body length} + \text{Sleeve length} + \text{allowance})}{10000000} \times ((1/2)\text{Chest} + \text{allowance}) \times 2 \times \text{GSM} \times 12$$

$$+ \text{Wastage}(\%) \frac{(91+41) \times 78 \times 2 \times 150 \times 12}{10000000}$$

$$+ 10(\%)$$

$$= 3.70 \times 10\%$$

$$= 4.07 \text{ kgs/dozen garments}$$

2. Grey fabric cost/dozen = Yarn cost/kg

$$\times \text{Fabric consumption/dozen}$$

$$= 3.4 \times 4.07$$

$$= \$ 13.84$$

TABLE 20.2

Garment Costing of T- Shirt

Fabric cost/dozen garments	$16.84
Printing cost/dozen garments	$4.50
Accessories cost/dozen garments	$1.50 (approx.)
CM cost/dozen of garments	$6.00 (approx.)
Sub total	$28.84
Transportation cost (0.5%)	$0.14
Clearing and loading cost (2%)	$0.57
Overhead cost (0.5%)	$0.14
Net cost	$29.69
Profit (15%)	$4.45
Net Free On Board price	$34.14
Freight (4%)	$1.36
Net C and F price	$35.50
Insurance (1%)	$0.35
Net CIF price/dozen	US$35.85
Net CIF price per piece	US$2.99

3. Actual fabric cost = Grey fabric cost + Knitting and Washing Cost

    + Dyeing Cost

    = 13.84 + 1.10 + 1.90

    = $ 16.84

The final cost of the garment can be arrived as shown in Table 20.2.

References

Anonymous. 2015a. http://docslide.us/documents/mer547387d4b4af9f40628b45a2.html. (accessed on September 11, 2015).

Anonymous. 2015b. http://www.scribd.com/doc/37249960/The-Knits (accessed on September 11, 2015).

Barbee, G. 1993. The ABCs of costing. *Bobbin* 34:64–7.

Bheda, R. 2002. *Managing Productivity of Apparel Industry*. CBI Publishers and Distributors, New Delhi.

Brown, P.K. and J. Rice. 2001. *Ready-to-Wear Apparel Analysis*. Prentice-Hall, Oxford.

Carr, H. and B. Latham. 2006. *The Technology of Clothing Manufacture*. Blackwell Science, Oxford.

Chuter, A.J. 1995. *Introduction to Clothing Production Management*. Blackwell Scientific Publications, Oxford, UK.

Clayton, M. 2008. *Ultimate Sewing Bible – A Complete Reference with Step-by-Step Techniques*. Collins & Brown, London.

Cooklin, G., S.G. Hayes and J. McLoughlin. 2006. *Introduction to Clothing Manufacture*. Wiley-Blackwell, London, UK.

Cooper, R. and R.S. Kaplan. 1988. Measure costs right: Make the right decisions. *Harvard Business Review* 66(5):96–103.

Easterling, C.R. and L. Ellen. 2012. *Merchandising Math for Retailing*. Addison-Wesley Longman Limited. Fairhurst, C. 2008. *Advances in Apparel Production*, The Textile Institute, Woodhead Publication, Cambridge, UK.

Hergeth, H. 2002. Target costing in the textile complex. *Journal of Textile Apparel Technology Management* 2(4):42–7.

Hogan, C. 2011. A framework for supply chains: Logistics operations in the Asia-Pacific region. *MHD Supply Chain Solution* 41(3):73–81.

Koshy, D.O. 2006. *Garment Exports – Winning Strategies*. Prentice-Hall of India Pvt Ltd, New Delhi, India.

Kothari, V.R. and S. Joshi. 2012. Fashion Merchandising: Garment Costing. http://www.textiletoday.com.bd/magazine/715 (accessed on March 27 2015).

Lowson, R. 2002. Apparel sourcing offshore: An optimal operational strategy? *Journal of Textile Institute* 93(3):15–24.

Pareek, V. 2013. Garment Costing Method http://textilelearner.blogspot.com/2013/08/methods-of-garments-costing-how-to.html#ixzz3VBd2IJgR (accessed on March 21, 2015).

Solinger, J. 1998. *Apparel Manufacturing Handbook-Analysis Principles and Practice*. Columbia Boblin Media Corp, New York, USA.

Somasekhar, B.V. and A. Rajmogili. 2002. *Textiles – Law & Policy*. PMR Publications (P) Ltd, Secunderabad, India.

Tyler, D.J. 1991. *Materials Management in Clothing Production*. BSP Professional Books, Oxford, UK.

Index

A

Abrasive strength, 22, 23
Absenteeism, 318
Absorption costing, 441–442
Accepted quality level (AQL), 5
Accessory stores department, 5
Accordion pleat, 226, 227
Achieved consumption, 340–341
Actual costs, 444
Adhesive paper dress form, 49–52
Adjustable zipper foot, 182
Air permeability, 22
Alternating presser foot, 172
Apparel industry, 1; *see also* Clothing
        industry; Textile industry
    apparel accessories, *see* Closures;
        Supporting materials
    apparel export merchandising,
        423–424
    apparel retail merchandising
        process, 433
Applied pockets, 258; *see also* Patch
        pocket
AQL, *see* Accepted quality level
        (AQL)
Assembly line system, 348; *see also*
        Modular production system
    PBS and UPS comparison, 356
    progressive bundle system, 348–351
    straight line system, 351, 352
    unit production systems, 351,
        353–355
Attachment preparation, 184; *see also*
        Guide attachments; Position
        attachments
    buttonhole foot, 184
    chain cutters, 185
    needle and stitch devices, 185
    pinking, 184
    pressing attachments, 184
    tape cutters, 185
    thread cutters, 185
Attrition rate, 319

Automatic rolling rack, 81
Automatic turntable spreader, 82
Available capacity, 390, 415, 449

B

Balance control, 401
Balancing efficiency, 405, 406; *see also*
        Line balancing
Band knife, 90
Barrel cuffs, 273
Barrel French cuffs, 276–277
Barrel shirt cuffs, 274
    double button barrel cuff,
        274, 275
    single button barrel cuff, 274
Basic Pitch Time (BPT), 369
Basic sleeve, 270
Basket weave, 14, 15
Bellows pockets, 265, 266
Bell sleeve, 271
Besom pockets, 264, 265
Bias grain, 18
Bill of material (BOM), 301, 432
Binder, 178, 179, 180
Bishop sleeve, 271
Blade pleating machine, 210
Blind
    loopers, 114
    stitch needle, 137
    tucks, 231
Blocking and relaying marker, 69
Block paper pattern, 40
Body measurement, 35–37
BOM, *see* Bill of material (BOM)
Bound seam (BS), 147
Box pleats, 224, 225
BS, *see* Bound seam (BS)
Bundle ticket, 380–381
Bursting strength, 23
Buttonhole, 288
    foot, 184
    scissors, 44
    stabiliser plate, 183, 184